Troubleshooting and Repairing Consumer Electronics

Troubleshooting and Repairing Consumer Electronics

Third Edition

Homer L. Davidson

McGraw-Hill

New York   Chicago   San Francisco   Lisbon   London   Madrid
Mexico City   Milan   New Delhi   San Juan   Seoul
Singapore   Sydney   Toronto

The McGraw·Hill Companies

Library of Congress Cataloging-in-Publication Data

Davidson, Homer L.
   Troubleshooting and repairing consumer electronics / Homer L.
Davidson.—3rd ed.
        p.   cm.
   Rev. ed. of: Troubleshooting and repairing consumer electronics without
a schematic. 2nd ed. c1997.
   Includes index.
   ISBN 0-07-142181-5
   1. Electronic apparatus and appliances—Maintenance and repair.
I. Davidson, Homer L. Troubleshooting and repairing consumer electronics
without a schematic.   II. Title.

TK7870.2.D387   2004
621.389'7—dc22   2003068886

1 2 3 4 5 6 7 8 9 0   DOC/DOC   0 1 9 8 7 6 5 4

ISBN 0-07-142181-5

*The sponsoring editor for this book was Scott L. Grillo, the editing supervisor
was Stephen M. Smith, and the production supervisor was Sherri Souffrance.
It was set in ITC Century Light following the EL1 design by Paul Scozzari of
McGraw-Hill Professional's Hightstown, N.J., composition unit. The art director
for the cover was Margaret Webster-Shapiro.*

Printed and bound by RR Donnelley.

McGraw-Hill books are available at special quantity discounts to use as premiums and
sales promotions, or for use in corporate training programs. For more information,
please write to the Director of Special Sales, McGraw-Hill Professional, Two Penn Plaza,
New York, NY 10121-2298. Or contact your local bookstore.

This book is printed on recycled, acid-free paper containing a
minimum of 50 percent recycled, de-inked fiber.

Contents

Introduction

The need to troubleshoot and repair consumer electronics without a schematic occurs every day in the life of a busy electronics technician. The technician who repairs all types of electronic equipment must make quick and practical repairs; otherwise, he or she will be out of the business within a few years. Yet even the largest and best-equipped establishment cannot have all the schematics required to service every piece of equipment that crosses the service bench. Sometimes, a repair center simply does not have room for them. Sometimes, schematics for import models are difficult to obtain; it might take weeks or months to get them, and the electronic product sits for days, in pieces, until the diagram arrives. Sometimes, a schematic is no longer available.

The more experienced and better informed the electronics technician is, the more productive in troubleshooting and repairing consumer electronics he or she will be. This book will help the beginning, intermediate, and experienced technician service and repair different types of consumer electronics without a schematic. Besides servicing tips and valuable information, practical case histories are found throughout the book.

Troubleshooting and Repairing Consumer Electronics begins with servicing methods. Chapter 2 shows how to locate, test, and repair the electronic product. Troubleshooting and repairing large and small amps are covered in Chap. 3, in which a list of required test equipment and a table of symptoms of problems and methods of servicing them can also be found. Chapter 4 is on servicing the auto or car radio receiver. Troubleshooting the cassette player is the subject of Chap. 5, along with various symptoms of problems and tips on solving them, as well as actual case histories.

Chapter 6 teaches how to repair the compact disc (CD) player. Troubleshooting and repairing the TV chassis is discussed in Chap. 7 as well as the many different circuits and troubleshooting tips. Servicing the power supplies found in many electronic components is covered in Chap. 8. Chapter 9, on servicing stereo audio circuits, examines most of the audio circuits located in many consumer electronics products.

Troubleshooting AM/FM/MPX stages is the subject of Chap. 10. VCR and TV/VCR combo repairs are covered in Chap. 11. Chapter 12 shows how to test remote control and infrared receiver circuits. The last chapter discusses the repair of the new DVD/CD players.

Don't push that electronic unit aside and wait for the correct schematic. Apply the methods in this book to turn out more repairs and fill up that cash register. Electronic products collecting dust provide no income. Troubleshooting and repairing consumer electronics can be fun and quite rewarding, even when the schematic is not available.

Homer L. Davidson

Acknowledgments

My thanks to Robert Douglas, who helped me correct messed-up programs that the stubborn computer and I created. A great deal of thanks to RCA Consumer Electronics and Radio Shack for service data. Special thanks to Nils Conrad Persson, editor of *Electronics Servicing and Technology* magazine, and electronic technicians Tom Rich from Tom's TV, Robert P. Saunders from Eagle Electronics, and Tom Krough from Krough Repair, who supplied troubleshooting tips and symptoms. Last but not least, thanks a lot to Michael B. Danish of Mike's Repair Service, Aberdeen Proving Ground, Maryland, for TV/VCR case histories.

1
CHAPTER

Servicing methods without a schematic

Many different electronic products are found within the consumer electronics field. Yesterday, the radio was king, and today we have several new electronic products, such as the TV/VCR combo, CD, DVD, DVD/VCR combo, and MP3 players. After the early radio receiver, along came the phonograph player constructed inside a radio/phonograph combination. Later on, the cassette player was introduced within the portable radio/cassette player. Now the new TV chassis also might include a VCR and a DVD player (Fig. 1-1). Today, the new DVD player is combined with a DVD/CD, DVD/VCR, or DVD/VCR/CD changer, and on it goes.

Servicing consumer electronic products can be a lot of fun if you let it be. Just as in solving a great murder mystery, the smart detective must have the skill, desire, and ability to find the culprit, so also the electronics technician must take the symptom and isolate, locate, remove, and replace the defective part.

The anticipation of locating a dead, open, or leaky component is like finding a needle in a haystack, so to speak. Your heart beats a little faster as you zoom in on the suspected part. Then great disappointment sets in when the suspected component is not the guilty party. So you try once again. Now victory is in sight.

Troubleshooting the TV chassis, VCR, CD/DVD player, and cassette player presents many rewards. Solving the intermittent or "tough dog" problem is a moment of triumph. However, if you can repair and service the electronic chassis without a schematic, you receive the highest honors of all. So let's begin.

1-1   The new portable TV also may include a VCR player.

Locating critical components

Damaged components, burned printed circuit (PC) boards, and pulled PC wiring indicate that the TV chassis has sustained lightning damage. Take a good look at the defective chassis. Visual inspection has repaired thousands of electronic chassis. Watch out for visual scars, such as a burned resistor, worn connections, arc-over parts, cracked boards, overheated components, and wear and tear.

For instance, firing or arcing inside the picture tube can indicate a cracked cathode-ray tube (CRT) gun assembly (Fig. 1-2). Arcing between parts on PC boards results from liquid spilled inside the chassis. Smoke curling up from overheated components signals trouble ahead. When sweat forms on a carbon resistor, an overheated circuit is nearby.

Cracked or blown resistors or capacitors show signs of higher voltage or defective parts. Arcing inside the flyback transformer indicates leaky high-voltage diodes. Overheated white and dark marks on the integrated circuit (IC) body might point out a leaky part. Cracks found in ceramic capacitors and resistors indicate overheated components. Too-hot-to-touch transistors or ICs might be leaky or shorted.

Several burned resistors might point to a leaky transistor or IC. Taking a close-up peek through a magnifying glass might solve the intermittent problem. Dark-brown areas around the component terminal leads on PC boards can indicate a poor connection or overheated part. Just spend a little time before jumping to conclusions. Look before isolating, and then take another peek at the collected symptoms.

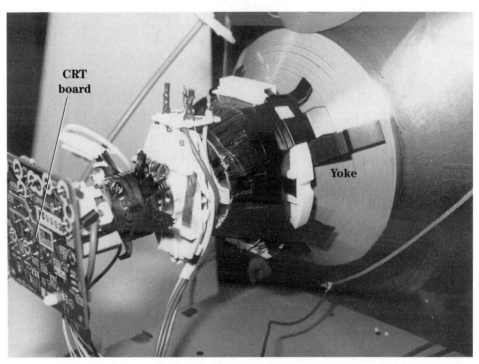

1-2 Sometimes the neck of the picture tube will crack inside the yoke area.

Just listen, man

Just listen and you can hear a defective component act up within the TV chassis. You should be able to hear the TV chassis with high voltage by listening with your ear close to the deflection yoke. Listen for arcing in the flyback or a tic-tic noise indicating horizontal problems (Fig. 1-3). Intermittent arcing noise within a high-voltage capacitor can indicate an internal break or loose terminal wire.

Check the main filter capacitors when extreme hum is heard in the speaker. A low hum can be caused by dried-up decoupling capacitors. Pickup hum might point toward a poor ground or an open base circuit. Hum with distorted audio might originate in the output transistors or IC components in the stereo circuits. A mushy speaker sound might be caused by a dropped or frozen voice coil.

Low frying and popping noises in large power amps might be caused by defective transistors or IC components. Especially check ceramic capacitors, transistors, and IC parts in the front end of stereo or cassette tape amplifiers. Loose grounds or soldered joints might cause microphonic noises in the input and output stages of the amplifier.

Hands on

Touching or feeling the overheated component might indicate a defective part. You can tell if the soldering iron is warm, cold, on the blink, or shut off by holding it near your

1-3   Carefully inspect the flyback for burned areas.

arm. Do not touch it. Locate the shorted yoke by removing it and feeling the inside area. Replace the yoke assembly if hot spots are found. Feel the outside winding of a power transformer or flyback, with the power off; heat might indicate shorted windings or overloading. You can feel the high voltage come up by placing your arm next to the picture tube; the hair rises up (Fig. 1-4).

Feel the speaker cone by pushing it in and out to determine if the voice coil is dragging. Loud speaker vibration can be felt with fingers on the speaker cone area. An intermittent voice-coil connection can be felt with the volume turned up and fingers pressed against the speaker cone. Touch can locate a lot of different problems in the electronic chassis.

Too, too bright

An extremely bright TV screen might indicate a defective component within the video or picture-tube circuits. An excessively bright raster without any control might result from a defective picture tube or CRT circuit or a change in boost voltage (Fig. 1-5). Usually B+ boost voltage is developed in the derived secondary circuits of the horizontal transformer. An open or burned isolation resistor or silicon diode or a dried-up filter capacitor can cause improper boost voltage. The excessively bright raster, followed by shutdown, might be caused by a defective picture tube.

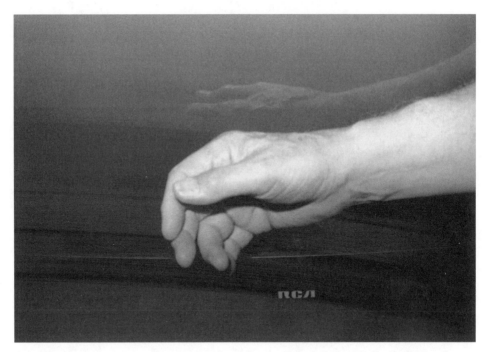

1-4 Place your arm next to the picture tube and notice if the hair rises, indicating high voltage.

High-voltage arc-over can be seen at the anode connection on the picture tube. If liquid is spilled on the CRT, arc-over between the anode and the aqueduct coating or ground spring can occur. Rubber damage on the anode button or socket can result in bright high-voltage arc-over. A pinched or cracked high-voltage anode cable results in bright arcing to parts or the chassis.

Often, when high voltage arcs over at the CRT anode socket, excessive high voltage is caused by a defective flyback or safety capacitor (Fig. 1-6). This capacitor, located across the damper diode or attached to the collector terminal, might open up, letting the horizontal output increase and resulting in uncontrollable high-voltage arc-over.

Audio pops and cracks

Sometimes you can find the defective component just by listening to the loudspeaker. Cracking or popping sounds in the speaker might be caused by a defective audio output transistor or IC. Isolate the output stage. Component replacement is the only answer. Low popping noises can be caused by defective ICs or transistors. Defective bypass capacitors within the audio input stages can also cause noise, hum, or buzz in the speaker. Use the external audio amplifier to locate the defective stage. Spray the component with coolant, and if the noise stops, replace the suspected component. Low frying or firing noises can be caused by a defective input transistor or IC.

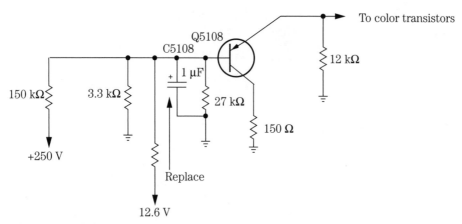

1-5 Excessive brightness with retrace lines caused by defective C5108 in a Zenith TV.

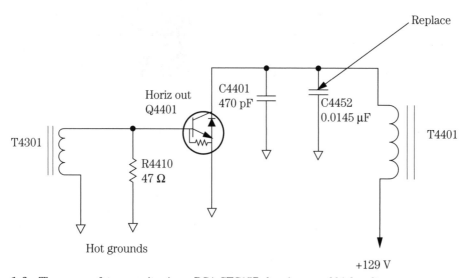

1-6 The open safety capacitor in an RCA CTC157 chassis caused high-voltage arc-over.

When switched, a dirty function switch might produce scratching or cracking noises in the speaker (Fig. 1-7). You might find one or two dead circuits with dirty switch contacts. Spray cleaning fluid inside the switch area, and work the switch control on and off to help clean the contacts. Replace the switch if it cannot be cleaned or has worn contacts.

High-voltage arc-over and popping noises are caused by excessive high voltage. When the TV chassis starts up or shuts down, a crackling noise can be heard. Of course, no problem exists with the yoke assembly expanding and collapsing with signal applied and turned off. This is a good crackling sound.

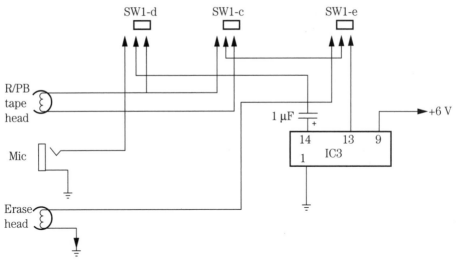

1-7 The bad or dirty function switch in the cassette player can cause noisy and intermittent operation.

Smoke and fire

Where there is smoke there is fire, or so the saying goes. This also applies to troubles that occur in the electronic chassis. Suspect a shorted power transformer with smoke and odors rising from the low-voltage circuits. Often, small power transformers with leaky silicon diodes or leaky filter capacitors might open the primary winding. In large power or switching transformers, found in high-voltage stereo amplifiers and TV chassis, the shorted transformer might burn or smoke (Fig. 1-8).

Remove all secondary leads from the power transformer to the power circuits, and let the transformer cool down. Again, turn the chassis on. If the transformer begins to smoke and run excessively warm, replace it. Remember, power transformers are rather expensive and sometimes difficult to obtain.

When liquid is spilled in an area where higher-than-normal voltage is found, firing on the PC board results. Smoke will curl upward and holes will be burned in the board if the unit is left on. Firing and arcing between components and wiring might melt down and destroy a few parts before the chassis is shut off. Arc-over between horizontal and vertical windings in the yoke assembly might be caused by liquid spilled into the yoke or breakdown within the yoke assembly.

Small burned spots on a board can be repaired by replacing burned or charred components. Spray on a cleaner/degreaser to wash away dirt from the parts and circuit board. Scrape off burned areas with a pocketknife or flat tool. Replace burned components with new parts. Recheck for overheating, arcing, and smoke after repairs. Spray on a light coat of premium preservative fluid and let dry.

Often the horizontal output transformer must be replaced when heavy arcing is found within the high-voltage windings or diodes. Check the flyback for burned areas.

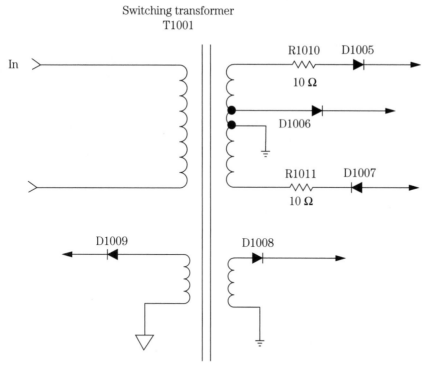

Switching transformer
T1001

1-8 A defective switching transformer in a Panasonic PV-M1327 TV/VCR combo produced a dead chassis.

Turn off the chassis and feel the transformer for overheating. A hot area represents shorted internal wiring. Replace the flyback with the original part number. You may be able to see a burned or arcing section of the flyback or no signs at all.

Under those eyes

You can locate many different components within consumer electronics products with your own eyes. Just take a peek at the chassis and notice any burned areas or overheated parts. Most symptoms are found by sight, sound, or touch. Of course, test equipment is needed to check voltage, resistance, current, waveforms, and continuity of the various components. Isolating and pinpointing the defective part are the most difficult jobs in troubleshooting. It only takes a few minutes to replace a defective part after locating it.

Sometimes we try too hard to locate a problem when the defective part is staring up at us, right under our eyes, so to speak. It is best to get away from the chassis when a defective part cannot be found in 1 hour. The longer you stay with it, the more dark and clouded the scene becomes.

A little common sense, electronics knowledge, the correct test equipment, and experience can help you locate the "tough dog." When the mind is free and clear, early in the morning, tackle that tough repair problem. Happy hunting.

Keep at it

After working on a chassis for several hours, you might want to give up. Why not try a different approach or method of troubleshooting? Each electronics technician has his or her own method of troubleshooting an electronic product and does not want to change. Have you tried asking fellow technicians if they have had the same problem? Help can be obtained from electronics distributors and service departments. Why not ask the manufacturer for help? Help is out there; just ask for it. Some manufacturers have service centers around the country where their electronic products can be serviced.

Don't give up the ship. Keep your nose to the grindstone, so to speak. Sometimes you wonder if this job is all worth it. You bet it is. There is no greater satisfaction than solving "tough dog" problems.

Same chassis, different model

The electronics manufacturers use several chassis that are the same, year in and year out. The cabinet and outside dressing are changed, but the insides are essentially the same. In fact, the same chassis might be found in other brands of TVs.

For instance, let's take the RCA CTC187 TV that has many different chassis, such as the CTC187AJ, CTC187BJ, and CTC187BH. However, you might find in a few different manufacturers that the circuits are the same with a different size picture tube. The BJ and BH chassis might have a different sound system than the CTC187AJ chassis. The horizontal and vertical controls and power supplies might be the same in all chassis. You can service all RCA CTC187 chassis with only one schematic (Fig. 1-9).

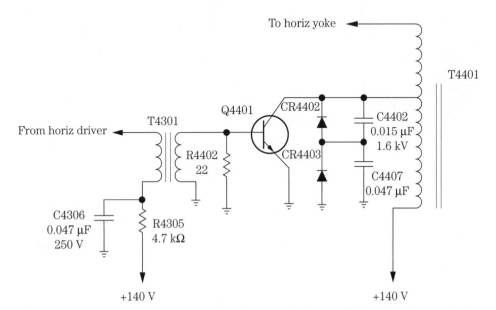

1-9   The horizontal output circuits are about the same in all color TV chassis as they are in the RCA CTC187.

Often the horizontal, vertical, and intermediate frequency (IF) circuits are the same. A different tuner might be used, but the IF sections are identical. The same color IC might be used in several dozen TV chassis. So why not try to use a schematic that is similar to the chassis found on your service bench instead of waiting for months for the exact diagram?

The same train tracks

Remember, the circuits found on the schematic of a manufacturer might be used in many different chassis. Some firms have their own pet circuits that are used in many different chassis, including other brands. In fact, several overseas companies make electronic units for other brands. These units might not look the same, but inside they are the same circuits. For instance, the horizontal circuits are about the same in all models.

When a breakdown occurs in one chassis, look for the same part to break down in other chassis. Sometimes the same component fails in many different chassis. You can make money on these types of repairs because they occur over and over. However, it is difficult to remember what parts break down in several hundred different chassis each year. Write it down.

The best place is to write it down is on the service schematic. Circle the defective component, draw a line out to the side, and write down the symptom. Circle the voltages that are measured when the part is defective, with proper waveforms (Fig. 1-10). Remember, the same part might break down in another chassis.

1-10 Circle the defective component and write the symptom on the schematic.

Service notes and troubleshooting tips by the manufacturer or service center can be added to the same chassis. If you are a warranty station for several brands of electronic products, the firm passes on troubleshooting guidelines that can help in servicing the "tough dog" chassis. And of course, manufacturers' service meetings and seminars are worth their weight in gold. Take a few hours away from the bench to gain valuable knowledge.

Lines down the road

While two yellow lines down the middle of the highway divide the coming and going traffic in half, the white lines along the outside edge of the asphalt show the edge of the pavement. Unwanted lines in the TV picture must be erased for good viewing. Sharp, jagged, or dotted lines across the picture can indicate outside interference or arcing in the TV chassis. Outside interference lines can be caused by motors, neon signs, and high-power lines arcing over (Fig. 1-11). Remove the antenna wire to determine if the lines are in the set or picked up by the antenna.

A dirty spark gap around the picture-tube socket can cause arc-over. Blow out the dust and dirt within the picture tube socket, or replace the socket. If one of these gaps continually arcs over, it can cause an extremely blurry picture, an out-of-focus picture, loss of color, a raster of one color, or chassis shutdown. Check for arcing under the anode high-voltage rubber cap assembly. Sometimes it is not hooked properly and comes loose, causing high-voltage arc-over.

Vertical lines to the left of the picture may result from Barkhausen or improper filtering in the automatic gain control (AGC) circuits (Fig. 1-12). In the early chassis, Barkhausen lines resulted from a horizontal output tube. Today, heavy-firing vertical lines can be caused by the horizontal output transistor or flyback transformer. Shunt small electrolytic capacitors in the AGC and sync circuits for vertical lines in the picture.

Identify part numbers

Sometimes you can locate the section with the defective component by looking at the part number stamped on the top. A large vertical output IC, as shown in Fig. 1-13, contains the vertical input and output circuits. Simply look up the part number stamped on

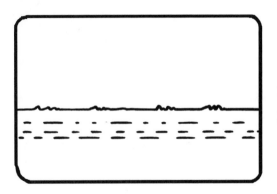

1-11
Remove the antenna cable to determine if the interference lines are in the TV or picked up.

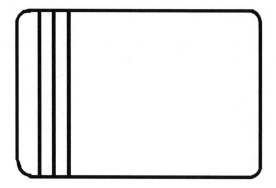

1-12
Black lines to the left of the TV
screen are caused by defective
parts in the chassis.

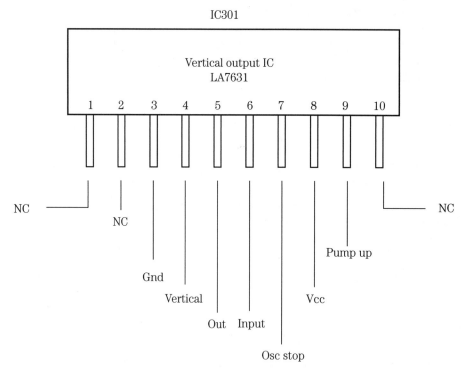

1-13 The vertical output IC301 in a Goldstar TV was replaced with a universal NTE1797 replacement.

the body area. The part number will tell you not only what circuit it works in but also the correct replacement.

Simply look up the part number in a semiconductor manual for the correct universal replacement. You can look at another chassis of the same brand, locate the same IC, and compare the voltage measurements with the other schematic. Not only will the part number indicate what circuit it operates in, but it also will provide the correct working voltage and terminal numbers.

Part numbers and letters found on ICs and transistors will help you to locate the correct circuit or component and sometimes the operating voltages. RCA, GE, Sylvania, Workman, and NTE semiconductor replacement manuals cover American, European, and Japanese solid-state devices. These manuals can be obtained from manufacturers, electronics stores, dealers, and mail-order firms.

If the exact schematic is not available, look on another schematic from the same manufacturer using the same part number. Compare the voltage measurements with those of the defective chassis. Sometimes the same part in other brands will have about the same voltage measurements.

Resistor color codes

You should be able to learn the color codes of the different resistors and by glancing at a suspected resistor read off the correct resistance. After many hours and weeks of servicing electronic chassis, the electronics technician can easily quote the resistance and tolerances by looking at the color stripes on a defective resistor. Sometimes if the part is burned or charred, the color may change and give a different resistance reading. Always replace the suspected resistor with one with the exact resistance and wattage rating (Table 1-1).

Table 1-1.   Fixed resistance.

FIXED			RESISTOR
First band	Second band	Third band	Fourth band
Color code	First band	Second band	Fourth band multiplier
Black		0	× 1 ohm
Brown	1	1	× 10 ohms
Red	2	2	× 100
Orange	3	3	× 1,000
Yellow	4	4	× 10,000
Green	5	5	× 100,000
Blue	6	6	× 1,000,000
Violet	7	7	× 10,000,000
Gray	8	8	
White	9	9	

Table 1-2.   Flat capacitor color code.

Numbers	Tolerance
Black: 0	Black: 20%
Brown: 1	Silver: 10%
Red: 2	Gold: 5%
Orange: 3	Red: 2%
Yellow: 4	Brown: 1%
Green: 5	
Violet: 7	
Gray: 8	
White: 9	

Flat capacitor color codes

The flat-type bypass capacitor might have only color bands on it with capacity and rated tolerance. This flat-type capacitor has no particular polarity and can be connected with any type of mounting. The first color at the top is the first number of the capacity. The second color is the next number, and the third color is the multiplier. The fourth color is the tolerance in percentage, and the last color is the working voltage (Table 1-2).

For instance, a 0.015-μF bypass capacitor will have a brown color at the top; next will be the green, yellow, black, and red. The yellow color is the multiplier, black is the percentage in tolerance, and red is the operating voltage (Fig. 1-14).

Foreign capacitors

Sometimes a foreign electronic product might have the capacity stamped on the chassis or listed on the schematic. All capacitors should be replaced with the same capacity and voltage or a higher working voltage. Check foreign capacitors against their American replacements as shown in Table 1-3.

Dogs of all kinds

The "tough dog" problem takes a lot of valuable service time while you scratch your head for the answer. For the headache, you can take an aspirin or Tylenol to relieve the pain. When servicing a "tough dog" problem in the TV, audio, or VCR chassis, nothing relieves the pressure or pain until the component culprit is located. A "tough dog" may end up with more than one trouble or component in the TV chassis.

"Tough dogs" are very difficult to locate without a schematic. In many cases the "tough dog" is right under our nose, so to speak. Isolate the symptom to a given section of the chassis, and take critical voltage and waveform readings. Often the difficult chassis may have an intermittent component that will only act up once a week or twice a month. These sets should stay in the home until the intermittent state acts up several times a day. Just wait, and they will begin to act up more often.

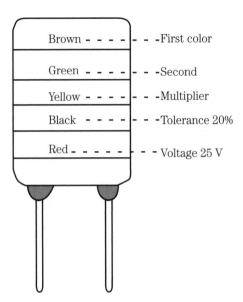

Brown - - - - - - - First color

Green - - - - - - -Second

Yellow - - - - - - -Multiplier

Black - - - - - - -Tolerance 20%

Red - - - - - - - Voltage 25 V

1-14
The flat 0.015-μF bypass capacitor
colors are marked on the body of
the capacitor.

Table 1-3.   Foreign capacitors.

Foreign (in nF)	U.S. (in μF)	U.S. (in pF)
0.001	0.000001	1
0.01	0.00001	10
0.1	0.0001	100
1	0.001	1,000
10	0.01	10,000
1,000	1	1,000,000
10,000	10	10,000,000
100,000	100	100,000,000

The intermittent chassis can result from poorly soldered connections, intermittent components, cracked boards, and poor terminal connections. Twist and push around on the board until you have located the intermittent section. Check all component soldered connections in that area. Sometimes a wholesale resoldering of the entire section may uncover the intermittent connection.

A large blob of solder may contain a poorly tinned component lead. Often a component lead that is not properly tinned or to which solder will not cling ends up a poor contact. Remove the large blob of solder with solder wick or a sucking tool. Remove the terminal lead. Clean it up, and tin the connections. Remount and resolder.

Foil wiring breakage around board standoffs can cause intermittent operation in a VCR. Inspect the PC wiring around where heavy components such as tuners, heat sinks, and transformers are mounted. If the unit is dropped in shipment or on delivery, these heavy components can cause cracked PC wiring (Fig. 1-15).

1-15   Check around transformers and heavy components for broken PC wiring in a VCR.

The surface-mounted components found in almost every consumer electronic product can cause an intermittent or dead service symptom. A poorly soldered end connection can result in a dead or intermittent circuit. A continuity check with the capacitor equivalent series resistance (ESR) meter can locate a high resistance or poor end connection of a surface-mounted device (SMD). Likewise, the ESR meter can locate broken or cracked PC boards (Fig. 1-16).

Another method is to spray coolant directly on the body of the component or parts that require an hour or so of heat-up time. Then apply hot air from an air gun or hair dryer and see if the part returns to the intermittent state. In the end, the "tough dog" symptom can cause high tension, headaches, and an "I do not know anything about electronics" feeling.

Cool nights and hot days

A coolant can help to locate thermal intermittence in transistors, ICs, resistors, capacitors, and SMD components. By applying several coats of coolant on a suspected part, sometimes one can make the intermittent component act up or return to its natural state. Coolant sprayed directly on transistors, ICs, capacitors, and SMD parts can locate the intermittent component. Choose a coolant that will not leave a residue or harm the plastics on PC boards (Fig. 1-17).

The no-residue cleaner or flux remover is especially effective in removing activated fluxes, dirt, and other ionic contaminants found on PC boards. Flux cleaner is ideal for

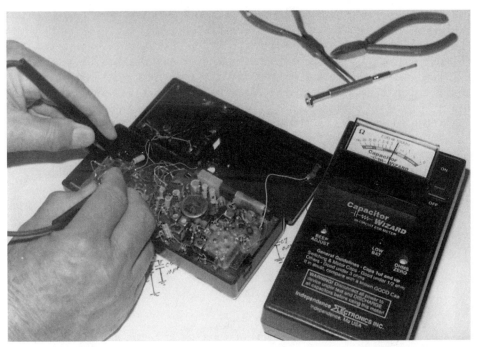

1-16   Use the ESR meter to check SMD part connections and cracked wiring.

1-17   Use the correct coolant to make the intermittent component act up.

use on electronic equipment, PC boards, relays, and soldered connections. Most flux spray cans leave no residue and will not harm plastics.

Choose a superwash spray that thoroughly removes dirt and grease from parts and assemblies. A can of wash can be used on tape heads, motors, relays, and PC boards.

Superwash head cleaner safely removes carbon deposits, oil, dirt, nicotine, dust, and other contaminants. The spray is ideal for sensitive magnetic and optical pickup heads. Clean cassette tape head and CD devices with head and disc cleaner. Choose a tuner chemical that cleans and lubricates corroded contacts. This spray will improve switch-type tuner performance and reliability. Choose a tuner spray that will not harm plastics and is not conductive.

Sometimes by applying heat to the body of an intermittent capacitor, transistor, or IC, the component might began to act up. A little heat from a soldering iron or gun close to the part can do the trick. Choose a portable hair dryer or heat gun to blow hot air directly on the suspected component. A heat gun with a flexible tube can blow hot air directly on the part without damage to other components. Applying heat and coolants to intermittent components can make them act up and solve a very tough intermittent symptom.

Double trouble

Here we are not talking about chewing gum when two or more components are defective within a TV, VCR, or cassette player. Often the electronic chassis has only one trouble symptom and one defective component. When you repair one problem and another appears, this is double trouble. Sometimes you may find more than two defective components in one section of a TV chassis.

For instance, the RCA CTC169 chassis was dead, and this was caused by a over-heated horizontal output transistor (Q4401). After replacing the output transistor, the chassis was still dead. A defective capacitor (C4402, 0.47 μF, 400 V) was located with the ESR meter. Finally, before the chassis came to life, IC4401 was replaced. All these components had to be replaced before the TV chassis was repaired (Fig. 1-18).

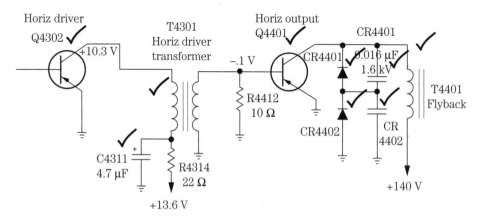

1-18 Check the indicated components within the RCA CTC169 chassis for horizontal problems.

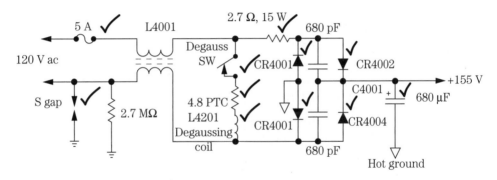

1-19 Check the indicated parts in an ac power supply for lightning damage.

Lightning damage

When a lightning- or storm-damaged chassis appears on the service bench, look for many damaged components. Power line outage conditions can damage any electronic product connected to the ac line. A direct hit can cause the unit to be totaled out. Often lightning damage comes in through the antenna or ac power cord and destroys components within these sections.

Look for burned or damaged antenna coupling capacitors and scorched cable wiring and terminals. A damaged ac receptacle and power cord and damaged line voltage regulator circuits in a TV chassis can result from lightning or power outage damage. Carefully check the entire PC wiring for pulled or lifted sections. After all repairs are completed, let the chassis run for several days before returning it to the cabinet (Fig. 1-19).

Down below

Do not forget to look underneath the chassis for defective boards, connections, and components. Look for possible overheated terminals, brown board areas, and poorly soldered connections on the PC board wiring side. These areas will show overheated parts and intermittent connections. Check the terminals on the component side to locate a possible defective component (Fig. 1-20).

Burned or stripped PC board wiring might indicate defective components, a power surge, or lightning damage. Several different blown-off areas of PC board wiring might indicate lightning damage. Often this means that the chassis is totaled and that complete set replacement is required.

Cracked or damaged boards might be visible by looking closely with a magnifying glass. A strong light above the board, with light showing through the board, can help locate fine cracks in PC board wiring. A magnifying circle light is ideal for PC board wiring inspection. Going from terminal to terminal with the low-ohm scale of a digital multimeter (DMM) might help to locate open PC board wiring. The ESR meter is great in locating broken PC wiring.

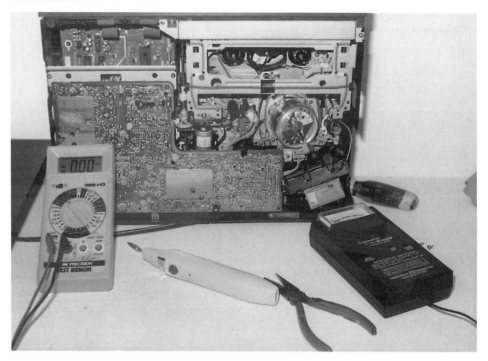

1-20 Try to locate the poor connections of a defective component on the PC wiring.

Outside the chassis

Often a separated PC board might be found mounted horizontally on a TV chassis and can show broken contacts where the board is connected to the main board chassis. In the latest high-powered car radios you will find several PC boards stacked on top of one another. You may have to remove several boards, soldered ground connections, and cable connections to get to the board that contains the defective component. A color symptom in a TV might result from a bad board or component on the CRT board outside the TV chassis (Fig. 1-21).

Check for defective components outside the regular chassis when all other tests are normal. For instance, a horizontal white line or no vertical sweep might be caused by an open winding or poor connection of the yoke assembly. Red missing from the raster might be caused by a defective color transistor or R-Y circuit located on the CRT socket board. Look outside the regular circuits when all other tests fail.

Do it today, not tomorrow

Do not set that "tough dog" product aside for several weeks before you attempt to repair it. Callbacks or electronic products returned for the second time can be put off until tomorrow, but tackle the "tough dog" problems at once. Often the repair time

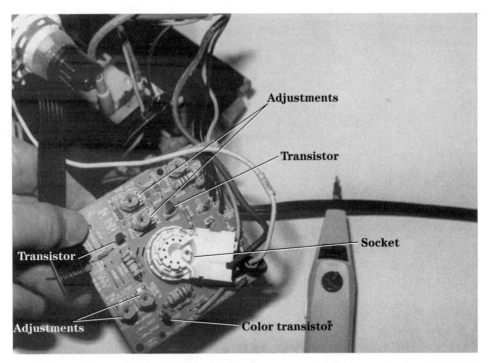

1-21 Check the indicated parts on a CRT board for missing colors and a poor black-and-white picture.

involved may be less than 1 hour. Don't put off today and do it tomorrow. It may never get done.

When you have repaired a boom-box cassette player for distorted sound and the unit is returned with another problem such as the cassette player is now eating tape, make another charge. Of course, the customer is angry because the repair lasted only a few days. Any electronic technician knows that the sound distortion problem is a long way away from the mechanical tape-eating symptom.

However, did you clean the tape heads, capstan, and pinch roller while the player was in for repair (Fig. 1-22)? Sometimes we forget to do preventive maintenance. A good cleanup and a thorough preventive maintenance check can save a lot of headaches. Besides a good cleanup, you should check the type of cassette and the condition of the suspected cassette, which may be defective. The customer is never wrong, but he or she may not know how cassettes operate or may mishandle the cassettes.

Intermittent problems

Intermittent service problems take a lot of service time to locate the defective component. The intermittent symptom might even turn into a "tough dog" problem. When TV audio cuts up and down, check the audio circuits on through to the loud speaker. An intermittent loss of color and intermittent shrinkage of vertical height at the top and bottom of the screen in a RCA CTC177 chassis might result from poorly soldered joints in

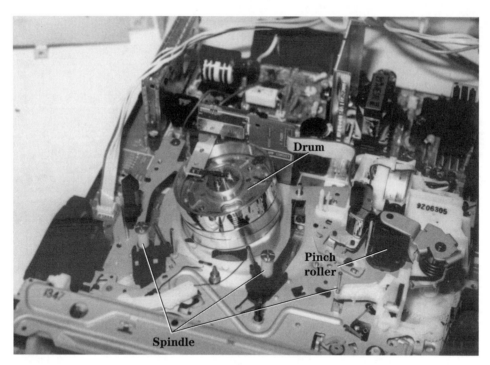

1-22   Clean the cassette player tape head and tape path with cleaning fluid.

the tuner. Although the color and vertical components are miles apart, so to speak, this service problem does occur.

The TV chassis that starts up and intermittently shuts off can be the result of poorly soldered joints on the horizontal driver transformer. A TV that intermittently shuts down but then turns back on and goes through auto programming can be the result of badly soldered joints on the shielded cover of the microprocessor. Intermittent brightness and height and audio volume that varies in a TV can be caused by badly soldered connections on an SMD resistor.

Intermittent symptoms must be isolated to a certain section within the electronic product. The intermittent component can be located with signal-tracing methods, voltage monitoring, and waveforms taken with an oscilloscope (Fig. 1-23). Isolate the defective part when the intermittence occurs by taking critical voltage tests. Spray coolant and apply heat to help make the intermittent part act up. Remember that the intermittent part might operate normally for days and the next hour start cutting up and down. Intermittence troubleshooting can be tough.

Time is precious

Time is one of the most important factors for the electronics technician. Time marches quickly when you are stuck with a difficult problem and no schematic. It might take months to receive the correct diagram and sometimes never on some older products. Wasted time is wasted dollars.

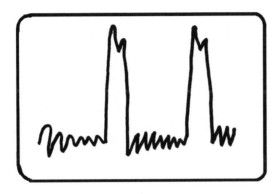

1-23
The horizontal output waveform
with the scope probe alongside
the flyback.

Servicing a radio or stereo chassis might require more time for a TV technician. Some TV technicians were not trained for audio-type service. Time can be wasted and never recovered when working on a "tough dog" problem. In the following pages I hope to reduce the time it takes you to troubleshoot and repair consumer electronic products without a schematic.

Test points

You will find many different tests points within a TV, CD player, or VCR player to take signal, waveform, and voltage measurements. If you can find the various test points in a given section, the circuits can be monitored for easy servicing. The test points might only consist of a lug, spot, or raised area that any test instrument can be connected to.

The test points within a CD player provide quick attachment of the scope for troubleshooting and making critical adjustments. Test points are found throughout camcorder circuits for adjustments, waveforms, and testing procedures. The test points in a TV chassis may be located in the video, radiofrequency/intermediate frequency (RF/IF), autokinne bias, and central processing unit (CPU) circuits (Fig. 1-24).

Say it isn't so

Have you ever checked the electronic chassis, ordered a new part, waited for the new replacement, installed it, turned on the switch, and the results are the same? This just cannot happen after installing a new IC component. Look at all the time lost in removing and installing the IC replacement. This can occur to the best electronics technician when all components tied to the terminal of the suspected IC are not checked thoroughly.

What if the voltages measured on each pin terminal of an IC were quite normal with a fairly normal input signal and no output waveform? Even the supply voltage (Vcc) was quite normal. The IC just has to be defective. You install a new IC after removing the old one and cleaning up each PC terminal connection. The results are the same—no output signal.

So you again make voltage, resistance, and waveform tests (Fig. 1-25). Everything appears normal. Maybe a piece of foil was cracked or broken. All terminal pins

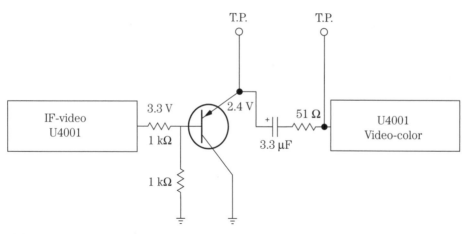

1-24 Test points are found throughout some TV circuits for signal and waveform tests.

1-25 Taking critical voltage tests on the bottom side of a suspected IC component.

are soldered up again. Resistance tests are made from each pin terminal to a common point on the same PC foil. All tests are good. What went wrong?

No doubt the new IC replacement is defective. However, this never happens because you test every component before installing it for this very reason. Of course, it is difficult to test the IC with so many terminals. So you just install it and take a chance. Nine times out of ten you never have to replace a replacement. These operations just happen in the life of an electronics technician. Always test each component before installation.

Ten points of servicing

Start with the symptoms provided by the electronic product. In a TV chassis, use the symptoms found on the TV screen and speaker. An insufficient vertical raster indicates trouble in the vertical section. Pulled width might be caused by the low-voltage power supply, horizontal circuits, or pincushion circuits. Intermittent sound might be caused by a defective audio IC.

Start with the symptom, and isolate the defective component. Break the circuits down, and list the probable circuits where the defective part may be located. Then make voltage, resistance, and solid-state tests in those circuits to locate the defective part.

To locate the defective section, you must know where each section is located on the chassis without the help of a schematic. For instance, if a tape player has weak and distorted sound, with normal AM and FM stereo reception, the audiofrequency (AF) and audio output circuits are normal. The defective component must be between the tape head and the stereo function switch. The most obvious things to check are a dirty tape head, electrolytic coupling capacitors, preamp transistors or ICs, and the supply voltage to the tape head circuits (Fig. 1-26).

Locate all large components on the chassis to find the defective circuits. Around large filter capacitors are the low-voltage power supply components. Look at the large transistors or IC audio output components for defective sound circuits. Vertical circuits are nearby when locating the vertical output transistors. The horizontal output transistors and regulators might be found on a large separate or chassis heat sink. Try to locate large parts in each section that point to the correct symptom.

After locating the defective circuits, take voltage and resistance measurements. Some technicians check solid-state parts first. The most accurate instrument for taking voltage and resistance measurements within a solid-state chassis is the digital multimeter (DMM) (Fig. 1-27). Voltage and resistance measurements of less than 1 V or 1 ohm are required within the solid-state chassis. Accurate base bias voltages (0.3 to 0.6 V) between base and emitter terminals indicate not only if the transistor is good but also if it is a germanium or silicon type.

Next, scope waveforms are needed to signal trace the various circuits. When the horizontal output circuits are functioning, the horizontal waveform can be traced from the vertical oscillator, amp, and output to the vertical yoke winding. Low distortion can be found in the stereo circuits with a sine or square wave at the input. Scope the audio stages for clipping and distortion.

After several tests, the defective part must be removed from the circuit. Desoldering equipment and solid mesh or wick material can be used to remove excess solder. Be careful when removing transistor pins so as not to pull or pop off thin PC board wiring connections. Too much heat might dislodge or destroy small PC board wiring. The controlled electronic soldering system or station is ideal for removing and replacing PC board–mounted components.

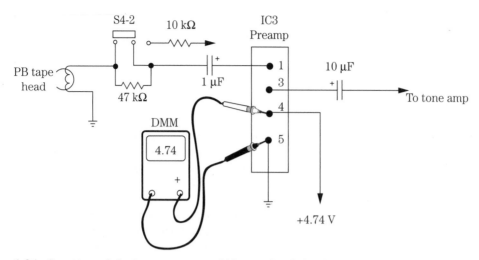

1-26 Locating a defective component within tape head circuits with critical voltage tests.

1-27 You can take critical voltage measurements with a digital multimeter (DMM).

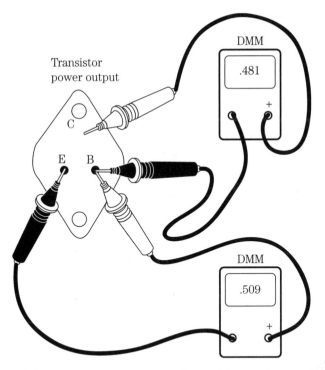

Transistor
power output

DMM
.481
+

DMM
.509
+

1-28 Checking an output transistor with a diode test of the DMM.

Check the suspected part after it is removed from the chassis. Sometimes the intermittent transistor may be normal after heat is applied and it is removed from the circuit. Replace it. Transistor, continuity, resistance, and special equipment tests may indicate a defective part after removal.

Test the new component before installing. Check transistors, capacitors, resistors, choke coils, diodes, and zener diodes before installation (Fig. 1-28). Make sure the parts are properly installed before soldering. By checking the part before installation, you may save a lot of valuable service time and confusion.

Install the new part with careful soldering of the terminals. Too much solder can overlap between terminals and short out the component when the chassis is turned on. Too much heat may damage a semiconductor. Clean out all solder and rosin flux between terminals.

Double-check all terminals. Are they all placed on the correct PC board wiring? Remember, the transistor or IC component can be soldered with reversed terminals. Check the wiring connections. Make sure that no cold joints or interconnected connections are made. Check and double-check before turning on the chassis.

The following steps must be taken to locate, isolate, repair, and replace a defective component:

1. Check the symptoms.
2. Isolate the circuits.
3. Locate the part.
4. Take voltage and resistance tests.

5. Perform scope tests.
6. Remove the defective part.
7. Check the defective part.
8. Test the new part.
9. Install the new part.
10. Clean up and double-check.

Safety measures

Excessive lightning or power outage damage can destroy a TV set, VCR, receiver, or cassette player. Extensive damage to the low-voltage circuits and tuner in a TV can total the entire chassis. When the 220-V neutral wire breaks before entering the home, it can place extra ac voltage on any electronic chassis. High–line voltage wires touching one another may result in high-power outage.

Always use an isolation transformer when servicing present-day TVs and CD and VCR players (Fig. 1-29). Most of these units are operated directly from the power line, and the chassis or some point in the chassis may be hot. Test instruments can be damaged as well as the product that you are working on. Besides, you might receive a shock that can cause damage to your body, or you might accidentally pull an expensive test instrument off the service bench.

Isolation transformer

1-29   Always use an isolation transformer when servicing electronic products.

When a TV set explodes and sets the house on fire, this can be very dangerous and fatal to the homeowner. If the fire spreads to the window curtains and to other rooms, someone can be trapped in the raging fire. Houses can be rebuilt, but a loved one is lost forever.

The electronics technician always should be careful when servicing an electronic chassis. Replace all safety-marked components with exact replacements. Never replace a flameproof resistor with a carbon resistor. Replace all line power supply and horizontal circuits with correct components. Make sure that cables and wires will not lie on hot components in the chassis. Inspect and replace a defective power cord. Always replace a fuse with one of the exact amperage.

Do not drop metal screws and small metal parts down into the maze of wiring and components and then forget to remove them before buttoning up the chassis. Make good, clean soldered connections. Botched wiring runs and poorly soldered connections may give the fire department and state commission an excuse to blame the electronics technician. In case of a death, you may lose everything you have worked for. Be careful out there.

2
CHAPTER

How to locate, test, and repair

After determining what the symptom of the electronic product is, you must locate the section that has the defective part before you can remove and replace that part. With a horizontal white line in a TV chassis, representing no vertical sweep, the defective component is found in the vertical section of the chassis. When the raster and picture are pulled in on the sides, you know the trouble lies in the horizontal output stages or insufficient voltage applied to the horizontal section from the low-voltage power supply.

If the TV chassis is dead and will not start up, check for an open fuse, badly soldered driver transformer, and problems within the low-voltage power supply. After determining what the symptom is, go directly to that section and make voltage, waveform, and component tests.

Remember, you don't have a schematic diagram to locate the suspected component with accurate voltage measurements. Make voltage, resistance, scope, continuity, and semiconductor tests to determine if a part is defective.

With only a DMM

One of the most useful test instruments in electronics servicing is the digital multimeter (DMM). The DMM can be used to make critical voltage, resistance, and current tests. The deluxe DMM can be used to test capacitors, diodes, and transistors. Besides checking out fixed diodes, the DMM diode test can be used quickly and efficiently to test each transistor right in the circuit. The DMM can determine the real low resistance of transistor bias resistors under 1 ohm. Besides taking critical dc voltages, the DMM can make accurate ac voltage measurements. Some DMMs also can check frequency, inductance, and capacity (Fig. 2-1).

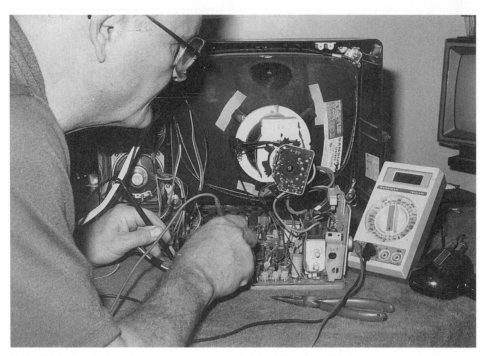

2-1 Taking critical resistance measurements with a DMM in a TV chassis.

Correct electronic test equipment

Besides the DMM, the oscilloscope is one of the most used instruments to test for correct waveforms on the service bench. You can easily locate a defective component with critical waveforms in many different consumer electronics products. A dual-trace oscilloscope is needed with at least a 35- or 40-MHz frequency bandwidth (higher in some cases). Knowing how to use the scope will save a lot of time on the service bench. Don't let the scope sit idle; use it. Besides regular hand tools, special and additional test equipment is listed in each chapter.

REQUIRED TEST EQUIPMENT FOR TV SERVICING

1. DMM
2. Oscilloscope (Fig. 2-2)
3. Semiconductor tester
4. Equivalent series resistance (ESR) meter
5. Audio signal generator
6. Sweep generator
7. Color-dot-bar generator
8. Radiofrequency (RF) signal generator
9. Frequency counter

2-2   The oscilloscope can quickly check critical waveforms within an electronic chassis.

10. Capacitance meter
11. Cathode-ray tube (CRT) tester and rejuvenator
12. High-voltage probe
13. Isolation transformer
14. Soldering station
15. Desoldering station

REQUIRED TEST EQUIPMENT FOR RADIOS AND CASSETTE PLAYERS

1. DMM (Fig. 2-3)
2. Volt-ohm-milliammeter (VOM)
3. Semiconductor tester
4. ESR meter
5. Signal tracer
6. Signal generator
7. Frequency counter
8. Test cassettes (1, 3, 10 kHz)
9. Soldering station
10. Isolation transformer
11. Dc power supply

2-3 Using a DMM to check resistance of an auto receiver–cassette player tape head.

REQUIRED TEST EQUIPMENT FOR CD PLAYER TROUBLESHOOTING

1. DMM
2. Semiconductor tester
3. ESR meter
4. Dual-trace oscilloscope (Fig. 2-4)
5. Optical power meter
6. Low-frequency audiofrequency (AF) oscillator
7. Signal generator
8. Frequency counter
9. Test discs
10. Several special tools, manufacturers' jigs, wrist straps, etc.
11. Variable isolation transformer

REQUIRED TEST EQUIPMENT FOR VCR REPAIR

1. DMM
2. National Television System Committee (NTSC) pattern
3. Ac millivoltmeter (RMS)

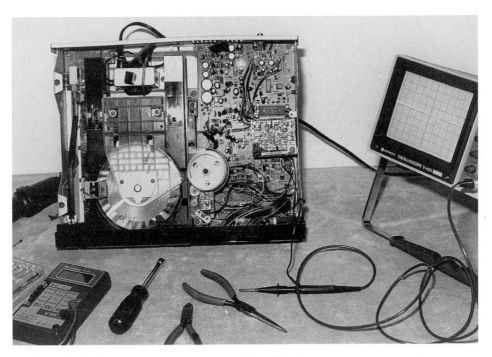

2-4 The dual-trace scope is ideal when servicing CD players.

4. Frequency counter
5. Dc power supply
6. Oscilloscope
7. Isolation transformer
8. Alignment table
9. Blank tape
10. Special manufacturers' jigs
11. Soldering station
12. ESR meter (Fig. 2-5)

Who's on first?

What is the first thing you should do? What is the symptom? You must know how the unit is acting up before troubleshooting the chassis. For instance, a dead TV might be caused by an open fuse, the low-voltage power supply, or the horizontal output circuits. A white line on the raster of a portable TV indicates a problem of insufficient sweep in the vertical section. Sides pulled in on the raster might indicate a defective low-voltage power supply, horizontal sweep, or pincushion circuits.

A dead symptom in stereo radio circuits might indicate a defective low-voltage power supply to the audio output transistor, integrated circuit (IC), or speaker. Excessive hum within a large stereo amplifier might be caused by defective filter capacitors. Pickup hum might point to worn input cables or circuits. No rotation of the disc in a CD

2-5 The ESR meter can quickly test capacitors and printed circuit wiring traces within a VCR.

player might be caused by a bad spindle motor or drive circuits. Use all the symptoms to locate the defective section within the electronic product.

Case histories

Keeping track of what occurs, such as an unusual symptom or a "tough dog," can help the next time around. A case history on a certain electronic product can help on another chassis with the same symptom. Case histories should be looked at after determining what section the symptoms point to. Keep all cases in a card file, on computer, or right on the schematic. Remember, the case history of one unit may help to solve problems in other chassis. Don't forget to write down or otherwise record all "tough dog" case histories.

A COMMON TV CASE HISTORY

An erratic or intermittent startup symptom with chassis shutdown in a TV chassis can be caused by poorly soldered joints on the horizontal driver transformer. Resolder all terminals on the driver transformer the first thing with no startup, intermittent startup, and shutdown problems. This same symptom occurs in just about every TV chassis (Fig. 2-6).

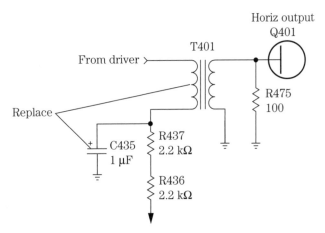

2-6 After replacing Q401, repeated failures of Q401 and IC501 (STR-30110) were caused by C435 and T401.

After replacing the horizontal output transistor for a dead symptom and the output transistor becomes shorted once again, solder up all terminals of the driver transformer. Check the resistance of the voltage resistor feeding the primary winding of the driver transformer. Replace the bypass or electrolytic capacitor at the transformer voltage resistor.

Suspect a defective flyback or poorly soldered connection on the horizontal output transformer. Scope for insufficient horizontal drive signal at the base terminal of the output transistor. Last but not least, check for a defective safety capacitor. Most of these service problems can happen in any TV chassis.

A COMMON RADIO-CASSETTE CASE HISTORY

Speed problems within the cassette player occur at one time or another in virtually all radio-cassette players. A stretched rubber drive belt or oil on the belt can cause improper speed. Worn rubber tires on the idler wheel can cause slow or erratic speed. A dry or worn spindle flywheel can cause improper speed. Check cassette speed problems with a test cassette (1 kHz) and a frequency counter to determine the correct speed.

A COMMON VCR CASE HISTORY

A defective capstan motor can cause many different problems within a VCR. A bad capstan motor or a frozen capstan bearing can cause erratic motor speed problems. A defective capstan motor with a flat shaft may not start up the next time. The bad capstan motor can develop slow speed rotation. A defective capstan motor also might cause improper or no fast-forward or rewind or shutdown problems. Just keep track of the unusual symptoms that can result from a defective capstan motor and write them down on the schematic.

Isolation

Look the chassis over for burned, damaged, or cracked components within the section that possibly can be the cause of the symptom. Inspect critical traces or printed circuit (PC) wiring on the bottom side of the chassis for overheated terminals and badly soldered spots. Usually a large blob of solder can cover up a poorly soldered terminal from a resistor or capacitor. Overheated parts can be isolated by touch. Touching the chassis and parts can indicate the intermittent component.

After isolating the potentially defective section, isolate the defective part with transistor, IC, resistance, scope, and continuity tests. Test the transistors in or out of the circuit with a transistor-diode test using a DMM or a beta transistor tester. Check all suspected ICs with signal in and out, voltage, and resistance measurements. Before you can pinpoint the defective part, you must locate it on the chassis without a schematic.

Location, location

After isolating the various symptoms, find the correct section in which the trouble might occur. Look for large components such as filter capacitors for the low-voltage power source, transistors mounted on separate heat sinks inside the chassis for vertical output transistors or ICs, and horizontal output transistors on separate heat sinks or metal chassis for the horizontal output circuits (Fig. 2-7).

2-7 Checking the voltage on the bottom pins of horizontal deflection IC in a TV chassis.

Another method for determining if you are in the correct section is to take the numbers located on the transistors and IC components and look them up in a semiconductor replacement manual. The manual tells you what pin goes where and what circuit it ties into on the chassis.

Look at the outline drawing for the correct pin connections. For instance, part number AN5435 lists as a color deflection signal processor IC in the RCA series replacement manual. The universal replacement is an SK9299 IC with pin 15 as the supply voltage pin (Vcc). The horizontal drive output is pin 6, and the vertical output is pin 9. Also, the SK9299 has a total of 18 pins. Double-check with the IC mounted on the chassis for the correct number of pins, the right section, and the correct universal replacement.

Let's say you have a Panasonic portable TV that comes in without a model number. The model number has been torn off, but it looks like a fairly new set, one that you have never worked on before. The symptom is a tic-tic sound with B+ present at the horizontal output transistor and no high voltage.

By supplying a horizontal pulse at the horizontal driver, you make the set run, indicating that the horizontal output stages and high-voltage and flyback circuits are normal. You think that the AN5435 IC is the horizontal/vertical countdown IC. But what pin is the voltage source (Vcc)? You look it up in the semiconductor replacement manual.

The AN5435 IC is a color TV deflection signal processor and can be replaced with an RCA SK9299 IC. Now you have located the correct IC and section where the trouble is. Before replacing the IC, take voltage measurements on all the IC pins. Usually the highest voltage is the supply pin (Vcc). Supply voltage to the deflection IC is usually from 10 to 25 V. In this case all pins are low in voltage. So you check the highest-voltage pin and trace it back to the low-voltage source. If the low-voltage source is supplied through flyback-derived secondary sources, you are out of luck. Look up the supply voltage pin on the outline drawing of the replacement SK9299 IC. Pin 6 ties to the supply voltage source.

Inject an external 10 or 12 V at pin 6 with the scope probe on the horizontal driver (pin 15). If the horizontal waveform is normal, the IC is good. Double-check the vertical output drive waveform at pin 9. If there is no waveform at either pin, suspect a defective IC or leaky components tied to the IC pins and common ground. Replace the IC with a universal replacement if the original part number is not available. Sometimes the IC's voltage and resistance measurements are normal, but there are still no waveforms. Replace the IC at once.

Check the voltage at pin 6. Low or no voltage can indicate a leaky IC, improper supply voltage, or no voltage from the flyback secondary source. Trace the PC board wiring back to the supply source to locate a defective isolation resistor, silicon diode, or small electrolytic capacitor. By simply checking the number on the IC or transistor you can determine the section where the trouble is located, the correct pin, and the total number of pins.

LOCATING CRITICAL TV COMPONENTS

When the front of a TV screen indicates that the trouble might be in the vertical section, such as an improper sweep, locate the vertical section on the TV chassis (Fig. 2-8). Today, most TV chassis use an IC output component on a heat sink for the vertical output circuits. The TV chassis can be broken up into the horizontal, vertical, tuner,

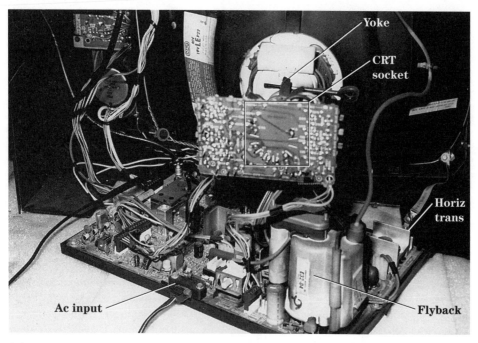

Labels on image: Yoke, CRT socket, Horiz trans, Flyback, Ac input

2-8 Try to locate the various components on the TV chassis.

intermediate frequency (IF) and video, audio, and CRT sections and high- and low-voltage power supplies. Remember that within a TV chassis, sometimes the horizontal section must operate before any voltages are fed to the secondary windings of the flyback to the other circuits.

LOCATING CRITICAL COMPONENTS ON A CASSETTE PLAYER

Take a quick peek at the cassette player, and go directly to the components that the symptom may suggest. The cassette motor and tape head are easily identified (Fig. 2-9). The preamp stage follows the tape head, and the audio output circuits are tied to the permanent magnet (PM) speaker.

If the tape player is running slowly, go directly to the cassette motor, belt, and revolving pulleys. A good cleanup can solve most speed problems. When the audio is dead in the right channel, locate the audio output IC on a heat sink, and make critical voltage and semiconductor tests.

LOCATING TV/VCR COMPONENTS

When both the TV and VCR are dead with no tape movement, go directly to the fuse block or low-voltage power supply. Determine if both units work from one power supply. Sometimes a separate power supply is found for the VCR circuits and another dc supply for the TV chassis. Then take critical voltage measurements.

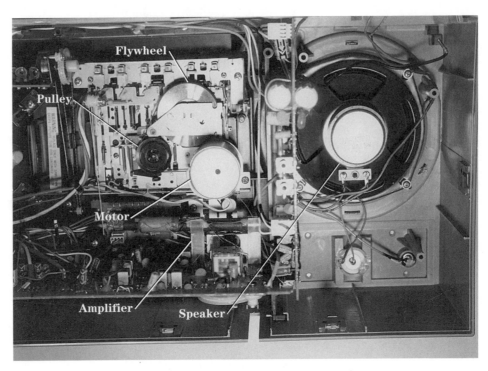

2-9   Locating the various parts on a radio-cassette player.

When the cylinder motor runs and then shuts down, suspect a bad cylinder motor or a cylinder motor drive IC. Take critical voltage measurements on the cylinder or drum motor terminals with no motor rotation (Fig. 2-10).

LOCATING CD COMPONENTS

Look over the CD chassis and try to locate the various CD parts. Go directly to the largest electrolytic capacitor when excessive hum is heard in the speakers with the volume turned down (Fig. 2-11). Often the main filter capacitor is from 2200 to 4300 μF capacity and is the largest part on the chassis. Clean up the lens assembly when the CD player starts skipping. Start at the digital-to-analog (D/A) IC output when one channel of audio is weak or intermittent. Take critical voltage measurements on the suspected transistor or IC component.

Out of bounds

Check outside the chassis for defective components. If the TV screen is all red when the color control is turned way up, check for defective components on the CRT board on the neck of the picture tube. Don't overlook defective components that are not on the main chassis. Look for defective transformers, power transistors, voltage regulators, ICs, capacitors, and separate metal heat sinks or chassis. Check for problems in circuits

2-10   Locate the drum or cylinder motor on the VCR section of a TV/VCR.

2-11   The CD player might be inside a boom-box player or a separate portable unit.

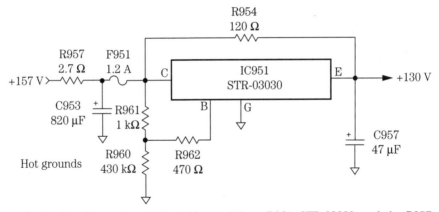

2-12 Besides blown fuse F951, bridge rectifiers, R961, STR-03030, and the R957 (2.7-ohm) resistor in a Mitsubishi CK-2602R TV were hit by lightning.

on separate PC boards that are soldered to the main PC board. Suspect poorly soldered connections from the horizontal PC board to the main chassis.

Lightning damage in a TV chassis

Sometimes lightning or storm damage may only blow the main fuse. When a TV is hit directly on the power line, you will find many parts damaged. A power outage also can cause damage to many components on a TV chassis. If the house fuse box is blown out of the cabinet with several light bulbs busted and the refrigerator will not run, suspect extreme damage to the TV and sound equipment.

Carefully inspect the antenna terminals and isolation capacitors on the tuner assembly of a TV. Notice if black and burned marks are found at the ac interlock and if several components are damaged in the power supply (Fig. 2-12). Peek at the PC board and notice any burned spots. Also look to see if any PC traces are blown or pulled up from the board. Remember that the low-voltage power supply must operate before any sections function. After all repairs are completed, let the chassis operate for several days before returning it to everyday service.

Burned and damaged

Burned or damaged parts are seen easily on a PC board. However, you might have to blow off dust to see the burned components. Sometimes you can smell burned or charred parts. The PC board might have to be repaired if it has been burned and damaged.

Large parts such as transformers, electrolytic capacitors, and power output transistors can be seen easily. Check for a part number on the body. It is difficult to determine the total resistance in charred resistors. If the color code cannot be seen, try to locate a schematic similar to the chassis on the bench.

In many cases the resistance may be close in value for several different known chassis in the same circuit. For instance, two charred boost-dropping voltage resistors were found in a Sony portable TV chassis. Because the correct schematic was not available, another Sony portable (KV-1747R) circuit was used that was quite similar. There was no doubt that R720 and R715 were damaged when the picture tube arced over with excessive high voltage. Make sure that the circuits are similar and have the correct operating voltage.

Ac input hot grounds

Today, the ac circuits have hot grounds above the common ground in TV chassis. Notice that the ac input degaussing circuit, silicon diodes, and main filter capacitor are in the hot ground circuits. When taking ac or dc voltage measurements in these circuits, make sure that the black voltage probe is on the hot ground side of the filter capacitor. To determine if the low-voltage circuits are functioning, take a dc voltage measurement across the main filter capacitor (Fig. 2-13). Do not take the voltage from the common chassis ground because it will not be correct.

TV SMPS hot grounds

Within the switching power supply or switch-mode power supply (SMPS), all components on the primary side are connected to the hot ground. Simply locate the switching transformer as the biggest part on the chassis. Take all critical voltage measurements on the switching regulator IC, transformer, and transistors on the hot ground (Fig. 2-14). Some schematics have a diamond-shaped or triangular symbol indicating the hot ground terminals. The secondary circuits of the switching transformer are connected to the common ground.

The common ground is where most voltages and critical measurements are connected to the common ground terminal of the DMM or scope. Common ground is a reference ground connection. The metal chassis of a TV or amplifier is the common ground. The main reason for using a common ground is to simplify the different circuits. The common ground is used as a return path for many different circuits. Most voltages and scope waveforms are taken with the test instrument black probe or ground clip connected to the common ground.

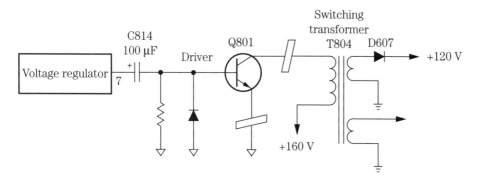

2-13   The hot ground is found in the ac input circuits of a Goldstar CMT-2612 TV up to the switching transformer T804.

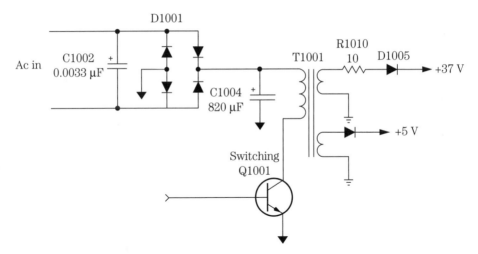

2-14 A shorted switching transistor (Q1001) caused a dead chassis in a Panasonic PV-M3217 on the hot ground side of T1001.

Running warm

A red hot or overly warm audio output transistor might show signs on the body from overheating with a cracked or burned bias resistor. Smoking resistors indicate a leaky transistor, IC, or capacitor. Power transformers running red hot might be caused by leaky silicon diodes or electrolytic capacitors. Red hot power output transistors and IC components might be leaky with burned bias resistors. Replace overheated IC components that have a chipped area or white and brown spots on the body.

Check the PC board where overheated parts are tied into the PC board wiring. Often, brown spots on the top, around heated terminals, indicate a poorly soldered joint. Clean the area and resolder. If the terminal lead will not take solder, cut off the bad terminal end, insert the lead through the PC board hole, and solder.

Red hot—do not touch

The power output IC within a deluxe receiver might indicate overheating with gray streaks on the body and a change in color. Check for burned bias resistors that might be operating quite warm. A dead high-powered amplifier with a loud hum can be caused by a red hot output transistor. Red hot output transistors in a car radio may have burned open the fuse and charred the hot wire to the receiver. When the horizontal output transistor in a TV chassis runs red hot, suspect insufficient drive voltage or a leaky output transistor. Suspect too-hot-to-touch transistors that are leaky or have improper bias voltage.

Check the transistor in the circuit for a leaky condition. You may find one transistor leaky and the other push-pull transistor open. Remove both transistors, and test them out of the circuit. Always replace both transistors when one is found to be leaky or shorted. Inspect the bias resistors for burned or charred areas. Check each bias resistor and diode while the transistors are out of the circuit.

Red hot audio output ICs indicate a leaky IC or excessive supply voltage. Most output ICs and transistors will run warm but not red hot. Check for a change in bias resistors and high-power supply voltage. Often, when a red hot IC or transistor is found, the bias resistors are burned or changed in value. Check for leaky directly driven transistors when the output transistor or IC runs red hot.

Same circuits, another chassis

Often the circuits within a TV, cassette, or CD player are the same as the ones you are working on. If you have a new TV chassis that has quit and you have no schematic, compare it with another chassis. Sometimes, even with a schematic, certain voltages are not listed, and with another chassis you can compare voltage and resistance readings. Of course, the chassis must be the same make and have the same chassis setup. Comparison tests can solve a difficult problem when a schematic is not handy.

Often the circuits within TVs, stereos, and radios are the same for several years. If you have a schematic for an older unit, use it. You also can compare the chassis with that of another brand. If the defective chassis has marked part numbers, use a schematic that is similar.

Take voltage measurements of a transistor from each terminal to common ground. With an NPN transistor, the positive lead will go to each terminal with the negative probe grounded. Just reverse the test leads for voltage measurements of a PNP transistor. The positive probe is at ground potential with a PNP transistor. Use a DMM for accurate low-voltage tests.

Most voltage tests on a suspected IC should be made with the positive lead at the IC terminals and the negative probe clipped to ground. Write down the voltage measurements on each terminal pin. Then determine if these voltages are correct. Always check the voltage supply pin first. If the voltage is very low, suspect a leaky IC (Fig. 2-15).

Remove the voltage supply terminal from the PC board wiring to see if the voltage increases. Simply unsolder the pin with solder wick and a hot iron. Flick the pin to make sure that it is loose and not touching the PC board wiring. Now take another voltage measurement. If the voltage has increased to normal or higher, replace the leaky IC.

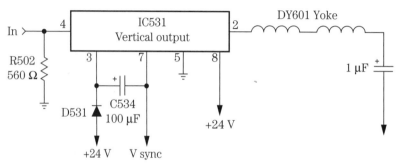

2-15 Only a horizontal white line caused by a defective IC531 was found in a Sharp 19SB60 TV.

When making high-voltage tests, such as on the anode terminal of a CRT, use either a high-voltage probe or a high-voltage probe attached to the vacuum-tube voltmeter (VTVM). High-voltage tests on microwave ovens should be made with a high-voltage probe or Magnameter. Make sure that the ground clip is attached to the chassis or you will receive a shock from the high-voltage probe.

Resistance in-circuit test measurements

Critical resistance measurements of high-ohm resistors in a circuit are not too reliable. Remove one end of the resistor and take another measurement (Fig. 2-16). Fairly normal resistance measurements of low-ohm resistors and coils are quite close in resistance. Always remove one end of a diode in the circuit for correct resistance measurements.

Resistance measurements with the diode test of the DMM across the transistor terminals are very accurate. When the same low resistance measurement is found between any two transistor terminals, the transistor is leaky or shorted. Likewise, when all voltages on the base, collector, and emitter are quite close, suspect a leaky transistor. Most failures in a transistor involve a short or a low ohm measurement between collector and emitter terminals.

When a low-ohm resistor or diode is found across a given component, the resistance measurement of the suspected part is not accurate. For instance, a resistance measurement of a diode within a relay solenoid circuit is not normal because both components have a low resistance. Remove one end of the diode for a correct measurement.

Critical resistance measurements on the terminals of a transistor or IC to common ground might locate the leaky part. Resistance measurements in transistor and diode tests with the DMM might determine if a suspected part is open or leaky. Critical resistance measurements on the emitter or bias resistors might indicate a leaky transistor or IC or a change in resistance.

Within an audio amplifier, accurate resistance measurements on the speaker output terminals, with the speaker removed, can indicate a defective transistor or IC or a change in resistance (Fig. 2-17). Safe resistance measurements can be made with the power off when an output IC runs hot. Comparable resistance measurements in the

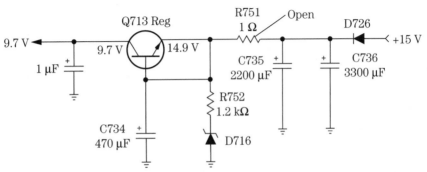

2-16 The open R751 resistor occurred with no voltage at the voltage regulator (Q713) in a Sanyo NK720K radio-cassette player.

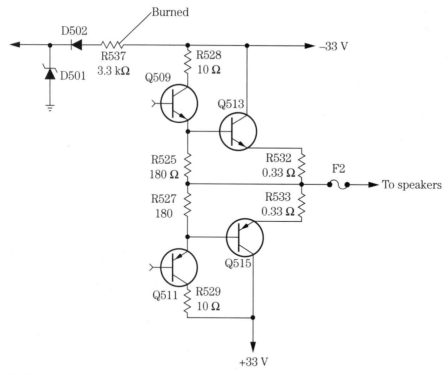

2-17   A critical voltage measurement of burned resistor R537 and leaky D502 caused a voltage to appear on the voice coil of speakers in a JC Penney tape deck.

Table 2-1.   The different resistance measurements from transistor terminals to ground shown in *Sams Photofacts*.

Item	E	B	C
Q201	9190 ohms	68 kilohms	5030 ohms
Q202	960 ohms	2650 ohms	470 ohms
Q203	351 ohms	640 ohms	360 ohms
Q204	0 ohms	1 kilohm	57 kilohms

stereo channels might help locate the defective part. Of course, a resistance or continuity check of each component might turn up a defective part. Very few schematics show the resistance measurements to ground, except *Sams Photofacts* (Table 2-1).

Continuity tests

A continuity or resistance check of a component is a quick way to determine if the part is open or leaky. Check the speaker terminals to see if the voice coil is open or grounded to the speaker magnet or framework. When taking speaker measurements with a DMM,

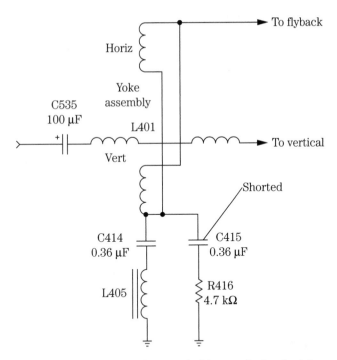

2-18 Both coils were fairly normal with a continuity check in a Samsung TC2540S TV with a leaky C415 that caused the relay to click and resulted in a dead chassis.

the resistance should be quite close to the speaker impedance. With a VOM, the resistance might be off 1 or 2 ohms.

Continuity tests on the tape head can determine if the winding is open. Place extra pressure on the tape head terminals if the winding is suspected of intermittent playback. The continuity test between the tape head terminal and the outside shield might help clear up the problem of weak or distorted tape player sound.

Continuity checks of any component can determine if the part is open. For instance, the resistance of a tape, VCR, or CD motor might not be known. A quick continunity test across the deflection yoke windings can determine if the yoke is open (Fig. 2-18). However, a continuity test across the terminals can indicate if the motor winding is open or if there is a broken connection. Likewise, check switches, coils, and transformer connections with a continuity test.

After installing a flyback transformer or IC component that solders directly to the PC board, continuity tests of the PC board wiring should be made. Check from the actual terminal pin to the next part tied to the same wiring to determine if the PC board wiring is intact. Sometimes the PC board wiring breaks at eyelets or terminal pins. Bridge the wiring and resolder.

Leaky or shorted transistor resistance tests

Resistance measurements across transistor terminals are not accurate with a VOM compared with a DMM. Within seconds you can check a suspected transistor be-

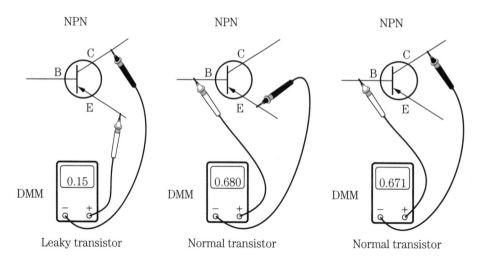

2-19   The leaky and normal transistor test with a DMM.

tween two elements with the DMM diode test. A low resistance measurement in both directions on any two terminals can indicate a shorted transistor or high leakage. If a shorted output transistor has a 0.15-ohm measurement between the collector and emitter terminals on the DMM, the output transistor is shorted. The transistor is leaky when the transistor has a 527-ohm reading in both directions. If the measurement is in one direction with the red probe on the base terminal, the transistor is likely normal (Fig. 2-19).

Comparable resistance measurements of transistors to the common ground can indicate a defective circuit. Take comparable resistance measurements if transistors are damaged directly after replacement, especially in directly connected circuits. Take a resistance measurement between the output speaker terminal and the ground with the speaker removed in the amplifier's left and right channels. Compare the resistance readings in both channels to locate a possible defective circuit.

Usually only one channel is defective in a stereo amplifier. When the resistance measurements are not the same at a given point, take a resistance measurement from each terminal of the output transistors and compare these readings. A normal circuit will have a similar resistance reading in both channels.

If the circuit is normal, the measurements of both channels will be within a few ohms of one another. Remember that this takes a lot of service time, and the DMM resistance reading must stop with each test. When the measurements are still way off, go to the next transistor and compare this resistance measurement. If by chance the next stage is real close in measurement, just back up a step to the defective stage. Now is the time to take critical voltage and semiconductor tests.

IC resistance tests

Rotate the DMM to the 2-kilohm scale on the meter, and take a resistance measurements from each IC terminal to the common ground. Double-check the resistance when

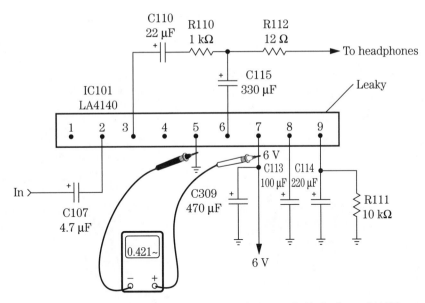

2-20 A leaky IC101 caused low ohm measurement at pin 7 of a Sanyo M4430 cassette player.

a low reading is found. Some terminals of an IC may be connected directly to the common ground. Simply check or trace it out. Now compare the same input and output terminals of a suspected IC in both channels. Usually only one channel is bad in a stereo amplifier circuit. When the resistance reading is way off in one channel, suspect the IC with the lowest measurement (Fig. 2-20).

Sams Photofacts

Most of the popular TV chassis with correct voltage measurements and waveforms are found in *Sams Photofacts*. Today, the *Sams Technical Publishing Annual Index* covers the TV chassis, TV/VCR combos, and VCR products. Besides a complete schematic, the service data might indicate grid trace location of parts, IC functions, important component information, adjustments, part lists, and a replacement chart. The replacement chart provides the latest semiconductor replacements for a given transistor, diode, or IC (Fig. 2-21).

The schematics might include the audio, CRT, power supply, system control, TV, and tuner diagrams. There are also schematic notes, service tips, stereo adjustments, test equipment, test jig hookups, troubleshooting, and a tuner circuit voltage chart. *Sams Photofacts* can be ordered from your local electronic distributor or directly from Sams:

Sams Technical Publishing
201 West 103d Street
Indianapolis, IN 46290-1097
1-800-428-SAMS

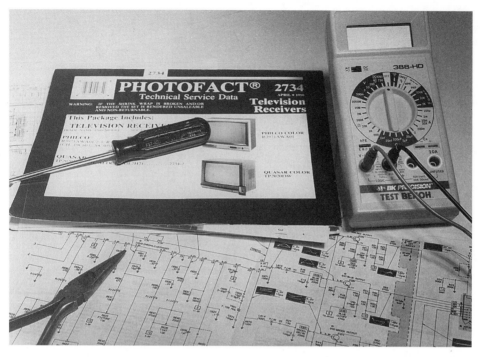

2-21   *Sams Photofacts* schematics can be used to check components in TVs and VCRs.

Electronic Servicing and Technology schematics

A TV or VCR schematic is found within each monthly issue of *Electronic Servicing and Technology* magazine. You can remove the schematic from the magazine and file it for future reference. *Electronic Servicing and Technology* is not on the newsstand and can be obtained only by subscription (Fig. 2-22). Besides containing a popular schematic, *Electronic Servicing and Technology* features many articles on electronic servicing of many different products.

Electronic Servicing and Technology
403 Main Street, 2d Floor
Port Washington, NY 11050
1-516-883-3382

The various waveforms

Take critical waveforms to determine if a signal is missing or present within a CD and cassette player, stereo amplifier, computer, or TV chassis. Critical waveforms taken within the horizontal section of the horizontal and vertical sections of a TV can quickly determine what stage is defective. A quick waveform on the deflection IC can indicate

2-22 You will find a schematic diagram of a TV or VCR in each issue of *Electronic Servicing and Technology* magazine.

if the horizontal oscillator stage is functioning (Fig. 2-23). Likewise, a vertical waveform taken at the input and output terminals of the vertical IC can show if the vertical section is performing.

A critical waveform taken at the output of the RF amplifier in a CD player indicates that the optical pickup and RF stage are normal. If the waveform is weak or missing, check the RF amp IC voltages and the pickup to determine what component is defective. Often, when the optical pickup has no output, the CD player may be totaled out and dumped, because replacing the optical pickup is quite expensive (Fig. 2-24).

Signal injection

The RF/mixer/IF circuits can be signal traced within the receiver circuits with the RF signal generator and the speaker as indicator. Signal trace the audio circuits of the amplifier with the audio signal generator. By injecting a signal at the beginning of each stage, you can quickly locate the dead circuit. The audio tone can be heard in the speaker. Then make critical voltage, resistance, and semiconductor tests to locate the defective component.

To signal trace the audio section of any amplifier, inject an audio signal (1 kHz) at the center terminal of the volume control. If a loud sound can be heard in the speaker, the output circuits are normal. Now proceed toward the input transistor or IC stages. When no signal can be heard when connected to the volume control, go from audio

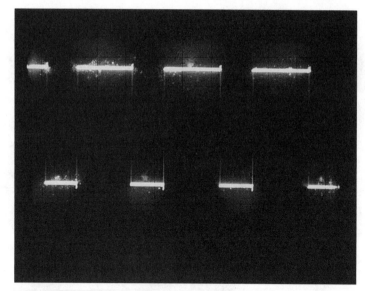

2-23 The horizontal deflection waveform on an IC pin terminal in a TV chassis.

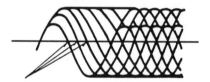

2-24
The encoding frequency modulation (EFM) waveform at the output of the RF amplifier indicating that the optical pickup and preamp are normal in a CD player.

stage to stage until a weaker audio signal is heard. Just back up a notch and take critical voltage measurements of the preceding circuit (Fig. 2-25).

When isolating a defective audio circuit within a TV, cassette player, or amplifier, start at the volume control by injecting a 1-kHz signal at the center tap of control, with no schematic. If the sound is normal, suspect the input audio signal circuit. Double-check the audio signal at both center tap terminals of the left and right stereo volume controls. Proceed toward the speakers with weak or no signal. Go from base to base of each audio transistor. Remember, as you proceed to the last set

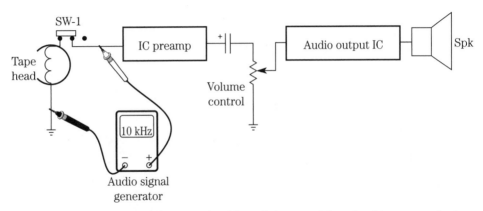

2-25 Injecting a 1- or 10-kHz audio signal into the preamp IC to signal trace cassette tape player circuits.

of transistors, the audio signal becomes higher. When the weak, distorted, or dead stage is found, take critical voltage and resistance measurements on transistors or IC components.

Troubleshooting with an external amp

The external amplifier can be taken from a large radio, phono amp, receiver, or a home-constructed amplifier. The audio amp can consist of a single IC amp or have two or more transistors within a self-enclosed speaker case (Fig. 2-26). The external audio amp can be used rather quickly to locate a defective component or circuit in the audio stages by injecting a 1-kHz audio signal at the input of the amplifier circuits from a generator and using the external audio amp to trace the audio signal through the audio circuits. Keep the audio signal generator as low as possible with an audible tone. This audio signal can be signal traced right up to the speaker terminals.

When checking for a loss of audio signal in a cassette player, insert a 1- or 3-kHz cassette and check the audio signal at the tape head ungrounded connection with the audio amp. Check the amplifier signal at the volume control. Test both channels for audio signal at the volume control and compare the signals. If one channel is lower than the other, suspect a defective preamp transistor or IC. Go from base to base of the preamp transistors or IC to locate the defective component. A preamp IC between the tape head and volume control is easy to locate on the PC board.

Transistor in-circuit tests

Transistors can be checked in or out of the circuit with the transistor tester or diode test of the DMM (Fig. 2-27). When checking transistors in the circuit, watch for choke coils, low-ohm resistors, and diodes within the circuit. Low or improper measurements may occur. Don't overlook a defective transistor in the circuit that might be open or leaky but when removed tests normal. Sometimes heat on the transistor terminals can restore the transistor. Replace it anyway.

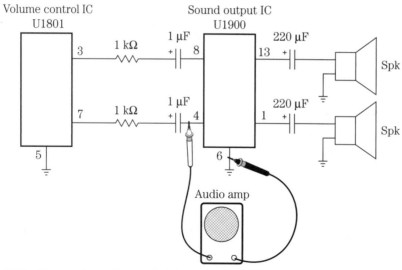

2-26   Tracing the audio signal with the external audio amp in a stereo audio amp circuit.

2-27   Checking the power output transistors in a portable TV with the diode tester of a DMM.

The suspected transistor can be checked in the circuit with voltage tests. Low collector voltage can indicate a leaky transistor. All close voltages on the three terminals of the transistor can indicate a leaky transistor. Higher-than-normal voltage on the collector terminal can indicate an open transistor. No emitter voltage can indicate that the transistor is open.

Take a voltage measurement between the base and emitter terminals for correct bias voltage. An NPN transistor should have a 0.6-V measurement, and a PNP transistor should have a 0.3-V reading. If an improper measurement is found of 1 ohm or more, the transistor is defective.

Accurate transistor tests can be made with the diode tester of the DMM. When checking an NPN transistor, place the red probe on the base terminal and the black probe on the collector terminal. If you get a low resistance reading, under 100 ohms, suspect leakage between the base and collector elements. The normal measurement is above 500 ohms.

Let's say that the measurement is 877 ohms. Place the black probe on the emitter terminal, and leave the red probe at the base terminal. The transistor is normal if a comparable measurement is noted (845 ohms).

When the reading is low between any two elements, the transistor is leaky between these elements. You might find that a transistor leaks between all three elements. Often a leaky transistor occurs between the collector and emitter terminals. When a junction test is high on one set of elements and not the other, it is likely that the transistor has a high resistance and poor junction; replace it.

When making transistor tests on a PNP transistor, the black probe stays on the base terminal while taking measurements with the red probe on the collector and emitter terminals. Remember, low resistance measurements can be caused by small coils, chokes, diodes, and low-ohm resistors in the base circuits. In this case remove the base terminal from the PC board wiring and take another test.

Digital transistor tests

Digital transistor chips (SMDs) are found in a camcorder, CD player, and VCR. The digital transistor might be an NPN or PNP type with internal base and bias resistors. R1 is found within the base circuit, whereas R2 is between base and the emitter terminals. A surface-mounted digital transistor may have one or two digital transistors within one component. A higher resistance measurement is noted when taking DMM diode tests.

NPN or PNP

Mostly NPN transistors are found in today's electronic chassis; however, be prepared for a few PNP types. When no schematic is handy and you don't know which terminal is the base or collector, try the diode test. Determine if the transistor has a part number on the body or marked terminals. Look up the part number in the replacement manual to tell if it is an NPN or PNP type, what its working voltage is, and what circuits it is found in.

If there are no part or terminal numbers on the transistor, try to locate the base terminal with the diode test. In a normal transistor the base is tested with the positive

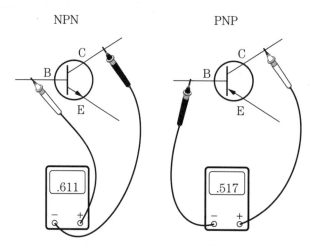

2-28   An NPN transistor has the positive terminal of the
DMM on the base terminal and the negative probe to the
base of the PNP transistor.

probe of the DMM, and the resistance is common to both the collector and emitter terminals. Usually a higher resistance measurement is from base terminal to emitter. For
instance, in testing an SK3710 horizontal deflection output transistor, the resistance,
with the red probe on the base terminal and the black on the collector, is 609 ohms. The
resistance measured from the base to the emitter terminal is higher at 667 ohms.

Also, this measurement indicates that the transistor is normal and an NPN type.
When the measurement is common with the red probe at the base terminal, it is an NPN
transistor (Fig. 2-28). Try checking any transistor out of the circuit with this method.

DMM diode tests

A fixed diode can be tested easily with the diode tester of the DMM. The open diode will
have no reading at all, even with reversed test leads. The leaky diode will show a low
ohm measurement in both directions. A normal diode will have a low ohm measurement
in only one direction and no reading with reversed test probes (Fig. 2-29). Always remove one end of the diode from the circuit for a good measurement. When a relay solenoid, coil, or low-ohm resistor is in the same circuit as the fixed diode, the in-circuit
measurement is not accurate. High-voltage diodes found in microwave ovens or black-
and-white TVs will not test accurately with the diode tester of the DMM.

The fixed diode has a line at one end indicating the positive or collector terminal,
whereas the other end of the diode is the anode terminal. Be sure to check for correct
polarity before installing the new diode. A shorted or leaky diode might show signs of
overheating or burned areas. Notice that the negative, or black, probe is at the positive
terminal of the diode for a normal reading. When the VOM is used, the positive probe
will be connected to the positive terminal for a normal measurement.

Zener, detection, and damper diodes must be checked in the same manner. The resistance of a normal RF diode (1N34 or 1N60) can be higher in one direction above 1
kilohm. All diodes checked on the resistance scale of a VOM will have a low resistance

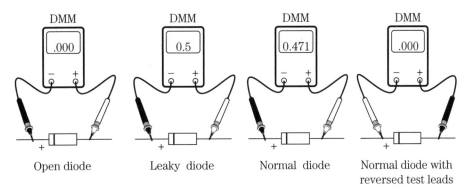

2-29 Checking a fixed diode with the diode tester of the DMM.

reading with the red probe on the cathode terminal and the black probe on the anode. Notice that this reading is just the reverse of a DMM diode measurement.

Safety transistor replacement

After removing and selecting the correct transistor replacement, install the transistor, and solder it with a heat sink at each terminal. A pair of long-nose pliers will do the trick. Before removing the suspected transistor, mark the three terminals on the PC board. You might find that the correct terminals are already marked on the PC board in white or dark letters.

If the transistor has no identification on the body area, the highest voltage test indicates the collector terminal of an NPN transistor. Likewise, the highest negative terminal is the collector terminal of a PNP transistor. The emitter terminal of an NPN transistor within the circuit has the lowest reading to common ground. A PNP transistor has the lowest reading to common ground and has the highest negative reading (see Table 2-2).

Make sure that the correct terminal is in the correct hole before soldering. Always use a heat sink to prevent damage to the new transistor. Grab the transistor terminal with a pair of long-nose pliers above the terminal to be soldered. If a heat sink is not used, sometimes the transistor will be ruined with too much heat from the soldering iron. Always make a cleanly soldered joint, but do not leave the iron on the heated junction too long.

Signal in and out tests of IC components

Check the voltage and resistance measurements of the suspected IC within the stereo audio circuits of a TV, receiver, or cassette player. Compare the defective audio circuit

Table 2-2. The different semiconductor components with correct terminal markings.

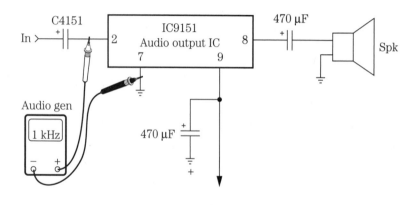

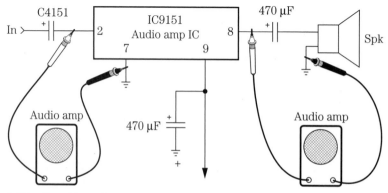

2-30 Signal in and out tests with a signal generator and an external audio amp.

with the normal stereo circuit. Then check IC components with signal in and out tests (Fig. 2-30). Inject the audio signal from an audio generator at the input, and use the speaker as an indicator. The audio signals can be checked with the external audio amp at the input and output terminals of the suspected IC. Audio signals found on the output IC can be heard on the output terminal if the IC is normal.

Critical voltage measurements on each IC terminal will indicate if the part is open, leaky, or shorted. Check the supply voltage pin (Vcc) for a higher voltage. Often, low supply voltage at the voltage source terminal indicates a leaky IC or improperly applied voltage. Take critical resistance measurements from each pin to the chassis ground to locate leaky parts tied to the ICs. Visually trace the pin wiring when a low measurement is found.

Voltage injection

The external power supply can be used to help locate defective circuits within the electronic chassis. Instead of using batteries on battery-operated electronic products, use the bench power supply while servicing a portable cassette or CD player. Within a TV chassis, inject a low-voltage source in the horizontal oscillator and deflection cir-

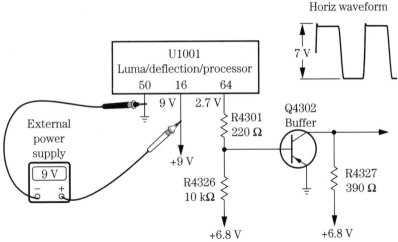

2-31 An external 9-V source was injected at pin 16 to start up the horizontal deflection IC.

cuit IC to make it operate. In many of the new TVs, the secondary voltage source must operate from the flyback because it provides operating voltage to many other circuits. This means that the horizontal circuits must operate before any secondary voltage is developed.

In an RCA CTC167, the screen was black with no high voltage. The horizontal output transistor was found to be shorted and was replaced with an NTE2331 universal replacement. There was still no horizontal sweep on the base terminal of Q4401. Since the horizontal secondary winding of the flyback (T4401) and the horizontal circuits must perform before any secondary voltage is developed, no supply voltage was found on pin 16 of the deflection IC (U1001).

An external 9-V source from the bench power supply was injected at pin 16 (Fig. 2-31). The scope was attached to pin 64 and ground of the deflection IC as a monitor. A 7-V horizontal waveform should be found at this terminal that feeds the horizontal buffer transistor. Replacing U1001 with an NTE1790 deflection IC solved the dead chassis.

Removing and replacing a defective IC

Before attempting to remove a defective IC, mark down where terminal 1 is on the PC board. Sometimes terminal 1 is marked in white on the board. The defective IC can be removed with the soldering iron and solder wick or a flat tool, tweezers, and an air-blowing unit. One of the cheapest methods is to use the soldering iron and solder mesh or wick material. Place the soldering iron tip over the solder wick, and place the mesh material over the IC terminals.

Pick up excess solder with the wick material. Slowly go down one side of the bottom of the IC terminals and pick up the excess solder. The excess solder must be removed several times. Then go down the other row of terminals to remove excess solder.

A surface-mounted IC can be removed with a thin tool or a pair of tweezers and a soldering iron (Fig. 2-32). Pick up the tip of the bent terminal, and pry it up as the iron

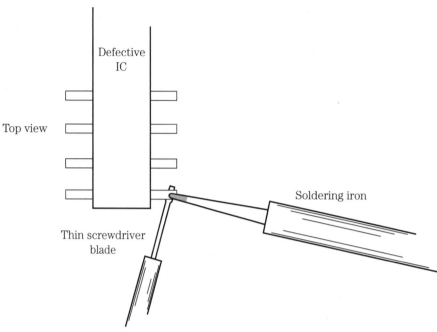

2-32   Remove a deflection IC by heating up and lifting the lug with a thin screwdriver blade.

heats up the joint. Use either a thin screwdriver blade or a pocket knife to free the soldered connection. Lift each terminal with the same method until the IC is free. After the IC has been removed, go over the traces or PC wiring with solder wick and a soldering iron to pick up any excess solder so that the surface-mounted part will lie flat on the PC board.

Make sure that the PC traces are clean and that the area is cleaned up before trying to replace the IC. Inspect each IC wiring connection to make sure that none are broken. A coat of flux might be added. Make sure that terminal 1 is in the right spot. Position the IC, and solder the two opposite pin terminals to hold the IC in place. Use a sharply pointed iron, and carefully solder each pin terminal.

Inspect each pin terminal with a magnifying glass to ensure a good soldered connection. Remove any solder bridges with the soldering iron and solder wick. Check with the low-ohm scale of the DMM from pin to corresponding trace or PC wiring for a good connection. Sometimes when too much heat is applied in removing the IC, the PC wiring can be damaged.

Here today and gone tomorrow

Intermittent operation can occur in any electronic circuit. Cracked boards can cause intermittent operation. Broken boards are difficult to replace. Sometimes placing extra pressure on sections of the board can cause the intermittent to act up. Place a strong light over the board, and use a magnifying glass to help to search for cracked wiring. Fine breaks can occur around heavy parts or chassis standoffs.

Badly soldered connections at terminal pins of components and ICs can produce intermittent sound in a TV chassis. Large blobs of solder on part terminals can cause a dead or intermittent chassis. Brown areas around terminals can mean poorly soldered connections or overheated parts. Sometimes resoldering the entire board section can solve this problem. Be careful not to overlap solder to another terminal or PC board wiring.

The intermittent chassis is the most difficult problem to repair and can be very time-consuming. Try to locate or isolate the intermittent with all the symptoms that are produced. Monitor the suspected section with voltage and scope tests. Within the audio system, use the speaker as an indicator. An oscilloscope can be used in areas where waveforms are found.

Coolant and heat sprayed on the suspected capacitor, transistor, or IC might make it act up or quit. Several applications might be needed before you get any results. Most intermittents are found by monitoring the chassis with a voltmeter and scope until the unit acts up.

Intermittent parts in an audio stereo chassis can be located by comparing the signal in the normal channel. Critical voltage measurements on each channel at a given point can uncover the intermittent part. Sometimes just moving the board or component can cause the sound to come and go. Check for intermittent coupling capacitors, transistors, and IC components in the stereo channels.

Overheated transistors or IC components in a TV chassis can produce intermittent symptoms. Intermittent horizontal circuits are very difficult to find because derived secondary voltages are required to make the circuits operate. Look for small things, such as poor connections, badly soldered joints, poorly soldered transistor pins, SMD end-part connections, defective or corroded sockets, and bad board connections.

INTERMITTENT RCA CTC169 CHASSIS

An RCA chassis might come on and stay on for a few minutes and then shut off or go into shutdown. Critical voltage measurements were made on the switching regulator when the chassis was intermittent. CR4102 was found to be leaky across the transformer (T4101) that ties to pin 2 of U4101. Also, CR4106 was replaced in series with R4117 and R4102 off pin 6 of regulator U4101. Replacing both diodes solved the intermittent RCA chassis.

INTERMITTENT FISHER RECEIVER AMP

The volume would cut up and down in the left channel of a Fisher RS853 receiver. The audio signal was normal at the volume control with the external amp attached. After several hours of signal tracing the intermittent channel, the intermittent sound was found to be caused by a defective speaker relay.

INTERMITTENT JC PENNEY CASSETTE PLAYER

Sometimes the right channel was intermittent when in the record mode. At other times a warbling recording was noted. The recording appeared normal when the record and play switches were moved. Cleaning up all contacts on the record/play switch solved the intermittent audio problem. If the record/play switch will not clean up, replace the defective switch.

Component replacement

Locating a defective component is more difficult; replacing the part is quite easy. Sometimes, however, the correct part is hard to find. Substituting another component might be the only answer. Electrolytic and bypass capacitors, resistors, and large-wattage resistors are easy to find. Special transistors, ICs, and diode parts can be replaced with universal replacements found in a semiconductor manual (Fig. 2-33). Transformers, coils, motors, and original replacement parts might be difficult to locate.

Always try to replace defective parts with the original component. Critical parts found in special safety areas should be original part numbers. It is difficult to substitute for the special tape motors found in VCRs and CD players. Tuners found in TV chassis or VCRs can be replaced with new units, or they can be sent in for repair at a manufacturer's service center.

Besides local wholesale electronics parts stores, manufacturers' service centers, and parts distributors, try obtaining certain parts from mail-order firms. It may take a few days, but such firms may have the part to fix that chassis that is collecting dust. Don't overlook the fellow technician down the street. Sometimes certain parts can be found right next door.

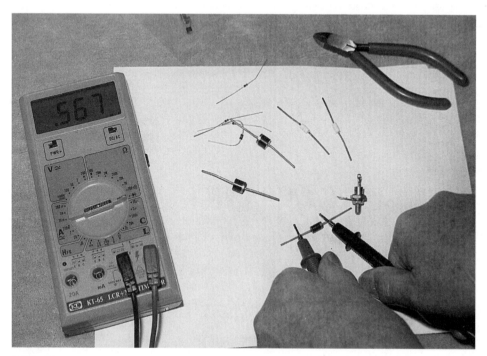

2-33 Be sure to test each part before installing.

2-34 Check the underside of a large receiver amp chassis for SMD parts and poor PC wiring.

Low and behold

Check down under the TV, VCR, or CD player chassis for surface-mounted parts. These components are found mounted and soldered directly on the PC wiring, whereas standard parts are found on the top side of the chassis board (Fig. 2-34). Extreme care must be exercised not to flex the PC board or drag the chassis across the work bench to prevent damaging the SMD components.

The surface-mounted components are soldered at each end to the foil of PC wiring. A surface-mounted resistor or capacitor may be flat or round. It may be found mounted anywhere under the TV chassis. Notice that several flat surface-mounted parts are even mounted between the large IC that is mounted on top with the pin terminals soldered to the PC wiring.

SMD components

Surface-mounted devices (SMDs) may consist of capacitors, resistors, diodes, transistors, ICs, and microprocessors. The SMD transistor looks like a three-legged bug on the PC wiring. You might find more than one electronic part in each SMD body. Sometimes it's rather difficult to find the difference between a fixed capacitor or resistor. Likewise, a feed-through SMD that is solid and connects two circuits together may look like a fixed resistor or capacitor. Do not confuse a shorted capacitor with a solid feed-through device. Check the SMD feed-through with the low-ohm scale of the DMM.

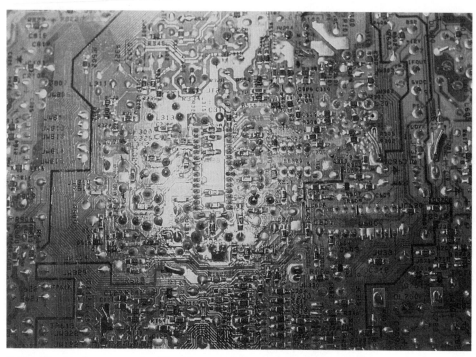

2-35   The black areas on the PC wiring are SMD parts of a color TV.

Check the top of the SMD part for markings that will indicate what the component actually is. An electrolytic SMD capacitor is stamped on top with the exact capacity and working voltage. A fixed resistor may contain a letter and a number on top of the component. An SMD resistor has three numbers stamped on top. The first number is the first digit, the second number is the second digit, and the third number is the multiplier. A 10,000-ohm resistor is indicated by numbers 103. The tie-through or solid SMD bar has 000 stamped on top. Do not mistake this for a shorted capacitor or low-ohm resistor (Fig. 2-35).

An SMD transistor has three terminals with the collector in the center at the top. The base terminal is at the bottom on the left corner, and the emitter terminal is at the right-hand corner. Use a sharp-pointed test probe to make critical tests on an SMD transistor and IC. Of course, the IC SMD component has many terminals, whereas the microprocessor has a lot more.

Flat or round

SMD parts may appear in either round or flat mounting devices. Only capacitors and resistors are found in the round form. All transistors, ICs, and microprocessors are found in the flat form. SMD parts have a habit of breaking at the soldered joint on the PC wiring, causing erratic or intermittent operation. Sometimes the large PC board will flex and bow, causing a break at the flat end of the SMD part (Fig. 2-36). Sometimes resoldering the joint is all that is needed if the SMD part is not cracked or broken.

2-36   Removing round SMD parts with a soldering iron and solder wick.

SMD components are tested in the very same way as those mounted on the top side of the chassis. To take resistance, voltage, and transistor measurements, sharpen the points of the test probes. If a schematic or parts layout is not available, it is very difficult to tell the difference between parts. Of course, a resistor will have resistance, whereas a capacitor should not. A feed-through SMD will have a dead short or zero resistance.

Removing SMD parts

Remove the solder at one end of the defective resistor or capacitor with a solder wick and a soldering iron. Sometimes the SMD part can be pried up with the iron applied at one end. Hold the SMD part with a pair of tweezers, and twist as the component is heated and removed (Fig. 2-37). Throw away all SMD parts that have been removed from the PC wiring.

To remove a transistor or diode SMD, melt the solder at one end, and lift the leads upward with a pair of tweezers. Do this to each terminal until all are removed. Some larger components may be glued underneath. Cementing the new SMD is not necessary. Remove jagged or overlapped solder from the PC wiring. Clean off lumps of solder from the wiring with the soldering iron and a solder wick.

Another method is to unsolder one terminal with the soldering iron and pry up the pin terminal with a sharp tool. Some electronics technicians use a butane-fueled soldering pen, found at most electronic distributors. The pen is held about ½ in from the

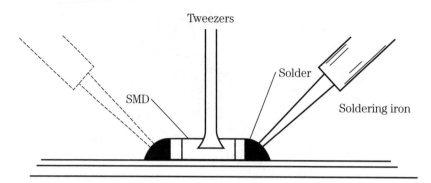

2-37 An SMD part can be removed with a soldering iron at each end by twisting the part with a pair of tweezers.

flat pin terminal, and the part is pried up with a small screw driver blade or sharp pointed tool.

With the gull-wing IC or microprocessor, many terminals can be removed by cutting each pin terminal near the body of the IC. Place a long resistor terminal alongside the body, and then cut each pin toward the IC body. When all pins are cut, lift the leadless body off the board. Remove each cut terminal with the soldering iron and tweezers. Be very careful not to cut into the foil pattern. Also, do not apply too much heat on one area to lift the foil from the PC board.

SMD replacement

Double-check the SMD part before attempting to remove and replace it (Fig. 2-38). Universal SMD fixed capacitors and resistors can be used to replace a defective component. These universal components can be purchased with a card of different values. Some SMD components come in sealed clear packages or on strips for easy removal. Always use direct factory replacements with dual resistors and diodes, transistors, ICs, and microprocessors.

Remove any solder left on the PC wiring with the soldering iron and solder wick. Clean the entire area with cleaning fluid, acetone, and a stiff brush. Make sure that the mounting area is clean and level.

When installing new SMD ICs and microprocessors, connect an electrostatic discharge strap to your arm, and clip it to the ground area of the chassis. Position the replacement SMD IC exactly in the correct place. Observe terminal number 1. Place solder flux over each SMD pin. Tack down each corner terminal of the SMD part with solder. Carefully solder each terminal pin, making certain that two separate terminals are not soldered together. Check the connections with the low resistance of the DMM.

ESD FCC ID numbers

Every VCR, cordless telephone, personal computer, and microwave oven must carry a Federal Communications Commission (FCC) ID number. The first three characters of

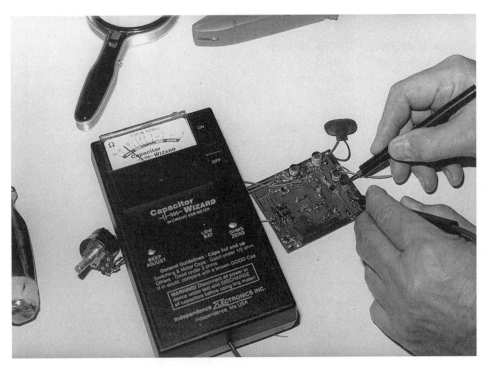

2-38   Double-check capacitors and PC wiring with the ESR meter.

that ID number identify the manufacturer of the product. When a unit comes in and the make is not known, look it up in Fig. 2-39 on p. 70. Often the standard life of a TV, VCR, stereo, or other consumer electronics product is between 7 and 10 years. Of course, there are many electronic components out there that are quite a few years older. You may find that parts and service literature are not available over the 10-year period. Check the correct prefix of the manufacturer, and contact the manufacturer for special components.

Cleanup

Blow out dust and dirt from on top and inside the electronic chassis. Clean up the radio dial numbers and frequency letters before placing the product on the finished bench. Wipe off the front of the TV screen, and wash the dial assembly and dirty knobs with window spray. Clean off the plastic or wood cabinet that may have color marks, dirt, and fingerprint smudges. Replace all chassis and back-cover screws. Make that electronic component shine like new before it is returned.

Code prefix	Manufacturer
A3D	NEC
A3L	Samsung
A7R	Orion
AAL	Phone Mate
AAO	Radio Shack
AAY	Midland International Corporation
ABL	Hitachi
ABW	JC Penney
ABY	Motorola
ACA	Yorx Electronics
ACB	Phonotronics
ACJ	Matsushita
ADF	Carterfone
ADT	Funai
AES	Uniden
AEZ	Sanyo
AFA	Fisher
AFL	Sharp
AFR	Curtis Mathes
AGI	Toshiba
AGV	Montgomery Ward
AHA	RCA
AIH	Litton Microwave Cooking Products
AIX	Sylvania
AJU	GE
AK8	Sony
AKC	Superscope Inc.
AKE	Marantz Co Inc.
ALA	Wells Gardner Electronics Corporation
ALI	Kenwood USA Corporation
ANV	Capetronic Int'l Corporation
API	Harman Kardon Inc.
ARR	AOC Int'l of America Inc.
ASH	Akai
ASI	Victor Company of Japan
ATA	Sharp
ATO	Zenith Electronics Corporation
ATP	Advent Corporation
BEJ	Goldstar
BGB	Mitsubishi
BOU	Philips
E0Z	Shintom

2-39 The FCC ID numbers and letters found on every VCR, cordless telephone, computer, and microwave oven. (*Courtesy of* Electronic Servicing and Technology *magazine.*)

3
CHAPTER

Troubleshooting and repairing audio amps, large and small

Troubleshooting the audio stages of any consumer product is fairly easy compared with TV, CD, and DVD chassis. Repairing audio circuits might take a few minutes longer without a schematic, however. Simply locate the audio output circuits by identifying the output transistors or integrated circuit (IC) on a heat sink. You can trace the audio circuits back from the speaker terminals or headphone jack. The audio input circuits can be traced from the volume control to the first or second audio amp. You may find most of the audio output circuits within one large IC. With a few test instruments, simplified instructions, and the test procedures found in the following pages, troubleshooting audio circuits can be easy.

List of audio test equipment

1. Digital multimeter (DMM)
2. Transistor tester
3. Capacitor tester or equivalent series resistance (ESR) meter
4. Signal injector
5. External audio amp
6. Frequency counter
7. Oscilloscope
8. Audio signal generator

Required test equipment

A good vacuum-tube voltmeter (VTVM), volt-ohm-milliammeter (VOM), or DMM can quickly locate a defective component with continuity, resistance, voltage, and current measurements (Fig. 3-1). You need a DMM that takes accurate low-voltage and resistance measurements. Besides voltage and resistance measurements, a DMM may have transistor, capacitor, diode, and frequency ranges. A low audible continuity buzzer is found in some models.

An ESR meter is ideal when locating a defective capacitor with equivalent series resistance. You can check the suspected electrolytic capacitor right in the circuit. An audio beeper sounds when a capacitor is good, eliminating the need to look at the meter pointer. Besides checking for a bad filter capacitor, the ESR meter can check small inductors, low-ohm resistors, and transistors. The meter is ideal in locating breaks in traces or badly soldered connections on the printed circuit board (Fig. 3-2).

Another important test instrument is a small audio amp. This tester is used to signal trace weak audio signals and distortion found within the audio stages. You can use a mono or stereo amp for signal tracing. In fact, you can build your own audio tester with only a few electronic components. With these two test instruments you can locate and repair most audio problems.

Test records, cassettes, and compact discs come in handy when signal tracing audio through the various circuits. A simple test speaker, cables, and test clips provide

3-1   A good VOM or DMM can quickly locate a defective part within an electronic chassis.

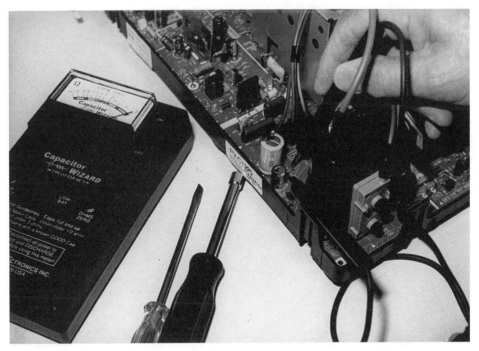

3-2 An ESR meter can locate a defective filter or electrolytic coupling capacitor.

quick audio connections. High-wattage, low-ohm resistors switched in parallel or series can provide speaker loading at the output terminals.

Optional test equipment includes a frequency counter, oscilloscope, capacity tester, audio signal generator, audio analyzer, audio watt meter, wow and flutter meters, and a dual stereo signal indicator. Additional audio test instruments are only required with certain audio problems (Fig. 3-3).

Suspect a defective coupling capacitor or open transistor or IC for a weak sound symptom. Double-check the audio output transistors and speaker coupling capacitors for weak or intermittent audio. A change in the bias resistors on a suspected transistor or IC can produce weak sound symptoms (Fig. 3-4). Simply locate the weak stage with the external audio amp and then take critical voltage and resistance measurements. Make critical low-ohm resistance measurements within the transistor output circuits to locate the defective component.

The intermittent, distorted, or noisy sound symptom is more difficult to locate. Check intermittent and noisy symptoms by going from stage to stage with the external audio amp or oscilloscope. Suspect poorly soldered or bad trace connections for intermittent sound problems. A low frying noise can be caused by a defective transistor or IC. Do not overlook bad ceramic bypass capacitors in the front end of the audio circuits for a low frying noise.

A low distorted sound is difficult to locate. Distorted sound in a speaker can result from leaky transistors or ICs. Use a sine and square wave generator with the scope as indicator to locate a low distorted sound. Double-check the audio output stages for ex-

3-3 A frequency counter, scope analyzer, and external power supply are optional test equipment.

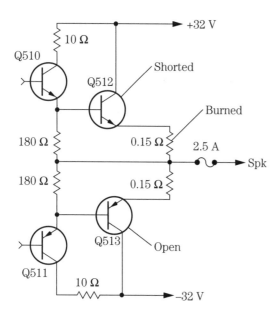

3-4
Double-check the bias resistors when a leaky, open, or shorted transistor is found.

treme distortion in the speaker. Distortion can be caused by leaky transistors or ICs, a change in resistance or burned bias resistors, and leaky coupling capacitors. Extreme distortion can occur in one or both channels. In a JVC AV-2749S TV, distorted audio in both channels was caused by a defective transistor Q963 (2SC1815).

Listen, listen

One of the greatest assets in electronics servicing is the electronics technician's own ears. You can hear a popping or frying noise caused by a defective transistor, IC, or ceramic capacitor. Your ears can hear the distorted, weak, or no sound symptom with the earphone or speakers. Often the frying noise is a constant low-level noise that might be heard in one or both audio channels. The popping noise can be caused by a defective output transistor or IC component. A bad power switch can produce a loud popping or arcing noise when pushed on with dirty or worn switching contacts.

A weak stage within an audio circuit can cause unbalanced channels. Often, the volume must be turned up on the other channel to compensate for a weaker channel. If a dual-volume control is used in the audio circuit, the weak stage must be located and repaired. Check for small defective electrolytic coupling capacitors and a change in bias resistors within the weak circuit (Fig. 3-5). The unbalanced channel can be checked with a scope or external amp and test cassette (3 kHz) within a cassette player.

The dead chassis

Servicing a low-powered audio amp is fairly easy even without a schematic. Check the batteries of the amplifier if it is powered by batteries, or check the ac power line otherwise. Go directly to the low-voltage power supply when the unit is operated from a power line. Measure the dc voltage across the large filter capacitor within the low-voltage circuits. Go to the audio output amp stages if normal voltage is found in the power supply. Usually, the two output transistors are mounted near the driver and audio output transformer in the early phono circuits. Clip a speaker across the suspected component to test it for no sound.

Only a whisper

Weak audio can be caused by bad coupling capacitors, transistors, and ICs and badly soldered connections. Inject an audio signal from the audio signal generator at the volume control to determine if the sound is normal. Go from stage to stage to locate the weak or bad component. Quickly check the transistors and ICs with in-circuit tests for leaky or open conditions. Check the suspected coupling capacitor by inserting the signal on one side and then the other to notice if the signal is there or lost. Go from base to collector of each transistor with the injected signal. Likewise, check the input and output terminals of the suspected IC for a weak signal. The signal should be greatly amplified at the output terminal.

A clicking test with a small screwdriver or test probe from the ungrounded side of a crystal cartridge can indicate a loud hum with the volume control wide open. Now go to the base terminal of the first transistor or IC and notice the hum level. Notice that the hum will become weaker as you proceed through the amplifier circuits going toward the

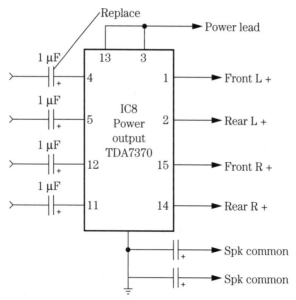

3-5 A defective coupling capacitor caused weak sound in the front left speaker of a car radio.

speaker. When the hum becomes loud after going past a coupling capacitor or transistor, you have located the defective audio stage (Fig. 3-6).

Repairing the audio amp

The audio amp found in a clock radio or a table model radio might consist of a few transistors or one large IC. Usually, a simple audio amp with transistors is found in directly driven transistor circuits. IC components are found in the latest sound circuits of AM/FM/MPX radios and cassette players. You may find a simple three-transistor circuit in an RCA TV chassis. Take critical voltage, semiconductor, and resistance tests of the bias resistors. Remove one end of each resistor for correct resistance measurements.

Distorted sound came from the speakers in an RCA CTC145 TV chassis. Voltage tests on both Q1202 and Q1203 were way off. A transistor test with the diode tester of a DMM indicated that Q1202 was open and Q1203 was shorted. Bias resistors R1209 and R1210 were still normal at 2.2 ohms. Replacing both output transistors solved the distorted sound problem (Fig. 3-7).

SPEAK UP, YOU ARE MUMBLING AGAIN

Test each transistor within the audio circuits with the diode tester of the DMM. Then take critical voltage measurements. Weak sound can result from bad coupling capacitors, a change in bias resistors, and defective transistors. Distorted sound can be

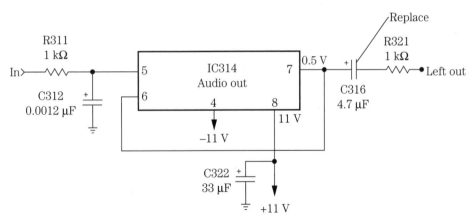

3-6 The defective C316 (4.7-μF) speaker coupling electrolytic caused weak sound in a Goldstar GCD-616 CD player.

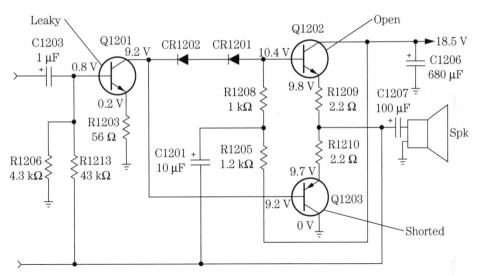

3-7 Check Q1202 for an open condition and Q1203 for a shorted condition with distortion in an RCA CTC145 chassis.

caused by leaky output transistors and ICs. Double-check the bias resistors when the output transistors are leaky or shorted.

Remember, when diodes or low-ohm resistors are found in the base circuits, the measurements may not be accurate. Remove the base terminal for accurate transistor tests. Very low voltage measurements on all transistors can indicate a leaky transistor.

Often, both output transistors are found close together, and they have separate heat sinks. If in doubt, locate the large coupling capacitor to the speaker, and then trace the positive end of the capacitor back to the output transistors. Another method is to eliminate the driver transistor by tracing the small coupling capacitor back to the base terminal.

Stereo IC amp repairs

A stereo IC amplifier might be located within an AM/FM/MPX receiver, a cassette or CD player, a boom box, a VCR, or a TV chassis. A simple stereo IC amp might have a couple of separate IC components within each stereo channel. You might find one dual IC for both audio channels. A DVD player may have a large IC as a digital-to-analog (D/A) converter that separates the audio into the two stereo channels. Two separate operational (OP) amps provide audio to each left and right channel with a muting switch in each stereo output channel (Fig. 3-8).

A deluxe TV stereo section might consist of a stereo IC decoder, an IC audio switch, and a stereo IC bus expander. One IC contains the left treble and bass with right treble and bass circuitry, and a separate left and right power output IC then connects to a woofer and treble speaker in each stereo output circuit.

The most simple IC stereo amplifier in an AM/FM/MPX receiver and tape player might consist of a dual-output IC. Locate the suspected IC on a metal heat sink, and take signal input and output tests. Take critical voltage and resistance measurements. The left side of an R. J. McDonald AM/FM/MPX receiver was distorted, and the right channel weak. Replacing IC104 solved the dual audio output problem (Fig. 3-9).

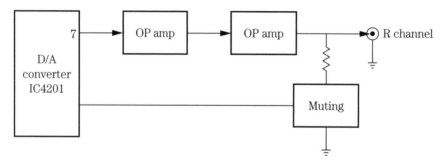

3-8 Block diagram of an IC stereo channel output circuit within a DVD player.

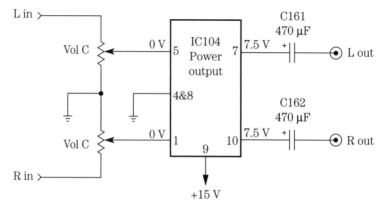

3-9 Replace IC104 for a distorted left and weak right channel in an R. J. McDonald AM/FM/MPX receiver.

3-10   Critical voltage and resistance tests on a single IC within a portable cassette player can locate the dead sound stage.

Repairing the cassette amp

Small cassette players may have three or four transistors, one large IC, or a combination of both. Cassette deck audio circuits might consist of several IC components or a combination of transistors and ICs. For instance, the equalizer or preamp tape head stages might consist of two transistors or one IC in a stereo audio circuit.

Look for a transistor within the recording bias oscillator circuits. Transistors can be used as audiofrequency (AF) or driver stages, whereas the latest stereo cassette decks might have several IC components. The audio output stage can be one large IC for both stereo channels or separate ICs (Fig. 3-10).

DISTORTED LEFT CHANNEL

Distorted audio can occur within either the right or left channel or both. If the right channel is normal and the left channel is distorted, go directly to the audio output IC. Try to locate the audio stages by looking for power ICs on heat sinks. You might find more than one heat sink, indicating separate audio output circuits. Sometimes both output ICs are mounted on one large metal heat sink or chassis.

You can compare voltage and resistance measurements within the stereo circuits against the defective channel. Often the left channel components are on the left side when looking from the front of the cassette deck. At other times they are not in line but are placed in a group of ICs, transistors, and decoupling electrolytic capacitors.

3-11 Replacing the Darlington IC output solved the distorted and weak audio in an AM/FM/MPX amplifier.

After locating the audio components on the printed circuit (PC) board, take critical voltage measurements or inject an audio signal into the audio circuits. Start at the center terminal of the volume control. Apply an external audio signal at each IC pin. When on the good channel, you will hear a loud tone in the speaker. The other IC output component is the distorted one (Fig. 3-11).

Distortion can be caused by a leaky IC, a change of resistance, or open or leaky bypass capacitors. Check each resistor tied to the power IC terminals, and compare them with the normal channel. Shunt each bypass or electrolytic capacitor or IC terminal to check for distortion. Check for high leakage across each capacitor terminal.

Noisy amp reception

Transistors, ICs, capacitors, and overheated resistors can cause noise in an audio amplifier. Notice if the sound is present when the volume control is turned down. If not, the noise is created in the front end of the audio circuits. Transistors and IC components can cause a low frying noise. Locate the noisy components by inserting a signal from a signal tracer or amp to the base terminal and then the collector of each audio transistor. When the noise is heard on the collector terminal and not on the base terminal of the same transistor, you have located the noisy transistor.

Likewise, signal trace the noise at the input and output terminals of ICs to locate a noisy IC. Sometimes applying coolant to the body of an IC can cause the noise to in-

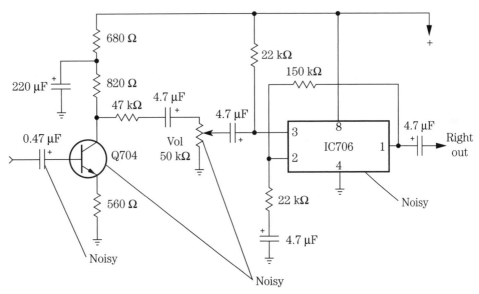

3-12 A noisy input electrolytic, transistor, volume control, or IC component can cause a frying or scratching noise in the speaker.

crease or go away. Heat applied from a hair dryer or heat blower also might increase the noise. Replacing the IC is the only answer (Fig. 3-12).

Warbling or squealing noises in playback mode can be caused by a dirty or worn function switch. Spray cleaning fluid into the switch area. Work the function switch back and forth to clean the contacts.

MAKE YOUR OWN SIGNAL CASSETTES

Signal tracing audio and adjustments within a cassette player can be accomplished with 1- and 10-kHz test cassettes. You can make your own test cassettes if you have an audio signal generator, cassette recorder/player, and blank cassettes. Blank cassettes can be picked up at most stores that sell cassette players.

Inject a 1-kHz signal from the generator to the tape head input or connect directly to the tape head. Connect the ground clip of the generator to the common ground of the tape head. Place the blank cassette into the recorder, and record both 1- and 10-kHz tones. Record the 1-kHz tone on one complete cassette and the 10-kHz tone on another one. Now you can signal trace a defective sound section in a cassette player and make the required adjustments with each signal cassette.

SIGNAL TRACING WITH A CASSETTE OR AUDIO GENERATOR

The 1- or 10-kHz test cassette can be used to signal trace a defective audio stage in a cassette player. Simply insert the test cassette and use the scope or external amp to signal

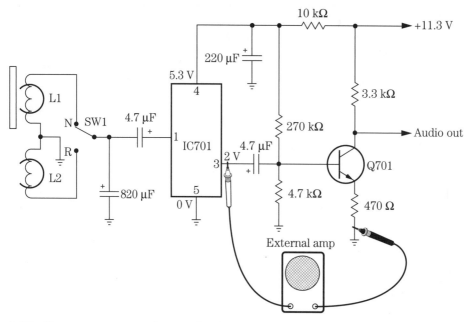

3-13 Use a test cassette and external audio amp to locate a defective audio stage.

trace the weak, distorted, or intermittent audio channel. Start at the volume control and work either way or at the tape head. Usually, the audio circuits are laid out with the audio transistor or IC preamps in line with the correct output circuits (Fig. 3-13).

By going from base to base of each audio AF or driver transistor, the volume should increase as you proceed through the audio output circuits. Check on each side of a coupling capacitor for weak audio reception. When the stage becomes weak or distorted, check the preceding circuits. Locate the base and collector pin terminals, and test for audio. When the signal is weak at the collector compared with the base terminal, suspect transistors or components within the circuit.

RADIO SHACK WEAK SOUND PROBLEM

In a Realistic portable AM/FM stereo double cassette player, the sound was very weak in the left channel. The external amp was connected to the left channel at the top of the volume control and double-checked against the right channel audio. Both stereo channels were normal at volume control sections VR3 and VR2. When the external amp was placed on pin 10 of the dual IC5, the sound was very weak compared with the right channel (pin 7). The C85 (1-μF) electrolytic coupling capacitor was replaced and solved the weak left channel problem (Fig. 3-14).

BLOWN SPEAKER FUSE

Go directly to the speaker fuse when a set or both speakers are dead. This can occur when the amplifier is turned up to a very high volume. Sometimes when the ampli-

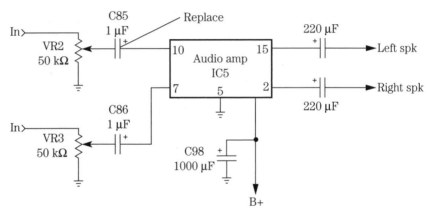

3-14 Weak sound in a Realistic double cassette player was caused by C85 (1 μF).

fier is pounding away at very high volume, the speaker fuse will open up. The speaker fuse will open up when failure in the output circuits becomes unbalanced and places dc voltage on the fuse or speaker terminals.

Check the speaker terminals for a low dc voltage before replacing the blown fuse. In high-volume amplifiers that do not have speaker fuse protection or an electrolytic coupling capacitor between the speaker and output transistors or ICs, a defective output stage can blow out the voice coil of a very expensive speaker. Connect a 7.5-ohm, 10-W resistor across each stereo speaker terminal when repairing the audio amp circuits. This speaker load resistor takes the place of a permanent magnet (PM) speaker. You do not damage a speaker with this method.

The speaker right channel fuse was found blown in a JC Penney 3223 high-powered stereo receiver. Power output transistors Q515 and Q514 became leaky and destroyed the 3-A fuse. Besides replacing the driver transistor and Q512, Q515, and Q514, both 0.33-ohm bias resistors were checked and replaced to eliminate the dc voltage placed on the FU3 fuse.

HUMMING ALONG

Suspect a bad filter capacitor or leaky and shorted output transistors or ICs when a loud hum is heard in the speakers. Rotate the volume control all the way down. Notice if the hum is still present. If so, go directly to the electrolytic filter capacitors. You may find one or two large upright filter capacitors mounted on the PC board. These large electrolytics have a tendency to dry up and produce a constant humming noise. Sometimes a connection inside the capacitor will break loose and cause a loud hum. Check the suspected capacitor with the ESR meter.

Shut the unit off and take a peek at the electronic chassis to find the largest filter capacitor (Fig. 3-15). Use test clips, and clip the negative terminal to the negative or outside terminal of the capacitor. Likewise, connect the positive or center terminal of the capacitor to the positive terminal of the suspected electrolytic. Do not clip the capacitor into the chassis with the power on. You might destroy some semiconductor components within the amplifier circuits.

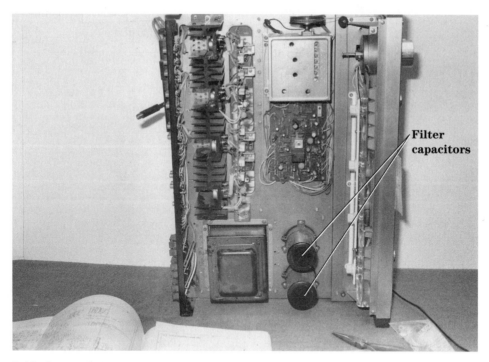

3-15 Locate the largest filter capacitors for speaker hum.

Choose an electrolytic capacitor with the same capacity and working voltage or one with a higher capacity and working voltage. If you do not, the capacitor might just blow up in your face. Excessive hum with both channels distorted was found in a Technics SUV7 amplifier and was caused by an STK805 audio output IC.

LOUD HUM, BLOWN FUSE

One of the main fuses was blown in a Pioneer SX-950 amplifier. Fuse FU2 (1 amp) was black inside the glass area. After replacing the fuse, a loud hum was heard, and the chassis was shut down. Since most line fuse faults are caused by leaky audio output transistors and ICs, one transistor was found leaky.

A quick test of the bridge rectifier circuits indicated normal positive 50.7 V and a high negative −64.5 V. Two voltage regulator transistors were found in the negative power source. Both transistors were tested in the circuit, and a 2SA720 transistor appeared open. The transistor was removed and tested again. After replacing Q1, the negative power source returned to −50.7 V.

ESR CAPACITOR TESTS

The ESR meter is ideal to check out and locate defective capacitors. The ESR meter can indicate a good or bad electrolytic capacitor. Make sure that the power is shut

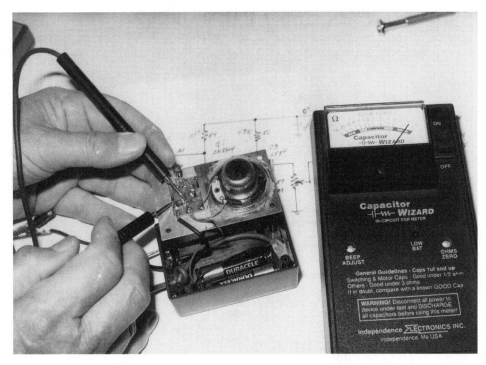

3-16   Check the electrolytic capacitors on the good scale of an ESR meter.

down in the electronic unit and that the electrolytic capacitor is discharged before connecting the meter. The equivalent series resistance (ESR) is the sum of all internal resistance within a capacitor. The ideal capacitor has 0 ohms (ESR). If the capacitor has 1 ohm of resistance, it generally is a good capacitor (Fig. 3-16).

Check the suspected capacitor by placing the test probes across the capacitor terminals. The ESR meter can check electrolytics in or out of the circuit. A good capacitor will measure below 1 ohm, whereas a bad capacitor will measure more than 10 ohms. The ESR meter will not test a shorted capacitor. Use the ohmmeter to check for shorts or leakage in an electrolytic capacitor.

Auto radio amp problems

In early auto radios, the tube was king; then along came the power output transistor, transistorized audio circuits, and IC components. Today, most auto receivers consist entirely of IC components. A new car radio might include both a cassette tape and CD player. The high-powered amplifier might be connected to the combination auto player.

Fifteen years ago transistors and IC components were found in deluxe AM/FM/MPX stereo car radio audio circuits. You might find one large IC as a preamp for the cassette tape head circuits, preamp IC audio, and power output transistors.

Today, IC components are found in most audio circuits of the common car radio. Besides IC components, processors and surface-mounted parts are found in car re-

3-17 Several boards must be removed to get at the different audio components in an auto receiver.

ceivers. In fact, several layers of PC boards must be removed before getting to the audio board (Fig. 3-17).

DEAD LEFT CHANNEL OF A SANYO FTC-26

The left channel was dead in a Sanyo auto receiver with the volume full on. Both the left channel cassette player and the AM/FM radio reception were dead. The external audio amp indicated no audio at the C713 (4.7-μF) coupling capacitor off of volume control VR704. The right channel was normal ahead of the volume control at C712. A resistance measurement across the control and sliding control of VR704 showed that the control was shorted. Replacing VR704 solved the problem (Fig. 3-18).

INTERMITTENT AND WEAK CHANNEL IN A FORD AUTO RECEIVER

The right channel in a Ford auto receiver was quite weak and intermittent at times. The external audio amp was clipped to the volume control as a monitor. When the volume went down in the radio speaker, the audio in the external amp was normal. Usually a driver transistor or coupling capacitor will go open, causing intermittent

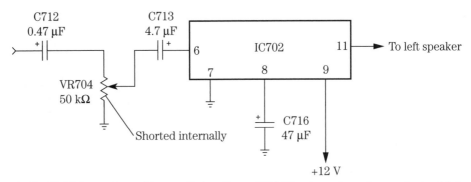

3-18 VR704 was shorted internally in a Sanyo FTC-26 auto radio and caused a dead left channel.

and weak reception. The audio output transistor or ICs can cause a weak and distorted symptom. Replacing a defective AF amp module (N3) solved the problem.

THE SPEAKER POPS AND CRACKS

When a car radio is turned on and a pop and cracking noise occurs, suspect a dirty or worn on/off switch. A loud pop and cracking noise can be caused by bad audio output transistors or IC components. Sometimes the music will pop and crack and become intermittent with a bad power output IC. Determine if the popping noise is before or after the volume control by monitoring with the external audio amp.

Very loud whistling and howling from the speaker can be caused by a defective filter capacitor. Shunt a similar capacitor across the suspected one and notice if the noise disappears. Tack the capacitor in with the power off. Howling can result from a red-hot leaky output IC. An annoying hum can be caused by leaky ICs or transistors. A leaky or dried up filter or decoupling electrolytic capacitor can cause a hum in one or both channels (Fig. 3-19).

The loud popping and cracking noise in a Craig T621 auto radio was caused by a defective output IC (UPC1185SH). This output IC was replaced with an ECG1293 universal replacement.

MOTOR BOATING

Replace both output transistors in the car radio with weak and distorted audio. Likewise, replace both leaky output transistors when you hear a motor-boating sound with the volume turned down. A motor-boating noise can be caused by poor grounding of the heat sink at the output IC. An open decoupling capacitor of 1000 µF can cause motor boating with a chirping sound. Replace the left power output IC when the channel is motor boating and the right channel is normal. A loud howling noise with no sound was the result of a red-hot IC303 in an Audiovox C979 car radio.

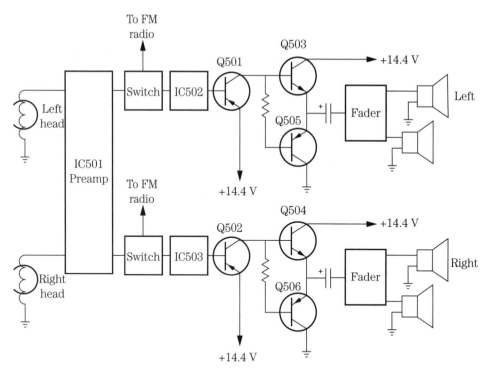

3-19   Block diagram of the sound system in an auto receiver.

Large stereo amp repairs

Most service problems within high-powered stereo amplifiers involve a weak channel, distortion, intermittent sound, or a dead amplifier section. Usually stereo amps
are easy to repair because one of the channels is normal and can be used as a reference. The channels are systematically laid out for location of components when servicing without a schematic. Voltage, resistance, and signal tests in the normal
channel can be compared with those in the defective channel.

Higher-wattage amplifiers may have transistors or ICs in the output circuits. These
components are located on a large heat sink or metal chassis. Often the bias resistors
and decoupling capacitors are found close by (Fig. 3-20). The AF, driver, and preamp
stages are found on a separate board or are mounted ahead of the power output components. The weak channel may be caused by a defective transistor, IC, bias resistor, or
coupling capacitor.

No left channel

Signal trace the audio at the volume control. If the signal is normal on both channels,
proceed to the output transistors. Inject an external audio signal into the base terminal of each output transistor or the input terminal of the IC and listen for a tone in

Power IC   IC heat shield

3-20 You may find power output ICs in the high-powered amp stages.

the speaker. The audio signal on the output transistor will be low when it is normal. The audio should be high at the input terminal of the power IC when it is normal.

Proceed toward the driver and preamp stages with the signal injection method. Likewise, the audio signal can be checked with a cassette tape or with the unit turned to the radio section in an AM/FM/MPX receiver. Use an external test amp for the latter method of signal tracing.

Check the left channel speaker for open connections. An open driver or AF transistors and IC preamps can cause a dead channel. Don't overlook open or dried-up electrolytic coupling capacitors. Leaky decoupling capacitors can open voltage supply resistors with no or improper voltage at the audio circuits. Often, leaky or shorted output transistors and IC components will have weak and distorted sound.

BOTH CHANNELS DISTORTED

Check the amplifier circuits for a common component when both channels are distorted in a stereo amp. A dual stereo power input or output IC can become leaky and cause distortion in both channels. Improper voltage from the power supply circuits can cause distortion. Either negative or positive voltage sources can cause distortion in both if one voltage is low or missing. Dirty or corroded function switches can apply incorrect voltage to the audio section, causing distortion.

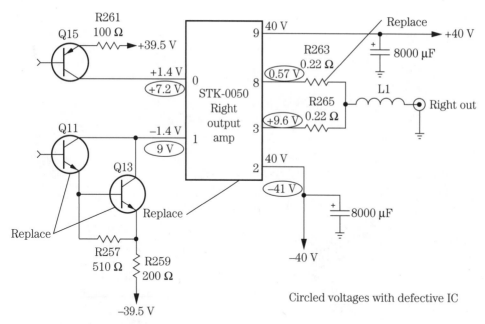

3-21   Leaky Q11 and Q13 keep destroying the power output IC (STK-0050) in a Pioneer amplifier.

REPEAT DAMAGE TO THE POWER OUTPUT IC

After a new power IC (STK-0050) was installed in a large Pioneer amplifier, the IC became warm, and extreme distortion was noted before the replacement was destroyed (Fig. 3-21). Dc voltage was found on the right speaker terminals. R263, a 0.22-ohm resistor, was found burned and was replaced. When the right channel IC was defective, the circled voltages (see Fig. 3-21) were found on the terminal pins.

All driver transistors connected to pins 0 and 1 were tested in the circuit. Q11 and Q13 indicated leakage. Both transistors were removed and tested. Since Q13 was directly shorted, both transistors were replaced. At the same time, bias resistors R257 (510 ohms) and R259 (200 ohms) were checked. R259 had changed resistance and was replaced. Zero voltage was found on the speaker load resistor (10 ohms, 10 W), indicating normal sound, after replacing the power IC (STK-0050) with an original replacement, Q11 and Q13, and resistors R259 and R263.

Troubleshooting CD player audio circuits

The stereo sound from a CD player originates at the D/A converter stage. Often the stereo sound is fed from the D/A converter to a dual IC audio amplifier. The audio output circuits might consist of a power output IC, left and right line output jacks, and a stereo headphone jack (Fig. 3-22). The CD audio signal may be switched in a combination cassette player, auto CD radio, or a boom-box player. In the high-powered amp, the CD music might be amplified several times before being heard at the speakers.

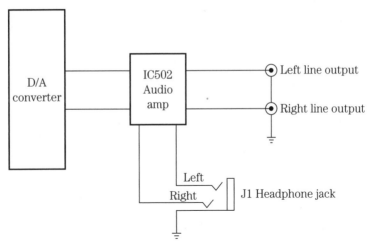

3-22 Block diagram of the audio circuits within a CD player.

Signal trace the audio from the output of the D/A converter IC through the sample hold IC and audio amp to the line output jack with an external audio amp. The stereo audio circuits begin at the output terminals of the D/A converter stage. Usually the audio amp in both channels is found in one IC component. Determine if the trouble occurs in the CD player or the sound amp connected to the line output jacks (Fig. 3-23).

If one channel is normal and the other is dead or distorted, use the good channel for voltage and reference measurements. Check the sound at points 1 through 4 (see Fig. 3-23) to locate the defective component. Check the input and output signals of each IC. When the signal becomes weak or distorted in the external amp, you have located the defective stage.

POOR MUTING

Do not overlook a defective mute switching transistor. The muting process takes place when mechanical action occurs in a CD player. The sound is cut out or switched to ground by a muting transistor. The emitter terminal of the muting transistor is at ground potential. Simply remove the emitter terminal and notice if the audio is now heard in the line output jacks. Check the suspected transistor with the diode tester of the DMM. If it is normal, suspect a defective system control IC when the audio is muted all the time.

In some deluxe CD players the muting transistor may control a switching relay. The servo IC may control the muting transistor and relay. The switching points of the relay open up the line output audio signal (Fig. 3-24). Suspect a defective transistor or relay when the muting system is not functioning.

DISTORTED AND NOISY HEADPHONES

In a Realistic 42-5029 CD player the right channel was distorted and had weak sound. The headphones were substituted with another pair, and indications were the

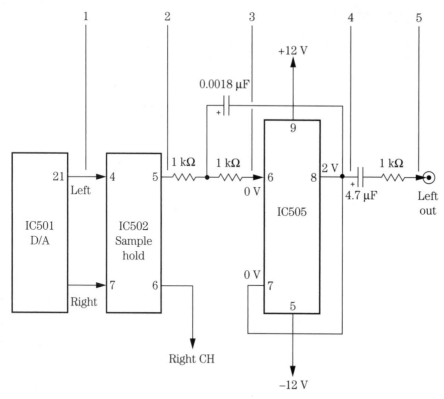

3-23 Signal trace the audio circuits within a CD player by the numbers.

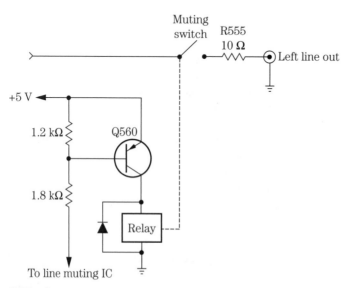

3-24 Suspect a transistorized relay assembly for poor muting within a CD player.

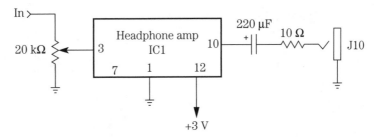

3-25 No headphone reception was caused by IC1 in a Realistic CD player.

same. The right headphone circuit was traced from the headphone jack back to a small 10-ohm resistor and a 220-μF coupling capacitor (Fig. 3-25). The electrolytic coupling capacitor was found connected to pin 10 of a dual-output IC. The voltage supply pin measured 1.7 V.

New batteries were installed, and the results were the same. Audio was signal traced into pin 3 of IC1, going to the volume control with distorted output at pin 10. The volume was up and normal in the left channel. IC1 was replaced, restoring normal CD headphone reception.

Servicing the TV audio amp

A TV audio amplifier consists of three or four transistors, a single IC, dual ICs, or two separate ICs in a stereo circuit. Sound symptoms in the audio output stages can include no sound, intermittent sound, weak sound, muffled sound, distorted sound, and noisy sound. Try to locate the sound discriminator coil and audio output IC. Often the IC is located close to the shielded discriminator coil assembly (Fig. 3-26). Several audio amp ICs in a TV chassis have their own heat sink on the PC board. Another method is to signal trace the speaker wires back to the audio output circuits.

HUM IN THE SPEAKERS

Most loud hum problems are caused by dried-up filter capacitors. A low hum can be caused by defective or open coupling capacitors. Poor grounds within the input sound stages can result in a low hum. A low hum with a squeaking noise might be caused by a defective output IC. Hum with a buzz in the audio might be caused by misalignment of the discriminator coil in early TV chassis. Check the surface-mounted transistor (Q1501) in an RCA F3570STFM TV with no audio or weak and distorted or severely distorted audio in both stereo channels (Fig. 3-27).

RCA TV WITH WEAK AUDIO

The audio was weak and distorted in both audio channels of an RCA CTC169 TV chassis. Check Q1501 for weak or severely distorted audio. A leaky Q1502 transistor can cause real loud distorted audio. Check both Q1501 and Q1502 in the stereo

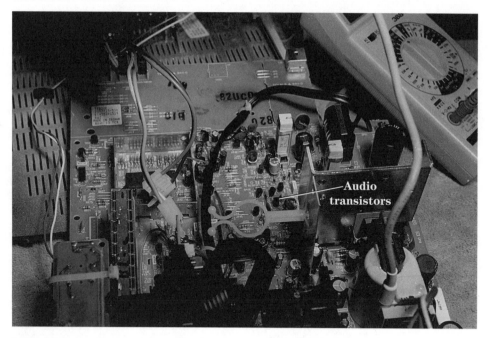

3-26 You might find three or more transistors within the sound output circuits of a small-screen TV.

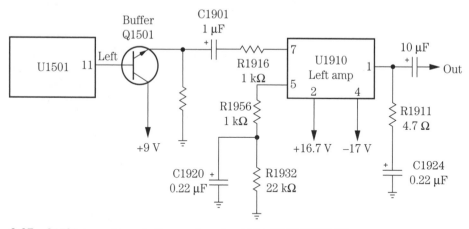

3-27 Q1501 caused hum in the speaker of an RCA F3570STFM TV.

sound channels with the diode tester of a DMM (Fig. 3-28). Be real careful not to damage the PC wiring traces when removing the surface-mounted transistors.

DEAD LEFT STEREO CHANNEL

Only a slight hum could be heard in an RCA CTC167 TV chassis. The right channel was fairly normal. The audio signal was traced through the left channel up to C1901

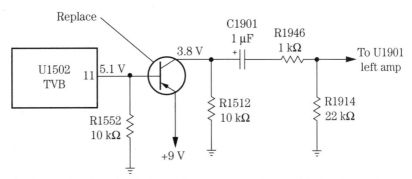

3-28 Replace both buffer transistors (Q1501 and Q1502) in the stereo channels of an RCA CTC169 TV chassis.

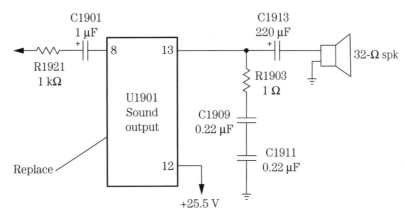

3-29 A slight hum with no audio was caused by sound output IC1900 in an RCA CTC167 TV chassis.

(1 μF) and was good. The sound was normal on both sides of C1901 and at pin 8 of the sound output IC (U1901). Only a slight hum could be heard at output pin 13. Replacing U1901 with a direct RCA replacement (181836) solved the dead left channel (Fig. 3-29).

POPPING AND SNAPPING NOISES

Look for defective audio output ICs or transistors when popping and snapping noises occur. Sometimes the popping noise occurs after the TV set has warmed up. Apply coolant on the suspected component to see if the noise disappears. Signal trace the noise at the input IC terminal or the base of the transistor with the external amp connected. If the noise is traced to the output component, replace it. Often, when the sound crackles and pops, check for a bad audio output IC.

The stereo light would flash on and off as the sound would pop and crack in a Goldstar NC-07X1 chassis. Critical voltage tests were made on the MPX/stereo IC601. Replacing IC601 removed the flashing stereo light and snapping noises (Fig. 3-30).

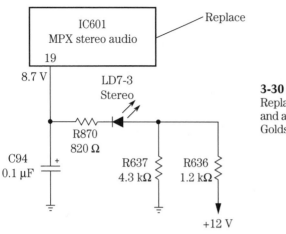

3-30
Replace IC601 with a popping noise and an intermittent stereo light in a Goldstar NC-07X1 chassis.

TOSHIBA INTERMITTENT SOUND

The sound was intermittent in a Toshiba TAC-7720 TV. The audio signal was monitored at terminal 2 of the IC (Q602) and was normal. The sound was normal up to C618, but on pin 9 the audio was intermittent at times at audio amp Q602. Remove and replace Q602. Replace the audio module for intermittent or no audio in the latest Toshiba CX32D80 TV.

DISTORTED AUDIO IN AN RCA CTC187 CHASSIS

The audio was distorted in both channels of this RCA model. A quick audio test at pins 59 and 60 of the T-chip (U1001) was fairly normal. The sound was normal up to the dual audio output IC (U1901). Extreme distortion was found at output terminals 3 and 10. Replacing U1901 with an original TDA7263 (part number 215526) solved the problem (Fig. 3-31).

WHITE RASTER WITH NO AUDIO

In a Panasonic CT-25ROK TV, the screen showed no picture and no sound with high voltage present. A voltage test on the line voltage regulator indicated that the B+ voltage had increased to +155 V. The normal working voltage of a STR30130 regulator is from 125 to 130 V. Replacing the STR30130 regulator and R802 (10 kilohms) restored the picture and audio (Fig. 3-32).

Extremely high-powered amps

A high-powered amplifier can have as much as 50 to over 2000 W of power to amplify pounding music. These high-powered amps may have flat gold contact areas for maximum current-handling conditions. Total amperage with individual amps

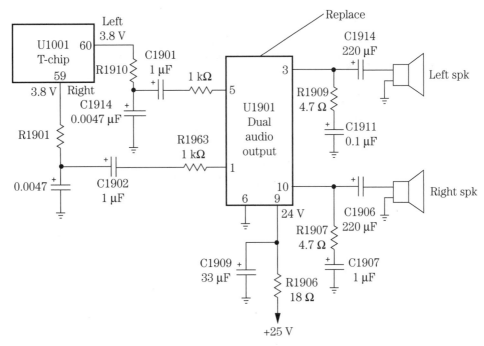

3-31 Replace dual audio output IC19011 for distortion in both channels of an RCA CTC187 chassis.

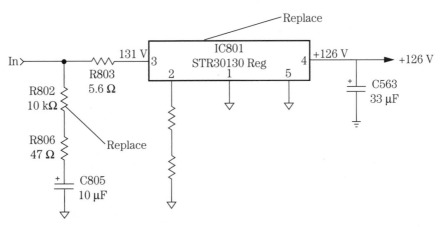

3-32 A white raster with no audio in a Panasonic CT-25ROK TV was caused by a line voltage regulator (IC801).

can exceed 1750 A. Each stereo section is lined up with gold-plated solid-copper bar strips down the center. The power output ICs and transistors may have extra heavy heat sinks or outside chassis heat sinks. Each stereo or mono amp might have its own power supply, removing channel-to-channel interaction and crosstalk (Fig. 3-33).

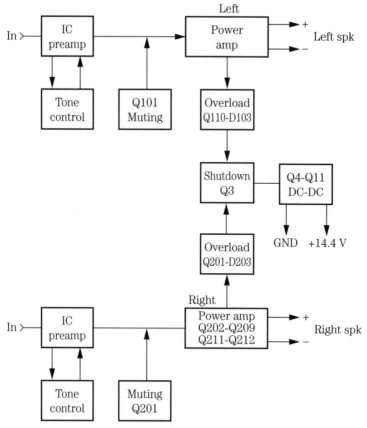

3-33   Block diagram of a 170-W power amp found connected to an auto radio receiver.

High-speed switching field-effect transistors (FETs) supply input power to the power toroids, keeping radiofrequency (RF) from entering or exiting the power amp supply. Fast-switching diodes may be found in some units. High-powered Darlington and triple Darlington output devices are found in some output stages.

Usually high-powered output stages are in line and easy to locate. Audio signal tracing methods can locate most defective components. Check for burned boards, flat bar connections, and burned toroids or resistors that might indicate leaky components. Compare the good channel with the defective one. Separate high-powered mono amps can be checked against a normal one.

Transistor and diode tests within the circuit can locate an open or leaky component. Actual voltage tests on power ICs can determine if an IC is leaky. Compare the voltage and resistance tests of a suspected component against the good channel. Substitution of high-powered ICs and Darlington transistors might be necessary after signal tracing the audio to the defective component.

3-34 A typical high-powered 2 × 350-W Pyle amplifier that connects to a 2-ohm speaker load.

Always replace high-powered amp transistors, ICs, Darlington transistors, FETs, and toroids with the exact part number. Remember, these are special parts.

A typical 150- to 200-W amplifier may have 2 ICs in the preamp circuits, a single transistor for muting, 12 power amp output transistors, and a single transistor in the overload and shutdown circuits within each audio channel. The high-powered Pyle PLA540 two-channel auto amplifier has a power RMS rating of 2 × 125 W in a 2-ohm load and a maximum power output of 2 × 350 W, and can be placed in a bridged mode to a single speaker at 1 × 700 W of power (Fig. 3-34).

DC VOLTAGE ON THE SPEAKER TERMINALS

When directly coupled power output stages are not functioning as they should, zero voltage is found at the speaker terminals. The power output transistors are connected directly together and are tied directly to the speaker output terminals or voice coils. If a power output transistor or diode becomes leaky or shorted within the power output circuits, the output circuits become unbalanced, and a dc voltage is found on the speaker voice coils. The voice coil begins to heat up and is found lodged against the pole magnet, eventually destroying the expensive speaker (Fig. 3-35).

Always check for a dc voltage on the speaker output terminals when no sound is heard in that channel. Check the output transistors for shorted or leaky conditions. You

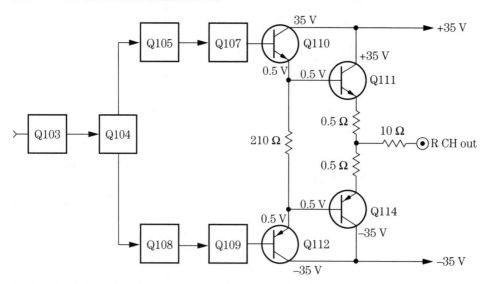

3-35 Check for a dc voltage on the speaker terminals when there is a damaged speaker in the right channel.

may find several shorted transistors and one or two that are open. Do not hook up a speaker to the output terminals until the dc voltage has been removed. Try to replace all damaged components with the same part number.

Check Table 3-1, power amp troubleshooting.

Table 3-1. Power amp troubleshooting.

Trouble	Remedy
Location—output	Locate the power output transistors and ICs on the large heat sinks.
Location—input	Check for the preamp transistors or IC connected to the tape heads or phono input. Check for a driver IC or transistor connected to the volume control.
Signal trace	Start at the volume control and work toward the front end or speaker.
Signal trace with test cassette	Insert cassette—signal trace at the volume control and output and input stages.
Signal trace with signal generator	Trace the 1-kHz signal from the input to the output speaker terminals. Go from base to base of each transistor and input to output terminal of the IC.
Locate the defective stage	When the signal stops, suspect the preceding stage.
Transistor tests	Check each transistor and IC where the signal stops. Test each transistor in and out of circuit.
IC tests	Signal trace the signal at input. If there is no output, take critical voltage measurements on each terminal. Take critical resistance measurements on each terminal.
Weak audio	Suspect defective coupling capacitors and bias resistors. Check for leaky or open transistors.
Distorted audio	Go directly to the audio output circuits. Check the transistors for open or leaky conditions. Check the IC voltages and resistance measurements. Suspect leaky coupling capacitors. Do not overlook an improper voltage source.
Weak and distorted sound	Check the audio output transistors and IC. Suspect bias resistors and bypass capacitors.
Intermittent sound	Suspect transistors, ICs, poor terminal connections, poor board connections, and cracked boards. Do not overlook the electrolytic coupling capacitors.
Too hot to touch	Suspect transistors and leaky output ICs. Test the transistors for leaky conditions. Look for burned components. Measure all bias resistors for a change in value.

4
CHAPTER

Servicing auto receivers

Today, the auto receiver has many new features, including a digital varactor and quartz tuning, AM/FM preset stations, Dolby B and C, auto reverse cassette player, CD and trunk changers, and a very high-powered amplifier. The car radio has made drastic changes since tubes and vibrators were king. Besides a cassette player, the auto receiver might have a CD changer control with the changer located in the car's trunk.

You will find many integrated circuit (IC) components doing double duty within the new auto receiver. One large IC might contain the front-end section, whereas one or two output ICs may be found in the audio circuits. Silent switching is now accomplished by using fixed diodes instead of a manual function switch. Several different printed circuit (PC) boards are found throughout the cramped spaces. Now a few surface-mounted devices (SMDs) can make the auto radio smaller physically (Fig. 4-1).

Required test instruments

The following equipment is needed to test car radio circuits:
1. Digital multimeter (DMM) or voltmeter
2. Transistor tester
3. Radiofrequency–intermediate frequency (RF-IF) signal generator
4. Sine- and square-wave generator
5. Oscilloscope
6. External amplifier
7. Frequency meter (optional)
8. Audio analyzer (optional)
9. Equivalent series resistance (ESR) meter

4-1 The early all-transistor car radio and cassette player had electronic components cramped on different PC boards.

The power supply

When the car radio is pulled from the dashboard, a bench power supply is required to provide a 12- to 14.4-V source in servicing the receiver. A low-priced power supply with a 13.8-V supply at 7 A can be purchased for less than $50. The variable 3- to 15-Vdc supply at 12 A may list for $150. When servicing those old car radios and high-powered amplifiers, choose a 30- to 40-A heavy-duty power supply. The variable regulated power supply with 0 to 24 Vdc is ideal for powering most electronic products (Fig. 4-2). Try to choose a variable power supply with voltage and current meters.

The front-end radio section

The front-end section of the latest auto radio might consist of an FM and AM tuner, an intermediate frequency/multiplex (IF/MPX) unit, an IF counter, a central processing unit (CPU), and analog switching. The AM tuner might have a radiofrequency (RF) transistor in the antenna circuit, RF and IF coils, ceramic stages, and oscillator circuits. Varactor diodes tune the RF and oscillator stages with a variable voltage (VT). One large IC might provide circuits for the RF, IF, and oscillator circuits (Fig. 4-3). The tuning voltage is controlled by a CPU IC. Locate the front-end circuits by themselves or from the antenna receptacle.

4-2 A bench power supply with a digital meter is a must item in servicing auto receivers on the service bench.

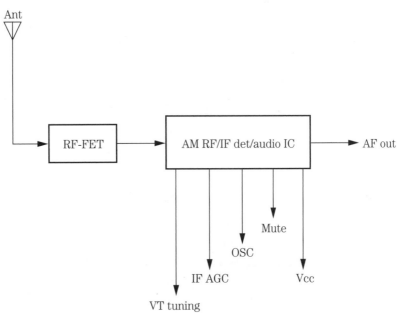

4-3 The front-end AM section of an auto receiver might have an RF FET amp and an RF/IF detector IC.

An RF transistor or a field-effect (FET) transistor amplifies the weak antenna signal to a tuned coil within an early car radio. Suspect a defective RF transistor when only one AM local signal can be heard. Measure the voltage on the RF transistor terminals. Test the RF transistor with a semiconductor test or the diode tester of the DMM. Likewise, test all RF and oscillator transistors with voltage and semiconductor tests.

The rear section

The rear section of a low-priced auto receiver may consist of an audio section, cassette player input circuits, CD player circuits, power supply, and AM/FM silent switching. A dual IC may be found in the cassette input circuits. The CD output circuits may feed to the stereo line output jacks or into a large power amp circuit (Fig. 4-4). You may find one large output IC with muting transistors that feed to left and right stereo fader speaker controls. Often the high-powered amplifier may contain 20 or more output transistors.

Start at the volume control center terminal or a test point on the external amp, using an audio signal generator to signal trace the audio circuits. The sine- or square-wave generator can determine if distortion is found in either stereo audio channel. Connect the signal generator to the RF or IF section to signal trace the front end. Injecting signals from a white noise generator can trace signals in both the RF and audio circuits.

4-4 An early auto radio might have a varactor tuning section and stereo audio circuits.

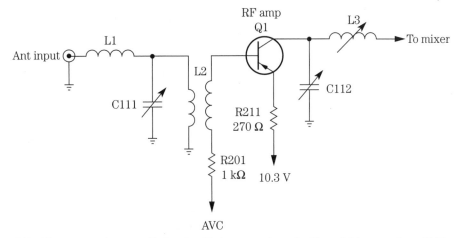

4-5   The very early car radio front-end stages are tuned with variable capacitors C111 and C112.

Variable-capacitor tuning

You might find in the early car radios a variable-capacitor tuning component. The antenna receptacle picks up the radio signal from the outside car antenna. The variable capacitor tunes the RF coil before the radio signal enters the base terminal of the RF transistor. Here the RF tuned signal is amplified and applied to the convertor stage. The converter or oscillator transistor selects the AM station and is combined with the RF signal within the converter stage. A variable capacitor tunes the RF, converter, and oscillator stages (Fig. 4-5).

Check the RF transistor for a weak or lost radio station. Sometimes only a strong local radio station can be heard when there is a defective RF stage. A broken antenna wire off the loading coil can cause a loud rushing noise with no tuned-in stations. Peek at the antenna coil for broken wiring. If a station fades out with a scratching noise, suspect that one or two plates of the capacitor are touching one another. Squirt a lubricating oil into the variable-capacitor rotor bearing when there is noisy tuning operation.

Inductance tuning

A variable-inductance or permeability tuner is the result of a powdered iron slug moving inside the RF, oscillator, and mixer coils. Permeability tuning was designed to take the place of ganged variable-capacitor tuning. These coils are wound with very fine wire, and as the tuning iron slug moves into the coil, a ganged tuning assembly moves the slugs in and out of the different coils.

Most problems with permeability tuning involve a dry gear assembly, broken wires on the coil winding, or poorly soldered connections, causing intermittent reception. A broken wire on the RF coil can cause no or only a local radio station to be tuned in. A broken slug inside the oscillator coil can cause stations to shift as the auto hits bumps. Poorly soldered connections or old soldered joints can produce intermittent reception (Fig. 4-6).

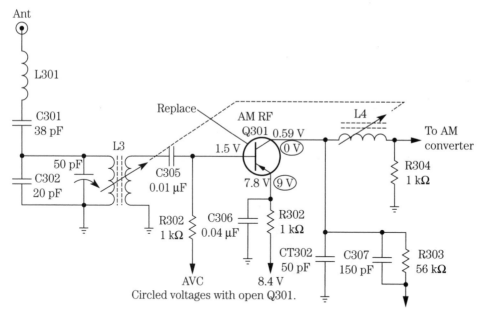

4-6 Only a local FM station could be heard in a Craig 3521 radio with open Q301.

Varactor tuning

Varactor tuning involves a varactor diode that changes capacity when a dc voltage (VT) is applied to it and is used to select broadcast stations. When the tuning system supplies voltage to the varactor diode, a different station is tuned in. Just by checking the different voltages on the varactor diode, you can determine if the tuning system or the RF, oscillator, and mixer stages contains the defective parts (Fig. 4-7).

In the Audiovox AV-215 with an FM station tuned in at 92.1 MHz, the control voltage is 2.1 V. This voltage increases as stations are tuned higher in the FM band. At 88 MHz, the voltage is 1.5 V, and at 108 MHz, the voltage is 5.5 V. In the AM band at 530 kHz, the voltage is 0.99 V, and at 1400 kHz, the tuning voltage is 5.83 V. This voltage will vary somewhat in different car radios.

When no tuning voltage is found at the varactor diodes or controlled transistors, check the CPU PLL processor. Measure the supply voltage (Vcc) on the supply pin. If voltage is supplied to the varactor diodes without any stations tuned in, suspect problems within the front-end circuits instead of the CPU quartz or digital synthesized circuits. Check the supply voltage on the suspected PLL or CPU processor.

Keeps blowing the fuse

Suspect shorted or leaky output transistors, ICs, a defective A lead capacitor, and several burned wires in the power harness when the fuse blows. The audio output transistor may be located on a separate heat sink or metal receiver framework. The output transistors in the high-powered amplifier are located on heat sinks with tran-

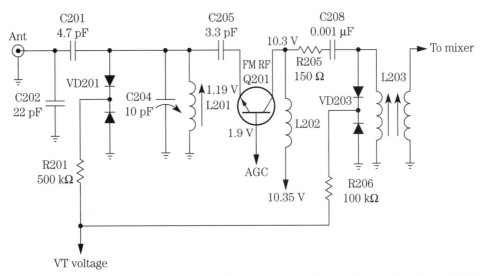

4-7 Improper varactor diode voltage (VT) caused the stations to fade out in a Ford D7AF (19A241AB) car radio.

sistors bolted to the large metal flange-type body. Check for a piece of mica insulation between the transistor and heat sink.

Locate the leaky transistor with in-circuit transistor tests. Remove the suspected transistor, and check out of circuit. While the output transistors are out of the circuit, check the bias resistors and diodes for leaky conditions. Double-check the directly coupled driver transistor for open or leaky conditions.

Too-hot-to-handle audio output IC components can cause the fuse to blow. When an oversize fuse has been inserted, the leaky output IC may burn the connecting foil of PC wiring and the A lead harness wires. Often the too-hot output IC may have gray or white burned marks on the body area. Check the IC circuits for a change in resistance or leaky bypass capacitor tied to the output IC terminals (Fig. 4-8).

PONTIAC 72FPB1 BURNED A LEAD

The 10-A fuse would blow at once when replaced in a Pontiac car radio. The A lead had been quite hot because the speaker wires were melted to the 13.8-V A lead. Power output transistor Q6 (DS-503) indicated a direct short between collector and emitter terminals. The audio output transformer wrapping had turned a brown color, indicating that heavy current was drawn through it. R63 was cracked and replaced. Replacing DS-503, R63 (0.68 ohm), and the 10-A fuse solved the blown-fuse symptom (Fig. 4-9).

Common auto radio failures

Blown fuses and leaky transistors and ICs cause most of the problems within car radios. The most common failures are the audio output transistors and ICs. Worn

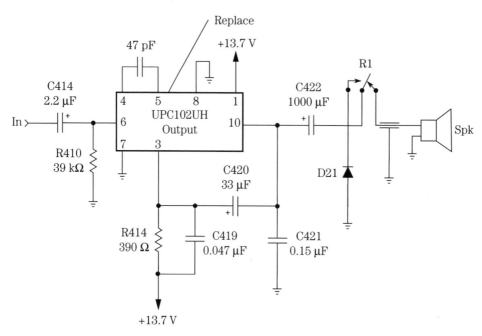

4-8 A leaky audio output IC in an Xtal XCB-23A car radio caused the fuse to open after replacement.

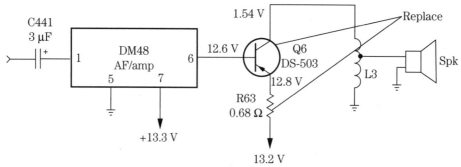

4-9 Replace both the DS-503 power output and fusible resistor R63 for a blown fuse and burned A lead.

or scratchy volume controls must be replaced after several years of wear and tear. Distorted, intermittent, and weak reception often occurs in the audio circuits. Fuse failure is caused by a leaky audio output component. Intermittent cassette audio also is caused by leads torn or worn off of the tape head connections.

Mechanical problems within the auto cassette player are speed, no-play, and eject problems. Speed problems result from worn or loose motor belts. Hold the capstan wheel, and if the motor stops rotating, the belt is good. Replace the drive belt when the motor pulley rotates inside the belt. A grinding noise can result from the capstan flywheel dragging against the bottom brace area. Install a new flywheel, or remove

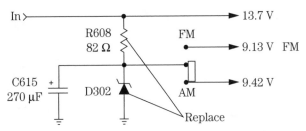

4-10 Weak AM and FM reception in an Automatic 6500 early car radio was caused by a leaky diode D302 and R608.

and tap the flywheel up into the correct position. Apply epoxy cement around the bottom of the capstan shaft and the body of the flywheel. Clean up the area with alcohol and a rag before applying the cement.

WEAK AM AND FM, AUTOMATIC 6500

Only a local AM station and a low rushing noise could be heard on an Automatic car radio. The tape player was normal. Critical voltage measurements were made on filter capacitors C615, C838, and C839. Hum was heard after touching a screwdriver blade to the center terminal of the volume control, indicating that the audio section was normal. The voltage check across electrolytic C615 was only 4.7 V that feeds to AM/FM selector switch SW904. Replacing the main filter capacitor C615, zener diode D302, and R608 solved the weak AM problem (Fig. 4-10).

WEAK AM RECEPTION

A defective antenna or bad receptacle, RF loading coil, RF coupling capacitor, and RF transistor and an extremely low supply voltage from the power supply can cause weak AM reception. In early car radios, the loading coil would shake loose, and the RF transistors produced most of the weak symptoms. Take a peek at the antenna receptacle and load coil for broken connections. Sometimes the RF coil has broken off and is found in the bottom of the car radio cabinet. Locate the RF coil, and trace out the circuit to the first RF or FET transistor. Take critical voltage tests on the RF transistor. An open or leaky RF transistor can cause weak AM reception.

When the RF stage is not functioning, placing the positive probe on the collector terminal of the RF transistor may bring in a local broadcast station. This means that both the mixer and the oscillator circuits are normal. Check the voltages of the transistors or IC components in the mixer and oscillator circuits to determine if the circuits are functioning. You might find that both AM and FM operate from the same component. Clean the AM/FM switch with cleaning spray.

The FM circuits consist of an oscillator and a mixer stage, as in the AM section, except at a much higher frequency (88 to 108 MHz). Most of the problems that occur in the FM stages are caused by leaky transistors, burned bias resistors, and improperly applied

4-11   Check for component markings on the PC board.

voltage. Check all three transistors within the FM circuits. Take critical voltage and resistance measurements.

If the AM section is normal, the FM problem must be in either the RF, oscillator, or mixer stages. Locate the FM RF transistor or IC from the antenna switch and small coils of wire. Notice that the FM coils are only a few turns of bare wire, whereas the AM coils have many turns on a form or rod. Parts in the front end can be located with markings on the PC board (Fig. 4-11).

A loud FM rush with the volume turned high can indicate a defective mixer or oscillator circuit. Sometimes a local station can be tuned in with a meter probe on the collector terminal of the RF FM transistor. Don't overlook a defective AM/FM switch.

WEAK CAR RADIO RECEPTION

Go directly to the antenna RF circuits when the signal is weak for either AM or FM reception. Locate the RF transistor tied to the coils and the AM/FM function switch. Take critical voltage measurements. If a schematic is not available or you cannot find a transistor number, use the operating voltage, and find it in the manual of universal replacement transistors.

Determine if the transistor is an NPN or PNP type. If the radio is fairly new, you can assume that it is an NPN transistor. With a positive voltage measurement on the collector terminal and emitter tied to ground, the transistor is an NPN type.

Improper supply voltage to the AM or FM circuits can produce weak reception. If the supply voltage is low, visually trace the voltage back to the isolation resistors and zener diode circuit. Usually the voltage is regulated by a zener diode or transistor.

INTERMITTENT AM RECEPTION

Determine if the intermittent signal is in the front or rear end of the car radio. If the cassette or CD player operates, you can assume that the audio circuits are normal. Go directly to the AM circuits if the FM reception is normal, indicating that the IF section is okay. Try to determine if the AM signal is intermittent within the front-end section.

If the AM and FM reception are intermittent and the cassette player is normal, you know that the trouble must result from the function switch to the IF section common to both bands. Usually when the AM, FM, and cassette player are intermittent, the trouble lies in the audiofrequency (AF) to the speaker. When both AM and FM stages are intermittent, either the IF circuits or the supply voltage contains the defective component (Fig. 4-12). Don't overlook broken or poor outside antenna connections.

Poorly soldered boards, bad transistors or ICs, poorly soldered terminal leads or eyelet connections, poor antenna connections, a defective antenna jack, or intermittent voltage applied to the front-end circuits can cause intermittent problems. Suspect a dirty or worn volume control with intermittent reception.

New FM circuits

The old FM front-end circuits consisted of an FM RF amp, an FM mixer, and FM oscillator transistors, whereas the latest FM circuits contain an FM tuner unit. The FM tuner might consist of a large IC that provides the RF, mixer, and oscillator circuits in one component. A PLL controller controls the AM and FM tuning voltages for the front-end tuner. The PLL circuits use digital locked-loop circuits to synthesize AM and FM oscillator circuits. Instead of IF coils, ceramic filters are used throughout the

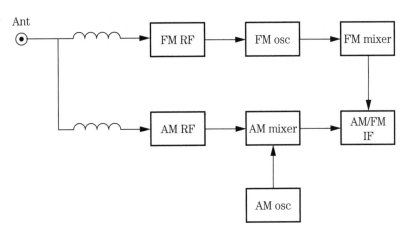

4-12 Determine if the intermittent AM reception is within the AM circuits when the FM, cassette, and CD player are working.

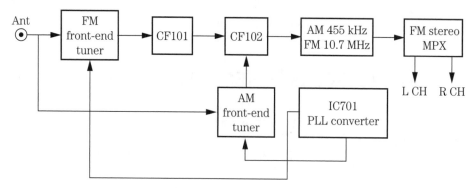

4-13 Block diagram of the AM/FM IF circuits within present-day auto receivers.

IF and detector stages. Usually the AM 455-kHz IF stages are combined with the 10.7-MHz FM IF circuits (Fig. 4-13).

WEAK FM, AUDIOVOX BLM CAR RADIO

A defective FM mixer transistor may produce only a rushing noise and no stations tuned in, the open or leaky FM oscillator transistor may not tune in a station, and the defective RF FM transistor may tune in only a local FM station or none at all. Place a test probe from the meter to the collector terminal of the RF FM transistor, and try to tune in a local FM station. Suspect the early RF FM FET transistor with no or very weak FM station. TR101 in an Audiovox BLM-105031 car radio was replaced with a 2SK49 transistor or with a universal NTE-312 or SK3834 replacement. Most transistors within the auto radio can be replaced with universal transistors without any problems (Fig. 4-14).

SMD COMPONENTS

You will find both conventional and surface-mounted components within the latest auto cassette/CD player receivers. The surface-mounted devices (SMDs) can be mounted on one side of the board, with regular parts on the other. Several different components are found within the cassette/tuner board (Fig. 4-15). Notice the very fine PC wiring that connects the surface-mounted processor to the circuit.

Surface-mounted transistors and IC components are used throughout the circuits. Besides surface-mounted ICs and transistors, diodes, capacitors, and resistors are used in many circuits. Notice that diodes can have two active tabs, and three-legged transistors can have different lead identifications. Power output ICs should be bolted to heavy heat sinks.

Digital transistors can be found in the audio preamp and the mechanism and system control. The digital transistor can have an internal resistor in the base circuit or another bias resistor between the base and emitter terminals. When these digital resistors are checked with a transistor tester or the diode tester of the DMM, the resistance is higher in the base terminal. Likewise, the measurement from base to emitter with the internal base-to-emitter resistor is different. Compare another similar digital transistor with the low base-to-resistor test before discarding the suspected leaky transistor.

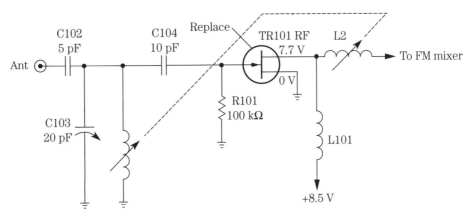

4-14 Weak AM reception was caused by a defective TR101 RF transistor in an Audiovox BLM-105031 car radio.

4-15 SMD transistors, ICs, and processors can be found on the auto receiver PC board.

Silent listening

In older car radios, the FM, AM, and cassette player are switched into the audio section with a manual switch. Today, you will find several diodes in each section that allow each section to be switched into the audio section. The AM and FM sections might be switched in with a separate AM/FM switch. The FM stereo and AM sections are switched in with diodes, and likewise, diodes are placed in the cassette player circuit when the power is applied to the cassette motor and audio circuits. No large manual function switch is found in the latest radios, cassette, or CD circuits.

Early car radio audio output circuits

Early car radios had power output tubes in the audio circuits that were replaced with a single audio output stage (see Fig. 4-9). The output transistor was replaced with a single or dual audio output IC. The dual audio IC might contain all or part of the audio circuits. Most of the audio stereo circuits are contained in one large IC (Fig. 4-16).

The FM multiplex (MPX) stereo sound circuits in a Delco 1980 series consisted of a driver IC U5 (DM116) capacitor coupled to a single power output IC in each stereo channel. The speaker fader control would shift the audio to the rear or front speakers. Remember that most Delco auto radios do not have the speaker return wire at ground potential. You can damage the output IC if a permanent magnet (PM) speaker is connected to the output IC and chassis ground.

DEAD, NO AUDIO, REALISTIC 12-2103

No sound with no lights lit up was seen in a Radio Shack 12-2103 in-dash AM/FM stereo cassette player with auto reverse. Both a 0.5- and an 8-A fuse were blown in the 12-V and memory B+ line. Since both channels were dead, critical voltage tests

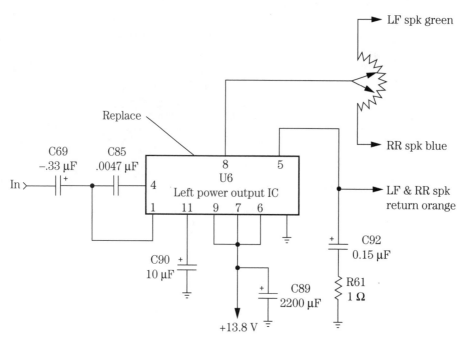

4-16 A leaky U6 power output IC in a Delco 1980 series resulted in weak and distorted music.

were made on the power output IC701. The power IC showed signs of overheating and would blow the main fuse. IC701 was replaced with a TA8210AH replacement.

WEAK AUDIO

Just about any component in a car radio can cause weak sound in the audio circuits. Try to isolate the weak section to the left or right channel. If both channels are weak, suspect an improper supply voltage on a common audio output IC. Weak audio can be caused by an open ground within the volume control. If sound is coming into the top and not at the center terminal, suspect a defective control.

Most weak audio conditions are caused by transistors, ICs, and coupling or bypass capacitors. If the left side is weak but not the right, isolate the weak stage with an external amp. Go from base to base on transistors or IC input and output terminals to determine where the signal is lost. Take critical voltage and resistance measurements when a suspected component is located.

DISTORTED LEFT CHANNEL

Any one or both channels can be distorted in an auto receiver. Most distortion problems are caused in the audio output circuits. Locate the power output transistor on the metal heat sink or metal chassis. Replace the leaky or shorted power output transistor or IC for a distorted left channel. Replace both leaky output transistors for a weak and distorted sound. A shorted decoupling capacitor in the power supply can cause dead or distorted audio. Check the suspected audio output transistors or ICs if they are running too warm to touch.

When the left channel is distorted and the right channel is normal, signal trace the left channel. Check for distortion at the volume control. Signal trace the rest of the audio circuit with an external amp to locate the distorted stage. Injection of a sine- or square-wave signal will indicate which stage is defective, with the scope as indicator. Clipped sine waves or rounded square waves indicate distortion.

Improper voltage found on the transistor or IC can be caused by a leaky zener diode or transistor regulator. Don't overlook shorted or leaky bypass capacitors and leaky decoupling electrolytic capacitors for insufficient voltage that can produce distortion. Overheated or red-hot output ICs definitely cause excessive distortion. Suspect the common output IC when both channels are distorted. Replace the leaky output IC (DM48) for left channel distortion in a Delco 1977 series car radio.

ONE OR BOTH CHANNELS ARE INTERMITTENT

Suspect a dual power output IC when both channels are intermittent. Sometimes a defective speaker coupling capacitor made by the same manufacturer can produce intermittent sound in all connected speakers. A shorted volume control can cause intermittent left and right channels. Broken foil or a broken coupling capacitor tied to the volume control can cause an intermittent channel. Weak and intermittent Ford D6TA-18806-AA car radio sound was caused by an intermittent AF amp module (N3).

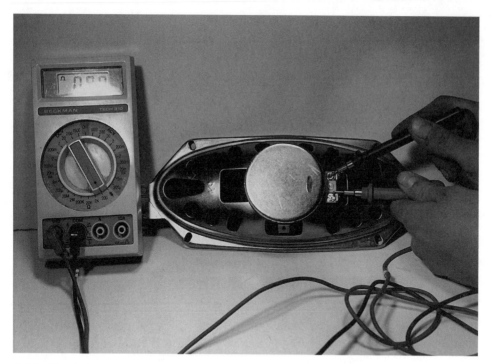

4-17 Check the suspected open speaker voice coil with the low-ohm scale of a DMM.

When the radio warms up and then cuts out, suspect a leaky power IC. An excessively worn volume control can cause the audio to cut out. A shorted volume control also can produce intermittent sound. Broken coupling capacitor leads or an intermittent break inside the IF coil can produce intermittent reception. Bad board connections, terminal leads, and speaker connections can cause intermittent audio.

Try to isolate the intermittent audio to the left or right channel. Locate the correct intermittent channel with an external amp connected to the volume control of the intermittent speaker. Sometimes when signal tracing, if the probe is touched to a transistor terminal, the sound returns to normal. Here, a defective transistor has been shocked back to life; replace it. Carefully go from stage to stage to locate the intermittent sound problem.

Suspect the dual audio IC when both channels are intermittent. If the IC seems normal, check for a voltage change at the low-voltage power supply. Check for a broken voice coil wire in one speaker. Loose speaker wires and cables can cause intermittent sound problems. Check all speakers for intermittent open or loose clip connections (Fig. 4-17).

NOISY RECEPTION

Suspect an intermittent power output transistor or IC when a car radio operates for 10 minutes, pops, and goes dead. A loud popping noise in the speaker can result from a 500-μF filter capacitor that has broken loose from the PC board. Replace the audio output IC when the radio produces a cracking sound when it is first turned on.

Suspect the AF transistor if there is noisy audio or the radio goes dead. You may have to replace both the driver and output transistors when there is a noisy left channel and an intermittent right channel. Replace the audio output transistor (Q6) when there is a pop and cracking noise in a Delco 01XPB4 car radio.

Check the output transistors and IC components for popping and crackling noises. If you hear a loud pop and the channel goes dead, suspect the output IC. Sometimes when stations are tuned in or the radio is first turned on, a noisy output IC will cause a crackling noise. Low frying noises can be caused by noisy transistors or ICs.

A motor-boating noise found in either channel can be caused by the power output IC. Suspect the dual output IC when very loud motor boating is heard with the volume turned down. A chirping noise can be caused by open or blown speaker coupling capacitors. Check for poor grounds in the output circuits or on the IC heat sink when motor-boating and buzzing noises are heard. Suspect the first AF transistor for microphonic noise.

Check the large filter capacitors or output ICs when only a hum is heard in the speaker. Overheated transistors or IC output components can produce hum in the audio. Low pickup hum can be caused by decoupling capacitors in the AF or driver circuits. Improper grounds in the volume control or AF transistor circuits can cause a pickup hum. Suspect a defective volume control for open or noisy conditions (Fig. 4-18).

Lower the volume control to determine if the noise or hum is in the audio or front-end section. Suspect a red-hot output IC for a loud howling noise. Shunt the main filter

4-18 Check the transistors, ICs, and various electronic components for noisy car radio reception.

capacitor for a loud or high-pitched noise. When a screeching noise is heard when stations are tuned in, check the small decoupling capacitors in the AF and driver circuits. Inspect shields and look for loose IC mounting screws when a squawking noise is heard after the set warms up and with the volume turned down.

Auto radio hookup

The battery and ignition power sources usually are fused within an auto receiver. When both a cassette and a CD player are found in a modern auto receiver, the two units are connected together with many cables and wires from each unit. You may find four different colored speaker wires from the unit with a red A lead in and a black ground wire. A simple car radio hookup might consist of only a left and right speaker cable, common ground for the speakers and radio returns, and a fused battery A lead that plugs into a socket at the rear of the auto radio (Fig. 4-19). Check Table 4-1, radio receiver troubleshooting.

High-powered amps

A 50- to 100-W amplifier might consist of a large high-powered IC or several transistors. A 100- to 2000-W high-powered stereo amp may have 24 transistors within the input and driver stages with a total of 32 or more transistors in the amp and power supply. Most high-powered amp power supplies have metal-oxide semiconductor field-effect transistors (MOSFETs) in the pulse width mode (PWM) and dc-dc converter circuits.

The root mean square (RMS) measures the power in watts that the amplifier can produce continually. A peak power rating is the maximum power that the amp can deliver during a brief moment of operation. RMS power at 2 ohms can tell you how much power the amp delivers into a 2-ohm speaker load (Fig. 4-20).

The high-powered amp within a radio, cassette player, or CD player may provide 15 to 50 W of peak power. You may find transistors and IC output components in these units. The really high-powered amp may use IC preamps, transistors, and many high-powered transistors in directly coupled output circuits. For instance, an RF Punch 500A2 amplifier has an RMS power of 250×2 W with a bridged RMS power of 500×1 W in a single-output speaker.

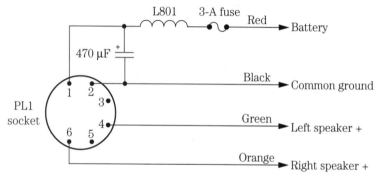

4-19 A very simple car radio hookup that plugs into a rear socket.

Table 4-1. Radio receiver troubleshooting.

Symptom	Circuit	Remedy
Cannot tune in stations	RF oscillator mixer circuits	Check the oscillator and mixer transistors and ICs. Check the voltage.
No tuning lights or digital channel	Synthesizer controller	Determine if the controller is tuning stations. Check for applied voltage on the varactor diodes.
No AM	AM converter	Test the AM converter transistor or IC. Take voltage tests.
Weak AM	AM antenna coil AM RF amp	Check for a broken core or winding. Check the RF FET. Perform critical voltage tests.
Distorted AM	AGC circuits	Check the resistors and capacitors in the AGC circuit.
Noisy AM	Transistors and ICs	Determine if there is noise in the radio. Remove the outside antennal plug. Check the AM converter and IF transistors.
No FM—normal AM	FM-RF oscillator and mixer circuits Dirty AM-FM SW	Check the FM voltage source. Check the FM FET. Spray down into the switch area, and clean the contacts.
Weak FM	Check RF-FM transistors and ICs	Test the RF FET. Perform critical voltage tests. Perform a complete FM alignment.
FM distortion	IF AGC circuits Matrix IC	Test the AGC transistor. Perform critical voltage tests. Replace the matrix IC. Perform critical voltage tests. Replace the matrix IC.
Noisy FM	IF-IC and FM matrix IC	Apply coolant and heat on the ICs and transistors.
FM hum	Power supply Voltage regulators	Check the filter capacitors. Sub the filter capacitors. Test the transistor and zener diode regulators.
No stereo operation	MPX circuits	Check the voltages on the MPX IC. Signal the stereo channels a scope or external audio amp.
Excessive stereo	MPX IC	Take critical voltage on the IC. Replace the MPX IC.
Stereo light does not light or stays on all the time	MPX IC or transistor light indicator	Check for a leaky transistor or IC. Check the LED or indicator light.

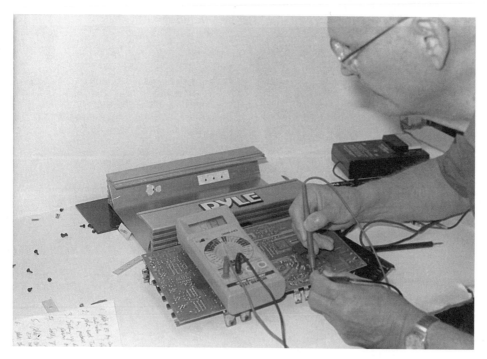

4-20   Checking the power output transistor in a high-powered amplifier.

In a typical high-powered auto system, you may find a separate amp for two tweeter speakers, another amp for midrange speakers, and a separate amp for four large subwoofer speakers. These high-powered amps and speakers normally are found in pickup trucks or large automobiles. The various 2- and 4-ohm speakers are connected to the two- and four-channel amps with left and right stereo channels.

Dismantling high-powered amps

You will find a heavy-duty heat sink that is part of the amplifier's cabinet to dissipate heat from the MOSFET power supply and high-powered audio output transistors. The molded heat sink is part of the outside cabinet or chassis. Remove the end pieces by removing several metal screws. Some of these covers have wires and sockets attached. Be very careful not to break loose cables and wires in removing the top metal cover.

Take a peek to see if the top cover will not come entirely loose until several power output transistors and MOSFET power supply transistor mounting screws are removed. Carefully pull up the transistors from their mounting insulators so that the chassis or cover can be removed. On some units the amp PC board will slide out of the bottom side of the metal chassis. After all repairs are made, make sure that the mica or clothlike insulators are replaced between the transistor and metal heat sink.

Locating the correct components

Look over the PC board for possible overheated or damaged parts. Overheated transistors may have gray or white marks on the body area with burned terminals. Check the body of each power IC for damage or chips blown out of the plastic body. Look for overheated and burned resistors that can be traced to leaky or shorted output transistors. High-powered output transistors are usually a little larger than regular TO-style transistors (Fig. 4-21).

Turn the PC board over and check for poor or burned traces and connections. Double-check the terminals of each power output transistor for burned or overheated connections on the PC board. You might find a poor connection on one of the high-powered transistors or MOSFETs in the power supply.

In a 50 × 2 W amplifier with power output transistors, the power supply might consist of two regular transistors and two high-powered MOSFETs tied to a toroid transformer. The dc-dc input voltage of 14.4 V comes from the car's battery with an operating output of +25 and −25 V. Look for the largest filter capacitor on the PC board to locate the various power supply components. Notice if four fixed diodes are nearby. Check for the dc-dc output voltage across the 1000-μF 50-V electrolytics.

The pulse width mode (PWM) and the dc-dc converter power supply of a 170-W amplifier might consist of a PWM IC, a couple of input transistors, and six MOSFETs

4-21   Locate the various parts on the high-powered amp chassis before checking the electronic components.

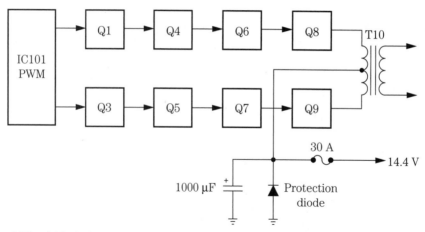

4-22 A block diagram of the dc-dc power supply in a 170-W high-powered amp.

tied to a toroid transformer. The secondary winding ties to four fixed silicon diodes or a bridged-type rectifier component. The input voltage is around 14.4 V, with an output voltage from the power supply at 34.5 V dc. The +34.5 V feeds directly to the six MOSFETs (Fig. 4-22).

Transistor high-powered amps

A 2 × 50 W auto amplifier might have four power output transistors in a directly coupled power output circuit. A negative −25 V and a positive +25 V are fed to the push-pull output transistors. Q16 might be a 2SA965 or a 2SB560 power driver transistor, whereas Q14 is a 2SC2235 or a 2SD438 PNP transistor. The final output transistor Q18 is a BD907 PNP, and Q20 is a BD908 NPN-type transistor. All high-powered output transistors should be replaced with the original part numbers (Fig. 4-23). If the originals are not available, try using universal replacement transistors.

In a 150- to 200-W amplifier, the output transistors operate at a higher 34.5 V and in a push-pull circuit. Both right and left stereo channel circuitry is the same, with two driver and two output transistors. Q206 and Q208 are driver transistors with NPN and PNP transistors, respectively. Q207 (NPN) and Q211 (PNP) are high-powered output transistors. There might be a total of 10 transistors in each high-powered output 150- to 200-W amplifier stereo channel (Fig. 4-24).

IC high-powered amps

A 15- to 50-W amplifier might contain transistors or IC components. You may find a separate power IC in each stereo channel. In many of the latest low- or medium-powered output circuits, a dual IC might be found. Audio muting takes place at the input of the dual-power IC. Usually the power IC is found on a large metal heat sink.

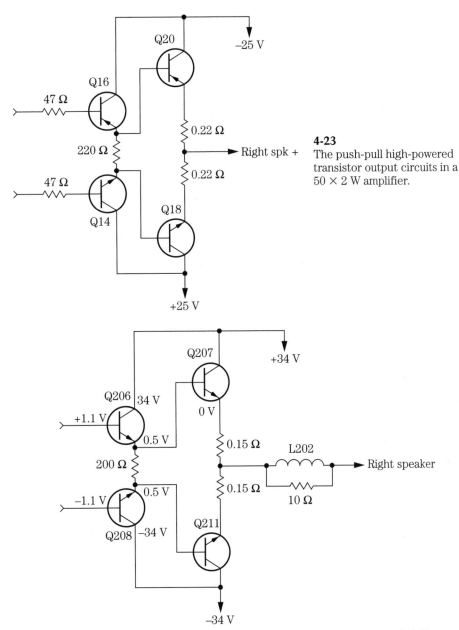

4-23
The push-pull high-powered transistor output circuits in a 50 × 2 W amplifier.

4-24 The right high-powered output transistor circuit in a 150- to 200-W amplifier.

In a 25-W Optimus 12-2103 AM/FM stereo cassette player with auto reverse, the left channel was weak and the right channel distorted. The audio signal was normal up to input terminals 2 and 7. Both outputs at pins 15 and 11 were weak and distorted on the external audio amp. IC701 was replaced with the original IC replacement (TA8210AH) (Fig. 4-25).

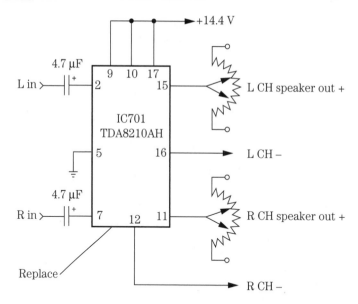

4-25 Replace IC701 for weak and distorted channels in an Optimus 25-W power output amplifier.

Troubleshooting high-powered power supplies

A simple power supply circuit is found in amplifiers under 50 W and might consist of the A lead, hash choke L101, a 5- to 10-A fuse, and several bypass capacitors. Some car radio power supplies have a polarity diode (D101) that protects the radio circuits if the battery is charged up backward or installed backward. D101 will blow the 8-A fuse if −14.4 V is applied to the A lead (Fig. 4-26).

You might find a transistor voltage regulator circuit, diode protection, and decoupling electrolytics in a low-voltage power supply. A bad electrolytic (1000 μF) on the voltage input line can cause some hum and motor-boating problems. An open decoupling capacitor (500 to 1000 μF) can produce motor boating and a chirping noise. A broken loose filter capacitor from the PC board can cause a loud popping noise. A dead Delco 51CFP1 car radio only hums with a leaky decoupling capacitor in the power supply.

Check for a shorted or leaky transistor in the high-powered power supply that can damage driver transistors and MOSFETs. A leaky or shorted MOSFET can cause one or two other transistors to be damaged because they are connected in a series-parallel power supply circuit. Quickly check all transistors with a semiconductor tester or the diode tester of the DMM. Likewise, check all diodes in the power supply with the diode tester of the DMM. Remove one end of the fixed diode for accurate measurements.

Discharge all electrolytics within the power supply before taking leakage or ESR tests. You can damage the ESR meter if electrolytics are not discharged.

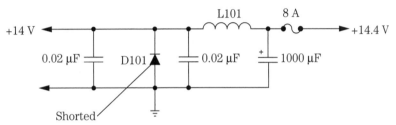

4-26   The repeated failure of the 8-A fuse can be caused by a defective protection diode (D101).

Check each electrolytic for shorted or leaky conditions with the 2-kilohm scale of the DMM.

Do not overlook a shorted silicon diode or bridged diodes in the secondary winding of the dc-dc power supply. Leaky or dried up electrolytics within the power supply can load down the dc-dc circuits and cause the high-voltage power supply to not function. Sometimes a shorted or leaky transistor in the amplifier can shut down the power supply and lower the dc output voltage. Check overloaded and shut-down transistors for leaky or shorted conditions.

If the 20- to 30-A fuse keeps blowing, suspect a shorted MOSFET, diodes, or electrolytic capacitors in the dc-dc power supply. Remember, if the polarity diode becomes shorted, the main fuse will keep blowing until the diode is replaced. Check for the right amp replacement fuse, silicon diodes, shorted filter capacitors, and MOSFETs when the fuse keeps blowing in the dc-dc power supply (Fig. 4-27).

Servicing high-powered amps

Check for a blown fuse if the high-powered amplifier is dead. Take critical voltage tests across the largest filter capacitors. Service the high-powered power supply if there is no or low voltage to the output transistors. Make sure that the dc-dc power supply is functioning before tackling the high-power output circuits. Make sure that a speaker load resistor (7.5 ohm, 10 W) is shunted across each stereo speaker channel. The dead channel might be caused by an open regulator transistor in the power supply.

The high-powered amp might have a dead right channel, weak left channel, distorted and weak audio, intermittent sound, a noisy channel, or hum in the speakers. The dead channel might be caused by a defective transistor or IC component. Quickly check all transistors with the diode tester of the DMM.

Remember that when one transistor is found open or leaky, look for another one or two damaged in the directly coupled audio output circuits. Take critical voltage measurements on each transistor, and compare with the same voltage in the normal stereo channel. Do not overlook a dead speaker caused by an open or blown speaker fuse or a defective speaker relay circuit.

A weak or distorted channel can be checked with an external audio amplifier. Always use an isolation transformer with the external audio amp to prevent damage to the

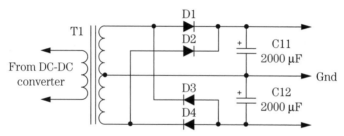

4-27 A shorted diode or leaky filter capacitor can shut down the dc-dc power supply.

test equipment or components within the amplifier. Extreme distortion symptoms generally are located within the high-powered output circuits or head unit. Leaky transistors, diodes, ICs, and coupling capacitors and bad bias resistors cause most distortion symptoms. The weak stage can result from leaky or open input and driver transistors and ICs, bad coupling capacitors, and a change in resistors. Compare the weak or distorted stage at the same point in the normal stereo channel. A shorted or leaky power output transistor can damage the emitter resistors.

When the audio signal is slightly weak or distorted, use the signal generator or external audio amp to locate the defective stage. Clip a 1-kHz audio signal from the generator to the inputs of the amp, and connect the external amp to check out each stage for the same amount of audio. Go from base to base of each transistor and notice the same volume setting in both channels. As you proceed through the amplifier circuits, the signal will get weaker in the external audio amp. Critical voltage and resistance measurements with transistor tests can locate most high-powered amplifier problems.

Bridged amp outputs

A high-powered amplifier might produce 700 to 2000 W bridged into a 4-ohm load. The two-channel high-powered amp can be bridged by connecting the positive terminal of the left channel to the positive terminal of the high-wattage speaker. The negative terminal of the right channel is connected to the negative terminal of a single speaker. A 2 × 85 W amp at 4 ohms with a maximum output of 2 × 350 W at 2 ohms can be bridged in a mono speaker at 1 × 700 W. The single-bridged speaker should be able to handle at least 700 to 1000 W of power (Fig. 4-28).

Two-channel amp hookup

A two-channel amp has a left and right channel in a stereo amplifier, whereas a four-channel amp has four separate speakers for each channel. Connect the positive lead of the left output to the positive terminal of the speaker. Likewise, connect the negative output of the left channel to the negative terminal on the speaker. One speaker is connected to the left side and another speaker to the right side (Fig. 4-29). Match

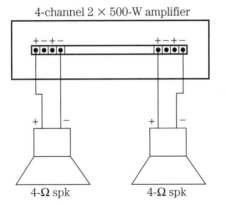

4-28
A bridged speaker hookup tied into a 2 × 500 W four-channel amp with a 4-ohm speaker load.

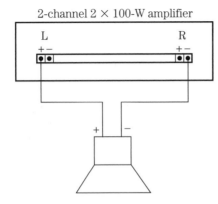

4-29
A mono speaker is connected to the left + channel and the negative terminal − to the right channel in a bridged speaker hookup.

the polarity of the speakers with the polarity of the amplifier for greater power and frequency response.

The two-channel amp can be bridged for a mono (1) PM speaker for higher wattage. For example, a 2 × 100 W two-channel amp can be bridged at the output terminals for a speaker output of 200 W. Just connect the positive terminal of the left channel to the positive terminal of the speaker and the negative terminal of the right channel to the negative terminal of the PM speaker.

Matching impedance

The impedance of the speaker should match the output impedance of the high-powered amplifier. A 4-ohm speaker should be connected to a 4-ohm amplifier and not to a 2-ohm impedance. Poor impedance matching can result in poor fidelity and power or volume. The speaker voice coils can be wired in series, just so the impedance of both units matches. For example, two 4-ohm speakers can be wired in series to match an 8-ohm impedance of the amplifier. When speakers are wired in parallel, the impedance is cut in half (Fig. 4-30). The early voice coil im-

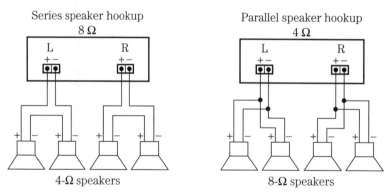

4-30   The 4-ohm speakers are connected in series to match an 8-ohm two-channel amplifier.

pedance was around 3.2 ohms, and the latest high-powered speakers impedance is 4, 8, and 16 ohms.

High-powered speaker hookup

Several different size speakers might be found in a high-powered amplifier system. A 500-W amp may feed a high-pass crossover network to a tweeter and midrange speaker in each channel. The same amp might have a low-pass level to a very large subwoofer speaker (Fig. 4-31).

Some head units are fed to a control equalizer unit, and this signal is fed to a separate 500×2 W amplifier. In turn, two large 10-in speakers are wired to the 500-W amplifier. The same control equalizer unit also might feed to a 75×2 W amplifier and in turn feeds a two-channel crossover network that feeds a tweeter and midrange speaker in each left and right stereo channel.

High-powered amplifiers can be built into a pickup truck as well as in an automobile. The deluxe speaker hookup might have a VCR, DVD, and screen monitor inside the pickup. The signal is sent to a 500-W amplifier that powers two different crossover units. The low-pass network is connected to two 15-in subwoofer speakers. The high and midrange power is fed to a tweeter and midrange speaker in the left and right output channels (Fig. 4-32).

The tweeter speakers, such as a really small PM speaker or a horn-type speaker, might be connected to the high-pass frequency network. The midrange speaker might have a 5- to 8-in speaker. A subwoofer can be a 10- to 18-in PM speaker for the bass frequency response of 20 to 500 kHz, whereas an 18-in speaker response is from 20 to 150 kHz.

Auto radio speakers

The speakers found in early car radios were 4, 5, 7, 8, or 6×9 in with a 3- to 4-ohm impedance. Along came the hybrid radio with 4×6-, 4×9-, 4×10-, 5-, 6-, 6×9-, or 7-in PM speakers. The all-transistor car radio might have 4×6-, 4×10-, 5×7-, 5-, 6-,

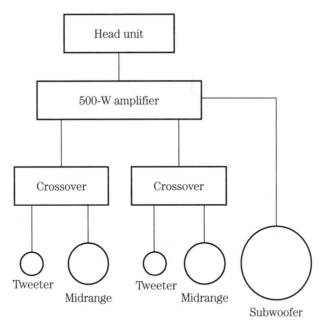

4-31 A 500-W high-powered amplifier tied to crossovers and the various speakers.

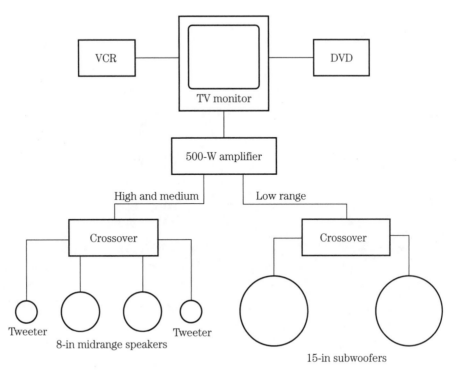

4-32 A front-end high-powered amp, VCR, DVD, and TV monitor with the various size speakers in a pickup truck.

6×9-, 6×10-, or 7-in speakers. The deluxe auto receiver might have a 4×10-, 5×7-, 6×9-, or 6×10-in speaker with a 4-, 6-, 8-, 16-, or 40-ohm voice coil.

The cones of early car radio speakers were constructed of paper and very light material. A high-powered subwoofer speaker has high temperature and composite cones, poly, polypropylene, poly and graphite linear cones that can create realistic sound. You might find a stiff Kelvar paper cone in subwoofer speakers. A surround speaker may use a silver-injected poly cone with a thermal paper and fiber-poly type of surround, and on it goes.

The voice coil is found on one end of the cone that moves through a magnetic field with a PM magnet. The early car radio replacement has an impedance of 3, 4, 8, 10, or 16 ohms. The voice coil impedance of a high-powered speaker is usually 2 or 4 ohms. Some auto speakers are found in very odd shapes and must be replaced with the original or they will not fit into the required space. Simply go to the car dealer and pick up a speaker that will fit in the same model of automobile.

The factory car rear-deck speaker openings might take a 4×6-, 5×7-, or 6×9- in speaker. Speakers mounted within the front deck or dash might be a $4^1/_2$-, $5^1/_4$-, 4×6-, or 4×10-in PM speakers. The different size speakers found in the rear wheel mount might be from 5-, $5^1/_4$-, $6^1/_2$-, 6×8-, or 6×9-in speaker openings. In high-powered amps, the very large speakers (12, 15, 18 in) might be mounted in columns or specially constructed baffles and mounted in the back or in the trunk area.

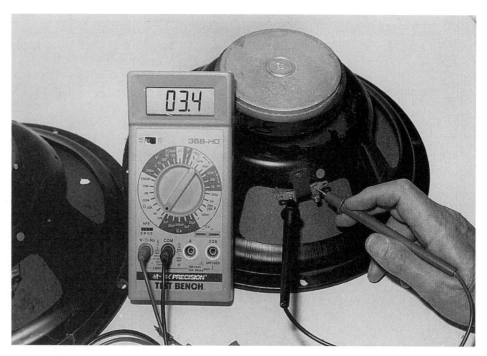

4-33 A 15-in 4-ohm speaker tests only 3.4 ohms on the DMM.

Speaker problems

Since an automobile operates in any kind of weather, auto speaker cones have a tendency to warp and drag. The defective speaker might have an open voice coil, bad voice coil connections, and drag against the metal pole piece. By pushing up and down on the speaker cone with the radio operating, you can locate an intermittent sound with a bad voice coil winding or poor connection. A mushy speaker cone drags against the center magnet pole piece. A dead speaker may have an open voice coil or a broken speaker connection.

Check for an open voice coil with the low-ohm range of the DMM. A volt-ohm-milliammeter (VOM) measurement of the voice coil will be greater than the speaker impedance. A DMM low-ohm test will show that the speaker impedance can be from 0.5 to 1 ohm less than the actual impedance. A 15-in high-powered speaker has an impedance of 4 ohms and a 3.4-ohm resistance measurement on the DMM (Fig. 4-33). Always have a speaker connected to the amplifier channels or a low-ohm resistor when servicing a high-powered amp. The large PM speaker can be damaged by applying too much volume or boom power from the amplifier.

Most speakers can be replaced from just about any electronic parts or mail-order dealer. The different sized auto speakers can be picked up at the auto dealerships. Very large midrange, woofer, and subwoofer speakers can be obtained from electronics part or mail-order dealers and Radio Shack. Do not overlook the electronics mail-order firms for speaker replacements and enclosures. Make sure that the speaker size, voice coil impedance, and weight of the magnet are the same as the speaker being replaced.

5
CHAPTER

Servicing cassette players

A cassette player may appear in tag-along, personal, portable, boom-box, microcassette, professional, cassette/CD portable, or double cassette form, as well as in the automobile (Fig. 5-1). Mono sound is found in small cassette players, whereas stereo channels exist in large portables, table models, and auto cassette players. Besides playback, recording of messages and music is why cassette players are so popular.

Although many service problems are the same in most models, the larger deluxe units and car tape players have additional features. When mechanical and electronic components work side by side, trouble surfaces. Most mechanical problems can be seen or located easily, whereas the audio channels must be isolated and located with test instruments.

Required test equipment

The following equipment is needed to service cassette players:
1. Digital multimeter (DMM) or voltmeter
2. Oscilloscope
3. Audio signal generator
4. Sine- and square-wave generator
5. Capacitor tester
6. Equivalent series resistance (ESR) meter
7. External amp
8. Audio analyzer (optional)
9. Wow and flutter meter (optional)
10. Test cassettes

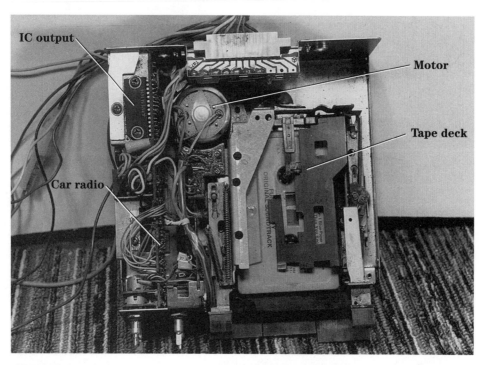

5-1 Radio cassette players are found in portable, boom-box, and home stereo systems and in automobiles.

Pocket cassette player layout

A pocket cassette player might include a tag-along for children to play their favorite music or a microcassette or professional player that can make news recordings. The small cassette player can be carried in a handbag or shirt or coat pocket for recording meetings or for gathering the latest news. A pocket cassette player can be used for fun activities or for recording professional meetings or special events (Fig. 5-2).

The portable cassette player

The portable cassette player may include tag-along, personal, boom-box, microcassette, and professional models. The portable cassette player is an electronic device that usually carries self-enclosed microphones for personal or professional recording. Most of the components within a portable cassette player are jammed together in a small cabinet (Fig. 5-3).

Auto radio/cassette player

The cassette player found in a car radio is used mostly to play music or various recordings of events. The cassette player is switched into operation as the cassette is inserted into the radio. You might find a combined cassette player and CD player

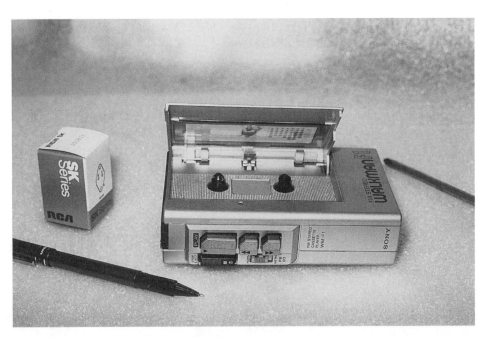

5-2 A personal or tag-along cassette player is small enough to be placed in your coat pocket.

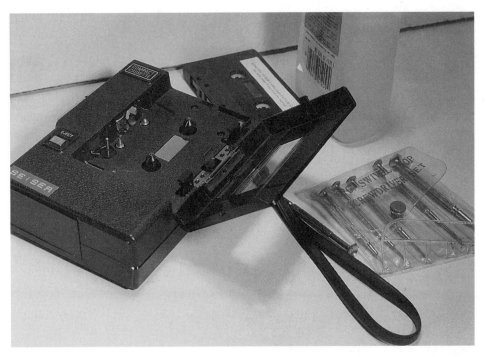

5-3 A pocket cassette can record events or live music.

5-4 Besides a cassette player and radio, an auto receiver might include a CD player.

in a deluxe auto receiver. A cassette player in an automobile is more difficult to service than a portable cassette player (Fig. 5-4).

Check out that cassette

Cassette player operation may consist of music, lessons, meetings, and various event recordings. Besides music, a cassette might include test material for servicing and alignment of the cassette player. The test cassette may be used to signal trace, align azimuth, or make speed adjustments on a cassette player. The minicassette or tag-along player may use a C-90 cassette.

The small C-90 cassette is quite fragile and never should be left out in the sun while in an automobile. Cassette tapes, both recorded and unrecorded, should not be stored in locations with high temperatures, high humidity, or direct sunlight. Keep cassettes away from magnetic sources such as high-powered speakers or TV sets.

A cracked cassette case should be thrown away. A defective cassette can run slow, began to drag, slow down, and pull out or "eat" tape. A worn or noisy tape should be discarded or thrown out. A defective tape may have poor recording qualities and poor frequency response. Remember that cassettes can be magnetized easily, resulting in very poor recordings in playback.

Slow speeds

Slow speed can result from a dry capstan bearing or pinch roller, tape wrapped around the pinch roller, a loose or oily belt, or a defective motor. Check for gummed-up or dry

5-5 Clean up all drive belts, idlers, and pulleys for slow speeds.

capstan and flywheel bearings. Simply remove the capstan, clean it with alcohol and a rag, lightly oil it with phono lube or light oil, and replace it (Fig. 5-5). Replace all oily, cracked, loose, or shiny belts.

Check the batteries when the tape slows down during battery operation. Try the machine using ac operation to determine if weak batteries are causing the slow speed. Sometimes batteries left in the player too long will pull down the motor speed during ac operation. Remove the batteries and check them in a battery tester or with a DMM.

Erratic or intermittent speeds can be caused by a dirty leaf switch, belt slippage, or a clogged pinch roller. Don't overlook a defective motor when slowdown or dragging occurs. Sometimes tapping the motor while it is operating will cause it to cut up and down. Replace a defective motor with the original part number.

Will not eject

Suspect the door latch, a defective plunger, or a loose pin when the cassette will not eject. Check for a defective eject button and no dc voltage to the solenoid plunger for no ejection of the cassette. The cassette player may not eject a tape when the door will not open because tape is wrapped around the capstan. Remove one end of a possible shorted diode across the solenoid winding, and check with the diode tester of the DMM. Poor switch contacts can cause the cassette to not eject from the player.

Readjust the cassette detector switch when a cassette is not accepted. Replace the main gear assembly when the player stops in the play mode and the cassette cannot be removed. The cassette would not eject in a Sony CFS212 player until the play lever was replaced (part number 334334201).

Dead, no sound

The dead channel might be caused by a bad off/on or leaf switch. A bad or defective cassette motor can cause a dead cassette player. A defective component in the left audio channel can cause a dead channel. Substitute another permanent magnet (PM) speaker for the suspected open voice coil in the speaker. A bad tape head can cause no playback or recording within the cassette player.

In a Sanyo M2566 cassette recorder, the sound was dead, but the motor operated with both batteries and ac power. No hum was heard when a screwdriver blade was touched to the center terminal of the volume control. Although the voltage was fairly normal on power IC1, no output signal or hum was heard. Replacing the IC1 (LA4104) power output with an SK3888 universal replacement fixed the dead portable cassette player (Fig. 5-6).

Poor tape motion

Notice if the motor belt is off when there is no tape motion. Replace with a flanged motor pulley if the belt keeps coming off as the motor starts up. Replace a loose belt or broken belt when it will not stay on the capstan flywheel or motor pulley. Does the motor rotate without any tape motion? Suspect an open motor winding or a flat shaft problem with no or poor motor rotation.

A dead or defective motor can cause slow tape motion. The erratic motor rotation might result from worn brushes or tongs in a dc motor. Check the motor terminals with an ohmmeter continuity test. No continuity indicates that the motor winding is open. A defective speed circuit can cause poor or no tape motion.

Check for a dry or frozen capstan or flywheel bearing for poor tape motion. Remove the flywheel, and clean out the bronze bearings. Clean off the capstan shaft with alcohol and a cloth. Apply light grease before replacing the capstan or flywheel. Make sure

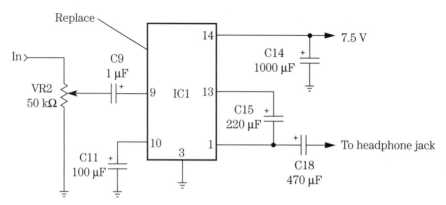

5-6   Replace IC1 in a Sanyo M2566 cassette recorder for no sound.

that the bottom keeper plate is not bent out of line, causing the flywheel to slow down. In a Sony FM31 cassette player, replace the motor belt and pulley if the belt keeps coming off the motor shaft.

No fast-forward or rewind

Notice if the tape motor is rotating when the radio circuits are functioning with no tape motion. Check for a worn or loose belt. Check for a dirty tape or radio switch assembly with no tape movement. Excessive tightness in the gear train assembly can cause no fast-forward or rewind. Suspect a capstan motor for no fast-forward or rewind. Replace the idler pin with poor torque in fast-forward and rewind (Fig. 5-7).

In fast-forward operation, the fast-forward roller or idler is pressed against the take-up reel at a faster rate of speed. Check for slippage on the idler drive wheel or pulley when the tape moves slowly in fast-forward mode. A good cleanup of pulleys and rubber tires on pulleys can solve a lot of speed and fast-forward problems. A bad clutch assembly can cause poor rewind and fast-forward modes. Check for oil on the rubber drive surfaces for erratic rewind. Clean the drive areas, idlers, and pulleys with alcohol and a cleaning stick. Replace rubber tires or resurface them if the pressure roller becomes stuck.

Check for a dry spindle and idler wheel bearing for poor rewind. Suspect a missing C washer on the supply reel shaft for poor rewind. Check for a binding capstan shaft if the tape is still slow in fast-forward. Remove the capstan or flywheel, clean the bearing,

5-7 Check the cassette motor, drive belt, and pulleys for no fast-forward or rewind.

and relube. Erratic fast-forward can be caused by a loose or oily belt and motor pulley. A dry idler wheel or pulley can cause erratic fast-forward. Check the fast-forward speed with a torque cassette. Suspect a defective zener diode in the power supply when neither auto nor manual operation will not reverse.

Suspect a bad motor if the cassette deck has two motors, one for forward and the other for rewind. A defective high-speed motor or bad motor circuit can cause the motor to not rewind or fast-forward. Take critical voltage measurements with the DMM across the motor terminals when the motor torque is not up to speed or rotation. Suspect a capstan motor or Q1607 in the power supply of a Pioneer CTF1250 for no fast-forward and rewind.

Tape everywhere

Check to see if the capstan or pinch roller is dirty or the take-up reel is not operating. The cassette player can "eat" or spill out tape when the take-up reel is not taking up excess tape. If the take-up reel rotates erratically or slowly, the tape can pull out. An out-of-shape pinch roller can cause the tape to spill out. Clean up the idler pulley and all tape contact surfaces in play and record modes. Replace the idler assembly if necessary; a malfunctioning one can cause insufficient tape tension and thus cause the tape to spill out.

Excess tape wrapped around the capstan drive can cause slow speeds and warbling sounds. Check for missing hub caps on the tops of the reels when there is pulling of tape. A sluggish take-up reel assembly can cause tape to pull out from the cassette. A gear-driven idler with missing teeth or dry bearings can cause the pulling of tape. Check for a missing washer or cam assembly for pulling of tape and loss of fast-forward motion.

Notice if the take-up reel is not operating or erratic when tape spills out of the cassette. Do not overlook a defective cassette. Try another one to see if it is the cassette or the tape player. If the take-up reel is operating normally, suspect a worn or uneven pinch roller. Clean the idler wheel with alcohol and a cloth. Check for any sticky substance on the capstan drive shaft. Remove all excess tape wound around the capstan and pinch roller assembly.

Jammed tape

Check for tape wrapped around the capstan when the cassette will not eject and must be pulled out of the cassette player. Try to rotate the tape backward with the capstan or flywheel. Reach in the back and locate the flywheel. Notice if the tape is loosening up as the flywheel is rotated backward. Often the cassette door will not open when there is a jammed tape in the player.

Sometimes the tape must be cut loose before the cassette can be removed. Now the cassette is damaged and should not be used again. Pull out all the loose tape from the capstan and pinch roller assembly. Make sure that all tape is removed from the pinch roller assembly or the tape will rotate slowly with clogged tape.

Clean the pinch roller assembly with alcohol and a cloth. Tape damage can result when the pinch roller is held against the capstan after the unit is turned off and may be caused by a defective zener diode in the power supply.

Squealing noises

The speaker in a cassette player may let out a loud squeal or warbling noise when the cassette function switch is pressed. A large cassette deck or player may have push buttons that operate many different functions. When the cassette switch is pressed and locked in with a loud squealing noise, suspect a dirty function switch. This switch can be a rotary or push button type that contains a stiff wire moving a flat function switch with many contacts.

Clean the function switch when there is extreme noise in the speakers. Spray cleaning fluid down in the dirty contacts, and rotate the switch back and forth to help clean the contacts. Push one button and then the cassette button to clean switch contacts. The silver contacts on sliding switches become tarnished and make poor switching contacts.

Too much oil

A case of too much oil might be caused by trying to correct the speed of a cassette player by squirting too much oil on the moving parts. Excess oil found on motor, drive belt, take-up, and supply reels can destroy tape motion rather than helping. Remove all excess oil with alcohol and a cloth.

Do not oil any components within a cassette player unless the bearing is frozen or squealing. Light oil sprayed on gears, take-up reels, and plastic gears can leak down onto moving parts and can cause extra slippage (Fig. 5-8). All oil spots must be wiped up.

5-8   Too much oil on the idlers, flywheel, and drive belt can cause slow speeds.

Do not apply oil or grease on top of the bearing of the capstan or flywheel if it is suspected of rotating at a slow speed. Remove the flywheel or capstan. Clean off old grease with alcohol and a rag. Apply a dab of phono or light grease at the capstan or flywheel bearing. Remount the capstan or flywheel, and wipe down the area with alcohol. Likewise, a drop of oil on the pinch roller is adequate. Wipe off oil or excess oxide on the rubber roller.

Most cassette motors do not need lubrication. They are lubricated at the factory. If one of the motor bearings begins to make a noise, a drop of light oil down into the bearing may help to stop the noise. Motors with worn bearings must be replaced.

Poor erase

The erase head is designed to erase any previous recording from a tape. A jumbled or distorted recording can be caused by a packed tape erase head with excessive tape oxide. The erase head may be out of place and may not touch the tape, and this can result in a messed-up recording. In larger tape players, the erase head may be moved up into place when recording begins and may not touch the moving tape. The erase head in other tape decks is out of the way in play mode and does not pivot up into position; this can cause poor erase. A ground wire broken off an erase head will result in no erasure of a previous recording.

The erase head in a battery-operated cassette player may operate from a dc voltage source, whereas in larger tape players a bias oscillator circuit is found. A defective bias oscillator circuit can produce damaged recordings. Check the oscillator waveform at the tape head with the oscilloscope to see if the bias circuit is functioning. Take critical voltage measurements for open or intermittent transformer windings. Clean the record/play switch for intermittent or noisy recordings (Fig. 5-9).

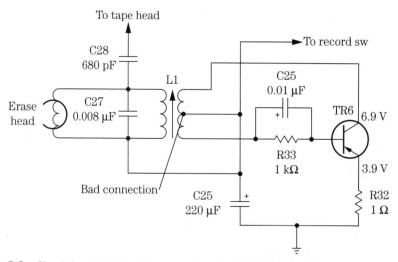

5-9   Check for a bad transformer connection causing intermittent recording in a JC Penney cassette player.

Poorly soldered connections

Blobs of solder on a printed circuit (PC) board can cause intermittent operation. Sometimes when the boards are run through a solder bath, several areas may contain excess solder. In replacing a component, the technician can apply too much solder, resulting in the shorting out of two or more traces or wiring. Extra care must be taken when replacing integrated circuits (ICs) or processors with gull-wing terminals. The PC wiring is very thin around these components.

Blobs of solder piled up at each end of a surface-mounted device (SMD) may result in a cold-soldered connection. Clean the old connection so that the SMD can lie flat, and melt solder to flow into the connection. Remember, too much soldering iron heat can destroy an SMD.

Dead left channel

Just about any electronic component from the tape head to the speaker can cause a dead left or right channel in a cassette player. A packed tape head with excess tape oxide can cause a dead or weak audio channel. Open coupling capacitors can produce a dead channel. Defective input transistors or ICs can cause a dead channel. If the left channel is dead in a cassette player and the AM/FM/MPX receiver is operating, check the input tape circuits for defective components. Improper voltage fed to the input transistors or ICs can cause a dead channel.

The left channel was dead in a Sanyo stereo component system GXT300. The AM/FM/MPX receiver operated fine, indicating that the trouble was in the tape head input circuits. With the volume wide open, the right channel at the tape head had a loud hum, and there was only a very low hum in the left tape head terminal. A loud hum was heard when the test probe was placed on terminal pin 1 of IC401. Tracing out from where pin 1 was located, a 3.3-μF electrolytic capacitor was tied to pin 1. Replacing C412 solved the dead left channel (Fig. 5-10).

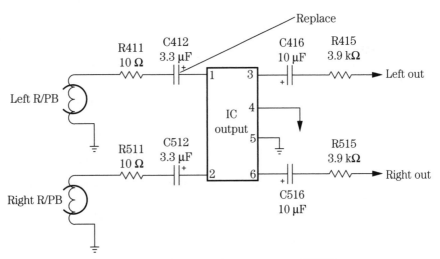

5-10 Replace C412 to fix a dead left channel in a Sanyo GXT300.

Intermittent operation

Intermittent tape rotation can be caused by a loose drive belt or pulley or a dry flywheel bearing or drive motor. A bad off and on switch can cause intermittent motor rotation. The motor might operate intermittently with a badly soldered connection on the pause control switch. A dirty or worn leaf switch can cause the cassette player to be intermittent. Suspect a dirty or worn function switch for intermittent cassette rotation. Look for a bad connection within the tape head to cause intermittent sound. Check for a defective automatic circuit when the cassette keeps shutting off.

Locate the left channel with the speaker or an external audio amp. Monitor the signal at the volume control to determine if the intermittent is in the front end or the output circuits. Most audio intermittents are caused by bad transistors, ICs, coupling capacitors, terminal connections, or board connections. Intermittent audio is the most difficult problem to locate and solve in the audio circuits and requires a lot of service time.

A Sharp RT1165 model would stop after a few seconds. After checking the motor and belts, a loose belt was suspected and replaced. See if the counter is turning when the tape runs. If so, you know that the belt driving the counter is now okay. Suspect the IC located behind the counter with a small magnet that rotates. Replacing the counter IC with part number VH1DN838A solved the intermittent rotation of a cassette player.

Under that magnifying glass

Sometimes it takes very sharp eyes and a lot of patience to trace out the PC wiring or cables to a given component. For instance, if one stereo channel is dead or weak and you do not know which components tie into that defective channel, start at the speaker terminals. Trace the speaker wires to a plug or soldered connection on the PC board. Since small stereo amplifiers have a speaker electrolytic coupling capacitor between the speakers and the output transistor or IC, trace it out.

By circuit tracing the PC wiring and locating large output ICs or power transistors, signal trace the defective channel with the external audio amp or scope. After locating the defective channel with IC or transistor output parts, signal trace the IC or transistors on the same side as the output component. The dead or weak channel can be compared with the normal one at several test points on the chassis. A magnifying glass and an ESR meter can help to locate broken connections, poor soldering, and cracks within the PC wiring (Fig. 5-11).

Locating and replacing bad transistors

By locating the number on the body of a transistor, you can determine what type it is, where it works, and its correct replacement by looking in the universal replacement manual (Fig. 5-12). Look for terminal markings on the bottom of the PC board to locate where each terminal connects. Mark each down on a separate piece of paper for reference.

Transistor styles include small, medium, and large power and special regulator types. The transistor number tells you what type of transistor it is and whether it is a

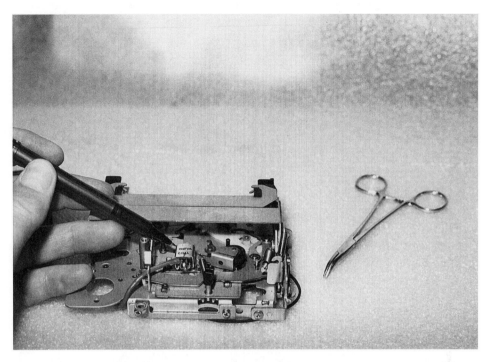

5-11 Remove some boards and sections of the cassette player before locating the defective component.

low-frequency, high-frequency, field-effect transistor (FET), silicon-controlled rectifier (SCR), triac, metal-oxide semiconductor field-effect transistor (MOSFET), or opto device:

- A—The type of transistor and the number indicates the active electrical connections plus one (Fig. 5-13)
- B—Either Japanese (EIAJ) or American (EIA)
- C—The transistor application and if it is an NPN or PNP type
- 2N—Transistor, FET, SCR, or triac
- 2SA—Transistor, high-frequency, PNP type
- 2CB—Transistor, low-frequency, PNP type
- 2SC—Transistor, high-frequency, NPN type
- 2SD—Transistor, low-frequency, NPN type
- 2SJ—FET, P channel
- 2SK—FET, N channel
- 3SK—MOSFET, N channel
- 3N—MOSFET, dual triacs
- 4N—Opto devices

By looking at the transistor number, you can determine if it operates at a certain frequency and whether it is a PNP or NPN type. The suffix letter after the last number indicates an improvement over another transistor. Note that a *B* or *C* suffix after the number indicates a much superior transistor than the letter *A*.

If no numbers are found on the transistor, measure the supply voltage on the transistor, and determine what type of a circuit the transistor works in, whether it is high or

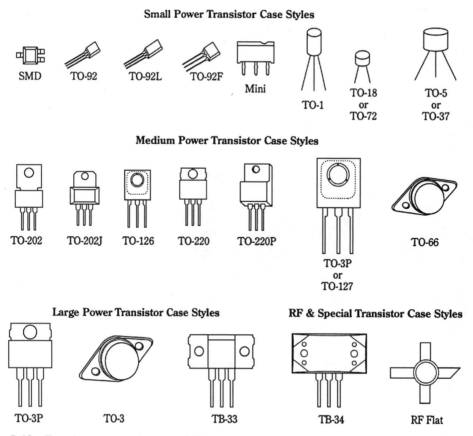

5-12 Transistors appear in many different sizes and shapes.

low frequency, and if it is of small, medium, or large power by the size of the transistor case. Look up the data in the universal replacement manual, and replace the questionable transistor. You also can look at a schematic of the same brand or a similar circuit to get the working transistor number.

Locating and replacing a bad IC

Locate a defective IC, and check the part number found on top of the IC. Look up the replacement in the universal replacement manual. Locate the input and output terminal pins. Signal trace the audio on the input terminal and then the output pin. The signal should be much greater on the output if the IC is normal.

The top numbers and letters on the IC indicate the manufacturer, the type of device, whether it is American or Japanese, improvements, and package style. Most manufacturers go by the following guidelines (Fig. 5-14):

- A—The manufacturer of the device when registered with the Japanese or American EIA numbers (see Fig. 5-14)

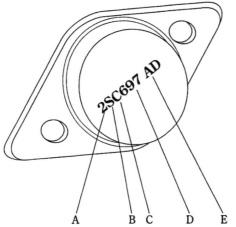

5-13
The 2SC indicates that the transistor is a high-frequency NPN type.

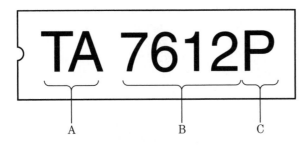

5-14
Locate the IC by the numbers on top, and look it up in the semiconductor manual.

- B—The type of device, root number, and that it is used primarily with American manufacturers
- C—Suffix letter that indicates an improvement (the letter *N* is superior to the letter *B;* the number can indicate the package style)
- C—Can or metal
- D—Dual in-line package (DIP)
- E—Flat package (SMD)
- K—TO-3 transistor package
- L—Single in-line package (SIP)
- P—Plastic package
- R—Reversed pin configuration
- S—Single in-line package (SIP)
- T—TO-220 transistor package
- V—Mounts vertically (ZIP)

Recording problems

Make sure that the cassette player is playing right before checking for recording conditions. Suspect a broken or dirty switch when the tape will play and not rewind. A worn tape head can prevent a recording in one or both channels. A jumbled recording and no erase can result from a ground wire off the erase head. The no record symptom may be caused by a defective switching IC. Clean the record/play switch if a loud cracking noise is heard when the player is switched to record mode.

A poor or garbled recording can be caused by not properly erasing the preceding recording, and this can be caused by a metal shield that is not grounded properly. Intermittent recording can be caused by a defective function switch. Check all small electrolytics (5 to 10 µF) within the input circuits for intermittent recording on the ESR meter. Check all bias controls when there is intermittent recording on the left or right channel. Clean the record switch when there are intermittent recordings on the right channel. An intermittent and warbling recording sound in the right channel can result from a worn and dirty record switch.

Replace a defective erase head for no record on the left channel. A packed record tape head can cause poor or no recording. The no record and only a rush of audio in the play mode may be caused by a broken internal connection in the tape head (Fig. 5-15). An extremely loud howl can result from a broken tape head connection with the volume wide open.

The packed tape head with excessive tape oxide can cause a weak recording in the right channel and no record in the left channel. Simply move the tape head with a wooden pencil or pen and notice if only a loud rush in the recording mode is caused by a broken internal tape head connection. Check for a record/play switch that is not fully engaged and needs alignment when a JC Penney 683-1795 cassette player will not record in the right channel.

In a Silver Marshall KMD-1022 cassette player, the left channel was intermittent, and the right channel was normal. Sometimes you could hear a low recording in the left channel. A loose and dirty VR101 (10-kilohm) thumbwheel control caused the intermittent left channel.

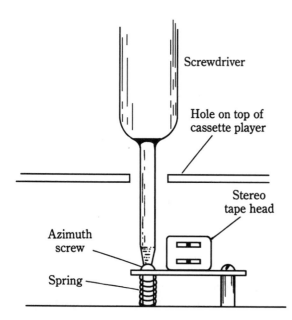

Screwdriver

Hole on top of cassette player

Stereo tape head

Azimuth screw

Spring

5-15
Check for broken connections on the tape head or a loose screw for poor playback and recording.

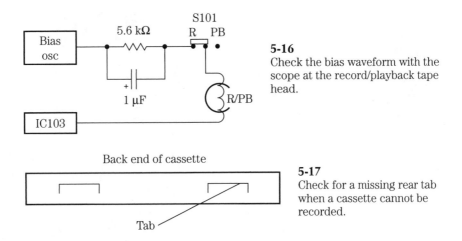

5-16
Check the bias waveform with the scope at the record/playback tape head.

5-17
Check for a missing rear tab when a cassette cannot be recorded.

Suspect a defective bias oscillator circuit when there is no waveform on one side of the recording head terminal (Fig. 5-16). Check for voltage on the bias oscillator transistor, an open transformer winding, and no supply voltage in record mode.

ACCIDENTAL ERASE

Check the back of the cassette to see if all tabs are still intact. If the tab is knocked out of the cassette, the cassette cannot be recorded again. The record button cannot be engaged if this tab is removed. To save a recording, remove the small tab at the rear of the cassette. Place a piece of tape over the tab opening when the cassette is needed for another recording (Fig. 5-17).

MESSED-UP AND POOR RECORDING

When both channels produce a messed-up, garbled, or jumbled recording, suspect the erase head circuits. Clean the erase head with alcohol. Sometimes the small gap can become packed with oxide dust, which results in no recording. Try another cassette. Check that the ground wire is connected to the erase head.

Check the bias oscillator for erratic or intermittent operation with poor recording symptoms. Poor oscillator transformer connections can produce intermittent recordings. Weak recording can be caused by a dirty record head or low supply voltage to the bias oscillator. Clean the play/record switch. If both record and play are weak or distorted, check the audio output and preamp circuits. If the record time indicator will not move, suspect a broken belt or jammed assembly.

NOISY PLAYER OPERATION

Check the capstan or flywheel bearings, idler pulley, and motor for squeaks or screeching noises. Lubricate with a light oil. A noisy cassette hub can be checked by changing to another cassette. An open head connection can cause a loud howling in

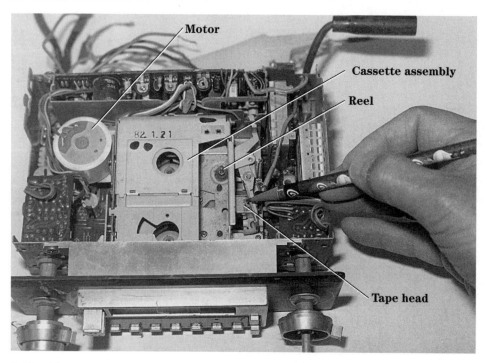

5-18   A loose motor bracket or noisy motor can cause pickup hum and static in the audio.

the speaker with the volume wide open. A dirty play/record button also can cause a howling noise.

When both channels have a low hum or frying noise, replace the output IC. If the noise continues with the volume turned down, suspect a noisy output IC or transistor. When the noise can be turned down with the volume control, check the preamp transistors, ICs, and bypass capacitors. Short the base to the emitter terminal of the transistors to locate the noisy stage. Don't overlook a defective motor (Fig. 5-18).

A hissing noise in a cassette player can be caused by a worn or dirty function switch. Noise in the sound can be caused by a noisy motor. Replace the electrolytic across the motor terminals with one of a higher capacity (220 to 470 μF) to eliminate a noisy motor. A bad filter capacitor (470 μF) on the audio output transistor can cause a popping sound. An intermittent and noisy left channel can result from a bad output IC. Motor boating in the sound can be caused by open decoupling and electrolytic capacitors in the voltage circuits.

NOISY RECORDING AND PLAYBACK

Suspect a broken wire on the tape head for a loud howl and rushing noise with no play or record mode. A bad internal connection of the tape head can cause the same intermittent loud and rushing noise (Fig. 5-19). A warbling sound and lost recording at the beginning of a tape can indicate a defective cassette. Check for a dirty or worn play/record switch when there is a loud howling sound. A squeaking noise as an idler

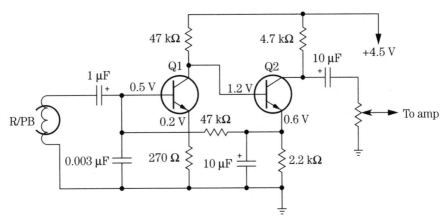

5-19 Check these components for noisy and frying noises in the front-end circuits.

wheel rotates indicates a dry wheel bearing. A loose motor bracket can cause a pickup hum and static in the cassette motor. In a Soundesign 5855 cassette player, the right channel was extremely noisy with the volume turned down after about 5 minutes of operation. Replacing the right output IC (STK435) repaired the noisy right channel.

INTERMITTENT RECORDING

The recording was intermittent in the left channel of a Sanyo M-X920 large portable tape player. The right channel was normal. The left channel playback of a cassette also was normal. A quick check of the C-Mic (microphone) input circuit seemed okay. When C801 (0.1 μF) was touched, the recording would cut in and out. After several hours of recording and playback, C801 was found to be intermittent and was replaced (Fig. 5-20).

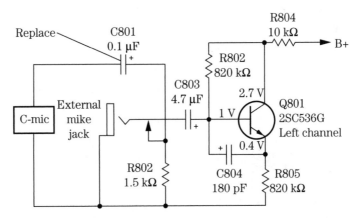

5-20 Replace the C801 (0.01-μF) microphone coupling capacitor for intermittent playback and recording.

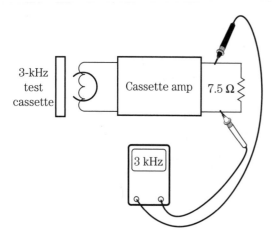

5-21
Check the cassette motor speed with a
3-kHz cassette and the frequency meter.

RUN, BABY, RUN

Defective motor-regulated circuits can cause the tape to run rather fast. High-speed
tape movement can result from a defective regulator circuit, a belt riding high on the
motor flange of a pulley, and a defective motor. Check the cassette motor speed by
inserting a 3-kHz tone cassette and connecting the frequency counter at the speaker
terminals. Make sure that the belt is riding in the motor pulley groove and not on
edge of pulley. Place the cassette player in play mode. Connect a 7.5-ohm resistor at
the earphone jack or speaker terminals (Fig. 5-21).

Measure the frequency at the frequency counter. If the reading is 3 kHz, the speed
is normal. A higher reading indicates faster speed, and a lower reading indicates slower
speeds. If the speed is rather low, suspect belt slippage, a loose belt, or a defective motor. Oil deposits on the capstan or flywheel can produce low speeds. Check at the end
of the motor to see if there is a screw speed adjustment. If not, check the motor speed
regulator for a control or screwdriver adjustment. Adjust the speed on the frequency
counter at 3 kHz.

The boom-box cassette player

The boom-box cassette player is found in an AM/FM/MPX radio receiver with several
big output speakers. A boom-box unit is much larger in size than a portable cassette
player. The boom-box player may have a double cassette deck, continuous playback,
a three-band graphic equalizer, and dubbing and soft eject features. The erase head
may be pivoted out of the way in playback mode and pulled into position in record
mode. The boom-box player may operate on batteries or from the power line or both.

The output circuits within a Radio Shack 14-731 boom-box player have a dual IC
output at 1.4 W of power. The left channel was weak and the right channel was distorted
on AM/FM/MPX reception, as well as on the cassette player. This meant that the trouble was in the audio output circuits because all three signals were weak and distorted.
The radio and cassette signals were normal at the volume controls with a clipped-in ex-

The boom-box cassette player

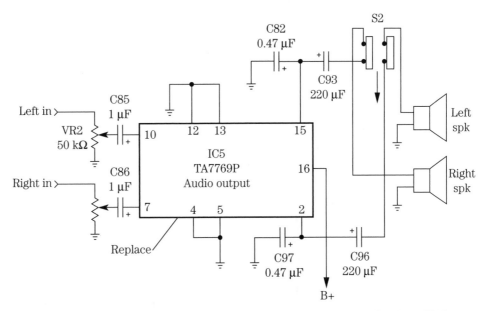

5-22 Replace IC5 for a weak right and distorted left channel in a Radio Shack 14-731 boom-box cassette player.

ternal audio amplifier. The weak and distorted channels were found at output terminals 2 and 15 of IC5. Replacing IC5 (TA7769P) with the original output IC resolved the weak and distorted output symptoms (Fig. 5-22).

You may find different service problems within the two separate decks of a boom-box cassette player. Replace the dummy erase head when deck A goes out of speed. Check the retainer washer when the left deck capstan slides down and causes no operation in the left deck. Suspect a defective IC402 in a Sony TCW255 cassette player with no audio in the B tape deck.

MOTOR PROBLEMS

The small cassette motor is designed to operate from a battery or the ac power line. The cassette motor may be intermittent, erratic, slow, or dead. The intermittent motor may start one time and be dead the next time. Then, when the motor terminal is rotated, the motor will begin to run. Erratic motor rotation is often caused by a worn or dirty commutator. Some of these motors are almost too small to take apart, but sometimes, with patience and care, the armature and wire tongs can be cleaned (Fig. 5-23). At other times, you can tap the outside metal belt of the motor, and it will resume speed.

Check the motor winding with a quick RX1 resistance measurement. No or high resistance measurements indicate an open winding or a dirty commutator. Most small dc motors will have a resistance of less than 10 ohms. Check for correct voltage applied across the motor terminals with slow speeds. A small motor operating from an ac power supply may operate from 12 to 18 V, whereas in the dc battery-operated cassette player, the voltage equals the total battery voltage.

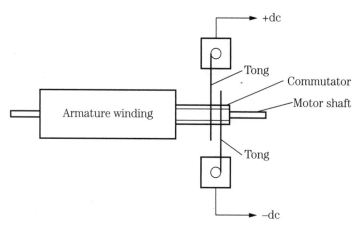

5-23 Clean the commutator and tongs for motor problems.

BOOM-BOX INPUT CIRCUITS

Boom-box input circuits may consist of transistors or IC components. Most of the later players have IC parts in the entire audio circuit. In a double-deck cassette player, tape deck 1 may only play the cassette and feeds directly into the input IC circuits. Tape deck 2 is switched into the input circuits in both play and record modes. Besides switching two different inputs into the IC, the input circuits also include switching the bias oscillator circuits in record mode.

Most problems within tape input circuits involve dirty tape heads and open or broken wires on the tape head. Suspect poor or intermittent switching when tape deck 2 is intermittent in play or record mode or will not play at all. Simply check the tape head continuity with the 2-kilohm scale of the DMM.

Spray cleaning fluid inside the tape head switch, and work the switch back and forth for a good cleanup. A bad record/play switch may cause intermittence and a warbling sound in one or both channels. Check tape head 1 with no playback in the right channel with tape deck 2 normal; this is caused by a ground wire off the tape head (Fig. 5-24).

CHECKING THE TAPE HEAD

A defective tape head may be packed with tape oxide, have an open winding or an intermittent internal connection, or be excessively worn. A worn tape head can result in a loss of high frequencies. A magnetized tape head can cause extra noise in a recording. An open head winding will produce a dead channel. Poorly soldered connections or broken internal connections can cause intermittent music. A grounded tape head winding to the outside metal shell may cause dead or weak audio.

Check the open or high-resistance head winding with the ohmmeter (Table 5-1). Inspect the cable wires, and resolder for intermittent conditions. These tape head wires tend to break where the cable connects to the head winding terminals. Make sure that the tape head is engaging the tape. Check for a loose screw, which would let the head swing out of line.

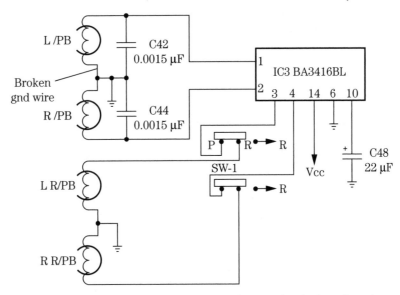

5-24 Check for a torn-off ground wire on the tape head when there is no playback in tape deck 1 of a Radio Shack SCR-55 cassette player.

Table 5-1. Tape-head resistance of several different cassette players.

Model	Tape-head resistance
GE-3-5808KA	225 ohms actual measurement
Panasonic RQ-L315	315 ohms actual measurement
Sony M440V	348 ohms actual measurement
Sony TCS 430	512 ohms actual measurement

Typical R/PB tape-head resistance: 200 to 830 ohms
Typical erase-head resistance: 200 ohms to 1 kilohm

The erase head is mounted ahead of the record/play head to erase any previous recording. Often the erase head has two leads and is excited by a dc voltage or by a bias oscillator circuit. Suspect the erase head when a recording is garbled or two different recordings are heard in the speakers. Measure the dc voltage across the tape head. If there is none, check the bias oscillator circuits. Simply trace the wiring from tape head to dc source or bias oscillator circuits. Check for an open erase head with the low-range ($R \times 10$) ohmmeter range (Fig. 5-25). The erase-head resistance can vary from 200 to 1000 ohms.

MORE POWER TO YOU

In large AM/FM/MPX receiver, cassette, and CD players, the audio section may consist of high-powered transistors or IC components. The high-powered amp may have an output of 1.5 to 10 W of power delivered into tweeters and midrange and large

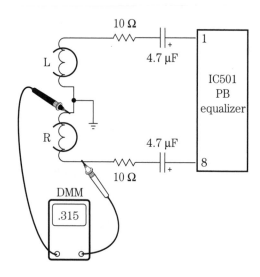

5-25
Check the tape head for open conditions
with the DMM.

subwoofer speakers. One large IC component may include most of the audio circuits for both right and left stereo channels (Fig. 5-26). Look for a large high-powered IC on a metal heat sink.

The audio from the radio, CD, and cassette player circuits is fed into an analog switch, a bass-treble board, and a high-powered IC that might include all the audio circuits. The audio output is amplified and coupled to a tweeter, subwoofer, or a very large PM speaker. Simply service the high-powered circuits as you would any audio amplifier circuits. Compare the dead or intermittent channel with the normal channel. Use the external audio amp and scope for signal tracing the signal in the audio circuits. Critical voltage and resistance measurements can weed out most defective audio amp components.

HEADPHONE CIRCUITS

The headphone jack may be connected to a separate headphone amplifier or directly from the audio output circuits of the output IC. A headphone jack may switch in the amplifier speakers when the headphone plug is removed. When the headphone plug is inserted, the speakers are disconnected. In some larger cassette players, the headphone jack may be switched into the amplifier circuits (Fig. 5-27).

A defective headphone jack may have dirty switching or contacts. Poor or dirty connections around the outside phone plug can cause erratic reception. Check for broken wires or poorly soldered joints at the back of the headphone jack for intermittent or dead sound. Intermittent sound in a Panasonic RX1940 model was caused by poorly soldered wires on the headphone terminals. Resolder the wires, and place a dab of glue on the connection to keep it intact.

The auto cassette player

The auto cassette player usually is found in an automobile AM/FM/MPX receiver. Besides the auto radio and cassette player, a CD player may be found in the same re-

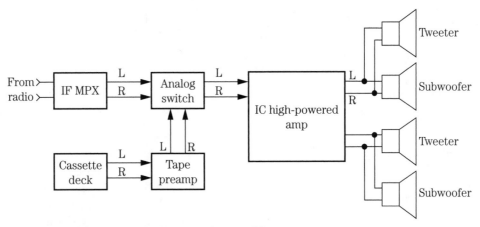

5-26   Block diagram for a high-powered output IC.

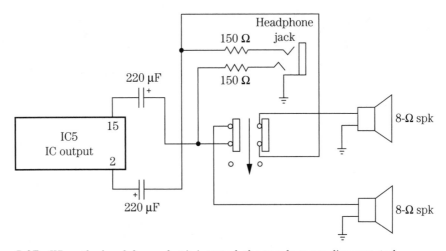

5-27   When the headphone plug is inserted, the speakers are disconnected.

ceiver. When either a cassette or a CD is inserted into the auto receiver, the radio and other player are disconnected by silent switching with silicon diodes. The AM/FM cassette receiver may have a flip-down, flip-up front panel. The AM/FM CD and cassette player may have a changer controller within the radio.

Besides a high-powered amplifier, the cassette receiver can control a CD player mounted in the trunk of the automobile (Fig. 5-28). A low-priced auto cassette/CD player and receiver may have from 1.2 to 10 W of output power, whereas a high-powered auto player may have from 25 to 50 W × four channels of power inside the audio system.

HIGH CASSETTE SPEED PROBLEMS

Excessive speeds may be caused by a defective cassette drive motor or a speed circuit. On some motors, you can adjust the speed from the rear end of the motor. Look

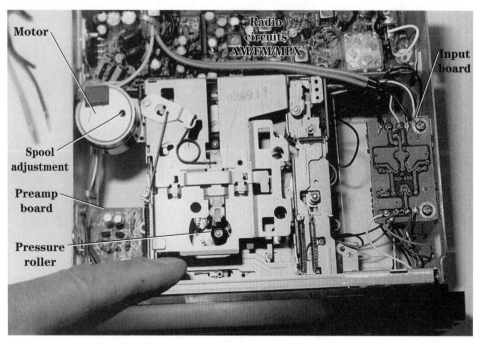

5-28 Inside view of the latest low-priced cassette car receiver.

for a small hole at the end of the motor. Also check the voltage applied to the motor terminals from the speed circuits. When the motor is running at a fast rate of speed, tap the end of the motor to see if it slows down. Replace a defective motor.

A leaky IC or transistor motor regulator circuit can cause high-speed problems. If the transistor or IC becomes open or leaky, the speed can increase or decrease depending on the dc voltage applied to the drive motor. Try to adjust the speed control in the regulator circuits. Suspect a defective circuit when the voltage will not change during adjustment (Fig. 5-29).

Check the motor speed by inserting a 3-kHz test cassette in the player and connecting a frequency counter at the speaker terminals. If the tape is running higher than 3 kHz at the frequency counter, readjust the speed control within the motor speed circuits or at the rear of the motor. The tape is running slow if the meter shows a frequency below 3 kHz (Fig. 5-30). Try to adjust the motor speed to exactly the 3-kHz frequency.

AUDIO COMPARISON TESTS

Suspect a defective audio stage when one channel is weaker than the other or will not balance with the balance control. The low or weak channel can be the result of dried-up electrolytic capacitors, poor transistor junctions and ICs, and a change in bias resistors. Do not overlook a bad electrolytic shorted across the emitter resistor of an audiofrequency (AF) or input transistor for weak sound. Check for a worn balance or fader control when the audio does not balance.

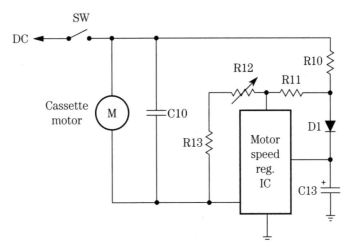

5-29 Adjust R12 to set the correct speed in a regulated cassette motor circuit.

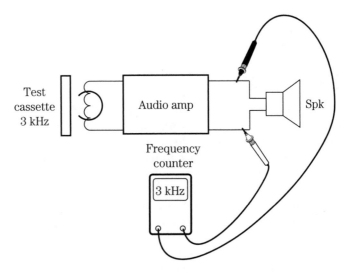

5-30 Readjust the speed control of the motor to read 3 kHz on the frequency meter.

Insert a 3- or 10-kHz test cassette, and notice the different amount of volume in either the left or the right channel. Start at number 1 in the audio circuit, and compare the volume in both channels. A bad or dirty tape head in either channel might cause a noisy and weak sound. Connect the external amp at spot number 2, and check for equal sound in both channels. Keep going through the audio channels until the weak stage shows up and can be located (Fig. 5-31).

Then check the audio on both sides of the electrolytic coupling capacitor. Check the sound on the base terminal of the AF transistor or IC to locate the weaker channel. Take critical voltage and resistance measurements after locating the defective stage. A lower supply voltage caused by a bad electrolytic decoupling capacitor or resistor network can cause the weaker sound stage.

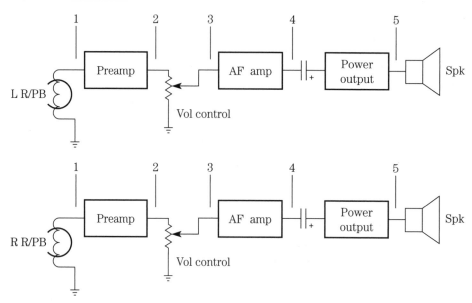

5-31 Compare the audio in each channel at the correct number in the tape audio circuit.

SILENT SWITCHING

In early cassette players and receivers, the outputs of both the radio and the cassette player normally would be switched into the audio output circuits. Today, silent switching is accomplished with silicon diodes in each path of the AM/FM/MPX radio, cassette, and CD player output circuits. Often the silent switching occurs before the volume controls in the audio circuits.

Two silicon diodes are found in each audio output of an AM/FM/MPX car radio and cassette player. D101 and D102 provide signal to each 4.7-μF electrolytic coupling capacitor of the volume controls within the radio circuits. The same audio is blocked by D103 and D104 from entering the cassette motor input circuits. D103 and D104 switch the audio from the cassette player to the same 4.7-μF capacitors, and the audio cannot enter the output circuits of the stereo radio circuits that are blocked by D101 and D102. All four diodes provide silent listening without any movement of parts (Fig. 5-32).

Suspect a shorted diode in the right channel of a radio when both the radio signal and cassette playback can be heard at the same time while the cassette is being played. Check each diode with the diode tester of the DMM. You may have to remove one end of the diode for an accurate test.

ACTUAL CAR RADIO/CASSETTE PLAYER PROBLEMS

A defective volume control in a Delco 01XPB4 car radio caused the volume to cut up and down. The intermittent left and right channels of a Sanyo FT1495 were caused by a shorted volume control. The intermittent right channel of a Craig 3135

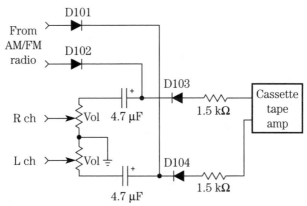

5-32 D101 through D104 provide silent switching in the audio circuits of a cassette player.

model was caused by broken foil or a broken trace at C210 coupled to the volume control. No volume was found in a Ford 5TPO model, and it was caused by an open volume control.

A loud popping sound was heard in an Automatic BTR-7284 car radio, and it was caused by a broken ground on a 500-μF electrolytic filter capacitor in the power supply. A Delco 51CFP1 model was dead, and only a hum was heard in the speakers, caused by a leaky decoupling capacitor C102 (250 μF). Motor boating and a chirping noise were heard in the speakers of a Plymouth 6BPD model, caused by an open decoupling capacitor (1000 μF at 50-V capacity).

A loud howling sound in the left channel with no audio was found in an Audiovox C979 model, and it was caused by a red-hot IC303. A loud buzzing and motor-boating sound in the speakers of a Soundesign 6024 model was caused by a poor ground on the heat sink. Replace the left power output amp IC for motor boating in the left channel of a Sanyo FTV98 car radio. The power amp IC was running warm in a Kraco KID-585H model, and the sound would cut out after a few minutes of operation.

HEAD AZIMUTH ADJUSTMENT

Improper adjustment of the tape head can result in distortion, muffled sound, and a loss of high frequencies. The head should be aligned horizontally for optimal sound reproduction. Check for a loose mounting screw that might cause muffled sound and a loss of audio. One side of the tape head is fastened with a small metal screw, and the other, with an adjustable spring.

Play a 10-kHz test cassette, and connect a frequency meter or volt-ohm-milliammeter (VOM) at the speaker terminals. Readjust the azimuth screw for a maximum reading on the meter. Likewise, a low ac voltage range on the VOM or vacuum-tube voltmeter (VTVM) can be used as an indicator. Adjust the head screw until a maximum measurement is found on the meter (Fig. 5-33). No highs and a muffled sound were heard in an Audiovox AV929 cassette player with a missing adjustment screw.

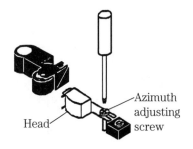

5-33
Adjust the azimuth screw for a maximum reading on a VOM or VTVM.

KEEPS REVERSING DIRECTIONS

In early car radio/cassette players, a magnetic switch would reverse the direction of the cassette motor when the tape had reached the end of the recording. With the tape coming to the end of the cassette, an automatic circuit mechanically reverses the direction of the cassette drive motor.

Some cassette players have a magnet mounted on the turntable or reel with a magnetic switch mounted below. When the tape stops rotating at the end of the recording, the magnetic switch reverses the direction of the tape and motor. Now the music is played in the reverse direction. Usually two tape heads and capstans and flywheels are found with this type of operation. Check for a broken wire on the magnetic switch when the auto reverse keeps reversing.

You also may find in early auto cassette players that the tape reversing process includes a commutator ring with springlike tongs that ride on the commutator rings. When the commutator stops rotating at the end of the cassette, the circuit changes the polarity of the cassette motor. Suspect dirty contacts on the commutator and tongs that cause the auto cassette to automatically keep reversing directions.

Clean the commutator rings with alcohol and a cleaning stick. Make sure that the tongs ride at the right spot on the commutator rings. When the tape was reversed in a Sanyo FT606 cassette player, no sound was heard in the right channel. Check for torn-off wires on the tape head as the tape is reversed in the other direction.

WOW AND FLUTTER

Usually wow and flutter occur during playback or in recording operations. A longer duration of the sound may be called a *wow* condition, whereas a short change in the audio may be called a *flutter* sound. Uneven rotation of the tape can result in wow and flutter.

Replace the capstan and flywheel for a wow or flutter sound. A binding capstan and flywheel can cause wow and flutter in the audio. A dirty or worn belt also can cause wow or flutter. The capstan may run slowly and drag, with many rubber belt particles found on the flywheel or pulleys. A dry or bent, out-of-line flywheel can cause wow and flutter. Replace the idler spring. A worn pinch roller with a bad idler spring can cause wow and flutter. A dry reel shaft also can cause wow and flutter during playback.

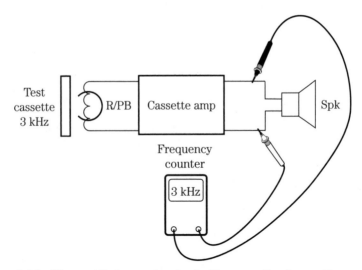

5-34 Wow and flutter can be checked in a cassette player with a 3- to 10-kHz frequency variation on the frequency counter.

A motor pulley that is way out of line on the shaft can cause wow and flutter. Do not overlook a defective cassette or a bad cassette motor for uneven sound in the playback modes. Check for a defective speed circuit for wow and flutter in the audio. A loud knocking noise with a bad idler spring can cause wow and flutter.

If a wow and flutter meter is not handy, check the tape deck for wow and flutter conditions with a frequency meter and a test cassette. Connect the frequency meter to a 7.5-ohm, 10-W resistor across the speaker terminals. Insert a 3-kHz cassette into the player, and notice any deviation in frequency in the meter. If the meter changes between 3 and 10 kHz of the test tape on the frequency counter, the cassette player definitely has a wow and flutter condition (Fig. 5-34). Audio flutter during playback was caused by a dry supply reel in a Panasonic RX1822 cassette player.

For additional cassette player troubleshooting tips, see Table 5-2 on p. 166.

Table 5-2. Cassette player troubleshooting.

Symptom	Cause	Remedy
Cassette cannot be inserted		Check cassette for damage. Check for foreign material inside player. Check if play or record button is depressed.
Record button cannot be depressed	No cassette loaded.	Load cassette.
	Cassette tab removed.	Place tape over cutout if you want to record on this cassette.
Playback button cannot be locked in	Tape is completely wound toward arrow direction.	Rewind tape with rewind button.
Tape does not move	Batteries are in backward.	Fix batteries.
	Weak batteries.	Test, and replace below 1.2 V.
	No ac.	Connect ac power adapter.
	Motor doesn't work.	Check voltage across motor. Continuity test motor winding to see if it is open.
No sound from speakers	Headphones plugged in.	Remove headphones plug.
	Volume is turned down.	Adjust location of volume control.
Fast tape speed	Incorrect speed control setting.	Readjust speed control.
Weak or distorted sound	Weak batteries.	Test, and replace below 1.2 V.
	Dirty heads.	Clean heads with alcohol and cleaning stick.
Poor recording	Weak batteries.	Test, and replace below 1.2 V.
	Dirty R/PB tape heads.	Clean up with alcohol and cloth.
Poor erase	Improper connection.	Check all head connections.
	Dirty erase head.	Clean erase head.
	No dc voltage.	Check dc voltage on head.
	No oscillator waveform.	Take scope waveform on head terminals.

6
CHAPTER

Troubleshooting
CD players

After World War II, the radio phonograph was included in the console radio, and now the CD player has taken its place. At first the CD player was found in a portable unit; then it progressed to the boom-box cassette player and advanced to the auto radio/cassette/CD player and to the CD changer, and now the CD player is located in the TV CD and DVD player. A portable and boom-box CD player may operate off batteries or the power line (Fig. 6-1).

Of course, a CD player in an automobile operates from the car battery, whereas a home system may include an AM/FM/MPX receiver with cassette and CD players. A CD changer may be found in a tabletop home system or in the trunk of an automobile. Today, a CD player may be found even in 13-, 19-, and 27-inch TV/VCR combo players.

Like a videocassette recorder, a CD player is difficult to service without a schematic, so try to locate a diagram to service such a player. However, there are many tests that can be done without a schematic. A block diagram will help in isolating the various circuits. The block diagram of each circuit may help you to understand how CD circuits perform.

Critical waveform, voltage, and audio tests may locate the defective circuit. Checking the laser diode, radiofrequency (RF), and encoding frequency modulation (EFM) signals will indicate if the laser pickup assembly is functioning. Voltage measurements and signal waveforms on the loading, slide, and disc motors can determine if the motor circuits are normal.

6-1 CD players can be found in portable, home, auto, and tag-along devices and inside a CD changer.

Laser diode circuits

The laser optical assembly is one of the most important sections of a CD player. If the laser diode or optical circuits cannot be repaired, the unit may be thrown out because the cost of replacing the optical assembly is rather expensive. Often the cost of replacing the whole optical assembly is too expensive. Of course, some components within the optical assembly can be replaced, if they are available. Make sure the RF signal from the pickup assembly is functioning before other repairs are made.

The laser circuit consists of photodetector diodes (A, B, C, and D), objective lens, focus and tracking coils, collimator lens, beam splitter, monitor, and a laser diode in some CD players. The tracking diodes are E and F, whereas the monitor diode is MD and the laser diode is listed as LD. The photodetector diodes sense the EFM signal from the disc. The tracking detector diodes (E and F) control the tracking, and the focus diode keeps the beam on the disc. The EFM signal from the photodetector diodes is applied to the RF amplifier circuits (Fig. 6-2).

The 8- to 14-bit (EFM) signal is a very complex encoding scheme used to transfer digital data to a form that can be placed on a disc. The EFM waveform is found at the output of the RF integrated circuit (IC) or transistors and is fed to the digital processor. When the error signal is not present at the servo IC, the chassis might shut down at once.

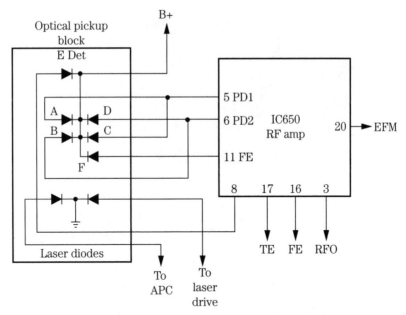

6-2   The laser circuits consist of photodetector diodes (*A, B, C,* and *D*), a monitor, and a laser diode.

The laser diode output can be checked with a laser light meter or infrared indicator. Measure the dc source applied to the laser diode. This might be a direct voltage or voltage applied from a laser driver transistor or IC. The laser diode output is fed directly to the RF amplifier IC or processor.

DIAMONDS ARE EXPENSIVE

The RF or eye pattern represents a diamond-shaped pattern taken from the output of the RF amplifier stage. When the eye pattern is not found at the RF amplifier IC or transistors, the CD player will shut down automatically. The laser optical assembly must be operating with a normal RF amplifier and an adequate low-voltage source feeding the front-end circuits. Scope the eye pattern (EFM) at the RF amplifier output (Fig. 6-3). This RF signal feeds the digital signal processor IC and servo circuits.

DO NOT LOOK AT ME

Do not look directly at the laser beam or lens at any time. The laser optical beam cannot be seen by the naked eye. Service technicians should avoid looking directly at the laser beam (Fig. 6-4). Keep a CD loaded on the disc platform all the time. Keep your eyes at least 25 in from the optical laser beam. Place a conductive mat under the test equipment and player while servicing it. Wear a wrist strap to leak off body charges to the chassis and ground. Do not forget to remove shorting or interlock devices after repairing the CD player. Take critical leakage tests. Replace the critical parts with originals.

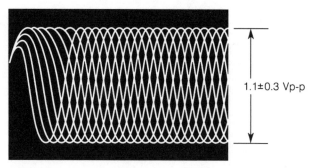

6-3 Check the eye pattern when the CD player is inoperative.

6-4 Keep your eyes away from looking directly at the lighted lens assembly.

RF AMP PROBLEMS

The RF amplifier receives the weak signal from the photodetector diodes and amplifies the digital signal as an RF pattern. Suspect a defective optical assembly when the RF signal is weak or missing. Measure the voltage applied to the photodetector diodes. Look for a test point on the printed circuit (PC) board for a quick scope test of the RF amplifier IC. An early RF amp might consist of several transistors, whereas the latest CD players contain IC components. The RF amp sends an EFM signal to the digital signal processor and a mirror signal to the servo mechanism.

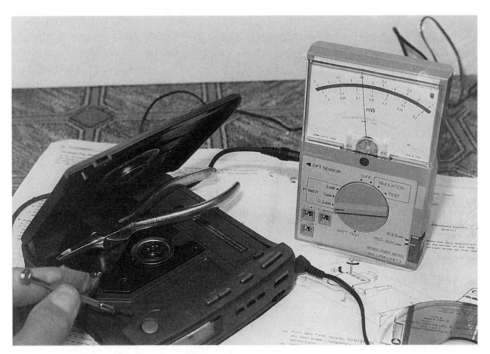

6-5   Check the laser diode lens assembly with a laser power meter.

Determine if the laser diode is emitting light with a laser power meter or infrared indicator strip. Place the power meter probe across the laser lens assembly. Do not look directly at the laser light. Move the probe back and forth to get the maximum reading on the meter. If the meter hand hits the peg, change the meter setting to the next highest measurement (Fig. 6-5). When the laser diode appears normal, place a disc on the disc mount and take an RF (EFM) scope waveform at the RF amp output.

Clean cigar or cigarette smoke residue and dust from the lens assembly. A dirty lens assembly can cause the disc to spin slowly and sometimes not play. Clean the lens with Windex. Wipe the lens assembly gently so as not to throw it out of line or bend the supporting spring. A dirty lens assembly can make a disc start and stop. Replace the laser scanner assembly when the CD rotates intermittently. A bad optical pickup assembly can cause static in the audio when the disc is playing. Usually the CD player will shut down at once with no EFM eye pattern.

Suspect a defective RF amp transistor or IC when no EFM pattern is found at the output terminal. Measure the supply voltage at the RF amp circuits. Improper or no voltage at the RF amp transistor or IC can result in no scope waveform. Test the RF transistor with a semiconductor tester or the diode tester of the DMM. Always be careful when working around the optical pickup assembly.

DIGITAL SIGNAL PROCESSOR

The signal processor IC usually contains the clock generator, EFM, data, LUCH, mute data, digital filter, timing control, oscillator, modulation, constant linear velocity

(CLV) servo, and servo system control. The RF output signal is fed to the input of the digital signal processor and the output to a digital-to-analog (D/A) converter circuit.

The EFM comparator changes the RF signal into a binary value. The binary-coded signal is then fed to pin 5 of the digital signal processor (IC3), where it is demodulated, a bit clock is generated, errors are detected and corrected, lost data are interpolated, and the subcode for track number/elapsed time is demodulated. A bit clock must be derived from the EFM signal by a variable crystal oscillator (VCO) so that the data in the EFM signal can be read.

The demodulated EFM signal is converted to digital data and stored in a random-access memory (RAM) (IC3). The data are read from the RAM and fed to the digital filter and then to a D/A converter (IC11) (Fig. 6-6).

A defective digital signal processor IC can cause a high-pitched noise in the audio while a CD is playing. A defective RAM IC can cause noise in the sound. Replace the RAM IC with a ticking noise while the CD is playing. A bad RAM IC can cause distorted sound in the speaker. Replace the RAM IC with a low-level noise in the audio. Check the crystal in the digital processor circuits when there is noise in the sound and no clock signal.

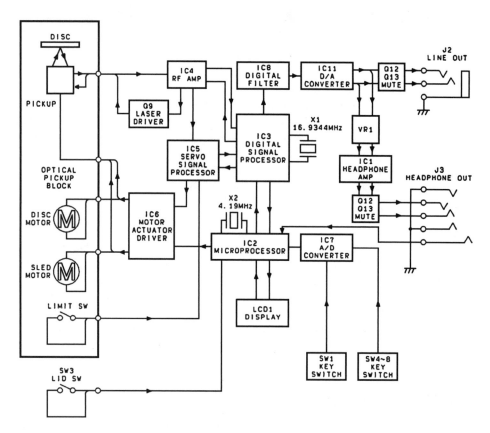

6-6   Block diagram of a portable CD player.

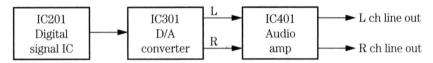

6-7 The digital-to-analog (D/A) converter circuit provides audio in the left and right channels.

Digital-to-analog (D/A) converter

The D/A converter receives a digital signal from a digital filter network or from the signal processor stage. The D/A converter (IC301) converts a digital signal to an analog signal (Fig. 6-7). The audio signal is then fed to the final output amplifier. One or two stages of amplification are found in most CD players. This audio signal is fed to a stereo line output jack or to a separate headphone amp. The line output voltage of most CD players is around 2 V.

Replace the D/A converter IC when there is no sound in the left channel. Check both the right and left output channels from the D/A converter for no sound in either channel. Replace a bad or noisy D/A converter IC when there is background noise during the quiet spaces of music. Distorted audio can result from a defective D/A converter IC. The D/A converter (IC203) in a Sharp DX100 CD player caused extreme distortion in both audio channels. No audio in the left channel of a Aiwa DXM77 CD player was caused by a low-pass filter (LPF101).

CD power supplies

A portable CD player may operate from a 3-V dc battery source or from an external power line cord adaptor. A power and change switch from the positive terminal of the battery is applied to the regulated supply as well as to the door switch. The door or cover of a portable CD player must be closed before the 3-V source is applied to the lens assembly. This protects the operator's eyes from looking down on the lens assembly (Fig. 6-8).

In a boom-box cassette/CD player, the low-voltage power circuits are quite simple, with bridge rectification and high filter input. The cassette motor and other circuits have a transistor-regulated voltage source, whereas the CD circuit has one or more transistor–zener diode regulators. A large input filter capacitor (2200 to 3300 µF) provides voltage to a CD regulator circuit. The output voltage may have a motor regulator transistor circuit that regulates the voltage to the cassette player and motor. The CD circuits may have several transistor-diode or IC regulators.

When higher dc voltage is needed within auto or home CD player systems, a dc-dc converter circuit provides a 5- to 10-V regulated source. Of course, an auto CD player operates from the car battery, whereas a home theater is plugged into an ac power line. You will find several transistors and zener-regulated circuits within the higher-voltage sources. Check the power supply voltage at the largest electrolytic filter capacitors.

Most power supply problems are related to leaky or dried-up filter capacitors, open fuses, and bad decoupling capacitors and transistor regulator circuits. A shorted or leaky filter capacitor can blow the fuse and damage the small power transformer if the

6-8 The top lid must be closed before a portable CD player will operate.

unit is not fused. Dried-up filter capacitors will cause hum in the sound at all times. Suspect a defective filter capacitor when hum is heard in the speakers with the volume turned down.

Leaky or shorted decoupling electrolytics can cause hum and low voltage to other circuits within a CD player. Simply clip in a known electrolytic across the suspected filter capacitor with the CD player turned off. If the hum disappears, replace the defective filter capacitor. Check all filter capacitors for no power to the CD circuits.

Do not overlook a defective regulator transistor or IC. Check the voltage in and out of the transistor or IC. An open transistor or IC regulator will have low or no dc output voltage. A leaky regulator transistor can cause a low- or high-voltage output depending on what elements in the transistor are leaky (Fig. 6-9). Besides voltage transistor regulators in the power supply, the servo and motor circuits also have transistor regulators. Intermittent sound and clicking of the protection relay were caused by poorly soldered contacts on the regulator transistors.

No audio and a dim liquid crystal display (LCD) in a Pioneer SXV500 CD player were caused by a change in resistance of R328 (8.2 kilohms), which supplies a 15-V regulator transistor Q312.

Suspect a shorted filter capacitor or leaky silicon diodes when no ac voltage is found at the power transformer (Fig. 6-10). Measure the ac voltage across the secondary winding of the power transformer. Pull the ac cord if no voltage is found across the secondary winding. Often, when a component in the input of a power transformer becomes shorted and the primary winding is not fused, the primary winding goes open because it has so many turns of a very fine wire. Quickly take a 2-kilohm ohmmeter continuity measure-

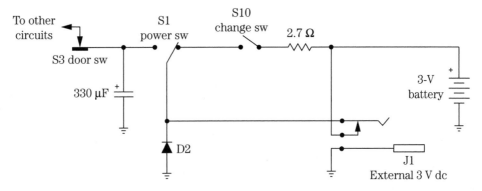

6-9 A portable CD player can be operated by batteries or an ac power line adaptor.

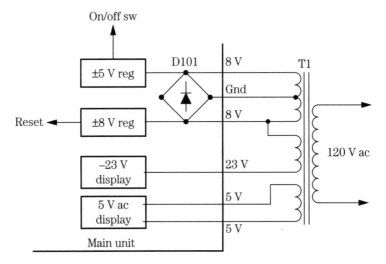

6-10 A defective 5-V regulator prevented operation of a Dennon DCM-560 CD player.

ment of the primary ac winding (Fig. 6-11). No low-ohm measurement indicates that the transformer primary winding is open; replace the transformer.

Replace the R305 and R306 (430-ohm) resistors in the regulator voltage circuit feeding the preamp ICs that caused one distorted channel and the other channel okay in a Pioneer SX780 CD player.

Check those electrolytics with an ESR meter

When hum is heard in the sound, check those electrolytics with an equivalent series resistance (ESR) meter. Discharge all terminals of the electrolytic by shorting them together. Make sure that no voltage is found on any capacitor terminals before mak-

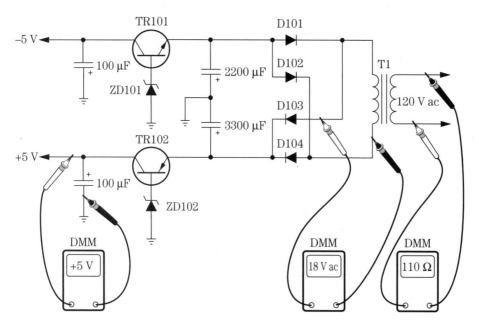

6-11   Check the primary winding for open conditions with the 2-kilohm ohmmeter scale and the ac voltage at the secondary winding of the transformer.

ing ESR tests. A charged voltage can destroy an ESR meter. Check across the capacitor's terminals with the ESR meter probes, and read the condition of the capacitor on the meter (Fig. 6-12).

A normal electrolytic will test in the green area of the meter. If the capacitor registers in the yellow or red area, replace it. A dried-up electrolytic will show up in the red or yellow area. Check the electrolytic for leakage or shorted conditions by taking a 2-kilohm measurement across each terminal to common ground.

Check all electrolytics within the power supply in the same manner. Do not overlook a decoupling electrolytic for a loss of capacity. The extremely low capacity of small electrolytics (1 to 10 µF) results in a tendency to lose even more capacity. Check all electrolytics at the various voltage sources within the power supply. A weak or low dc voltage source can be caused by a defective electrolytic at that voltage source (Fig. 6-13). A Sony CDXA20 CD player was dead with no power, and this was caused by a leaky 330-µF, 6.3-V electrolytic.

Focus amp problems

The focus coil keeps the lens assembly focused at all times and is driven by a focus driver transistor or IC. The focus driver receives the focus commands from a servo IC, whereas the servo IC is driven by a signal from the RF amp and signal processor. In some CD players, you may find that one servo driver IC controls the focus and

6-12 Check all electrolytics in the power supply with an ESR meter.

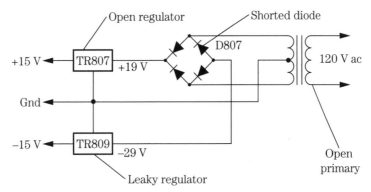

6-13 Various components can cause failure in a low-voltage power supply.

tracking coils and the SLED and spindle motors. The focus and tracking coils are located within the optical pickup assembly (Fig. 6-14).

Take a peek at the focus and tracking coils when the CD player is first turned on. Both coil assemblies will move and search, indicating that the circuits and coils are normal. Take a waveform test across the focus coil when the CD player is turned on. Check the supply at the driver transistor or IC. Measure the voltage across the focus coil winding. Turn the set off and take a continuity ohmmeter test of each coil for an open coil winding.

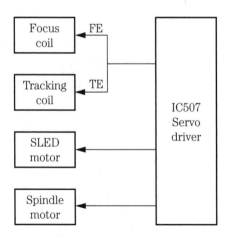

6-14
The focus and tracking coils may be driven by one large servo driver IC507.

Battery focus coil check

The focus and tracking coils can be checked by applying a low dc voltage across the coil windings. Remove the ground lead from the focus coil, and insert the battery voltage. Notice if the focus coil moves when 1.5 to 3 V is applied across the coil. Now reverse the battery polarity and notice if the coil moves in another direction. Test each focus and tracking coil in the same manner. You can assume that both coils are normal if they move when a battery source is applied to them.

Tracking amp problems

The tracking circuits are designed to keep the optical lens assembly on the right track as the CD spins. Usually, the tracking coil is operated by the same driver and servo IC. Take a low-ohm continuity measurement across the tracking coil to determine if it is open or has a poorly soldered connection. A tracking coil waveform across the tracking coil indicates that the tracking coil and driver IC or transistor are normal. Locate the tracking coil assembly wires, and trace them back to the driver transistor or IC. Take a supply voltage measurement on the driver IC when the tracking coil is not moving (Fig. 6-15).

Besides a defective tracking coil or driver component, a laser head groove gear may be split, causing improper tracking. A bad ribbon cable to the laser head assembly from the main PC board can cause the player not to play the outer track on the CD. Mistracking and skipping can be caused by a bad worm gear. In an Onkyo DX530 model, improper tracking was caused by a split gear at the laser head asssembly.

Loading motor problems

The loading motor moves the tray outward and loads the CD into the CD player. In a home entertainment CD player, the loading motor moves the tray out, the CD is loaded, and the tray motor pulls the tray back into the CD player. In most cases the loading motor is driven by a motor driver transistor or IC from a microprocessor or

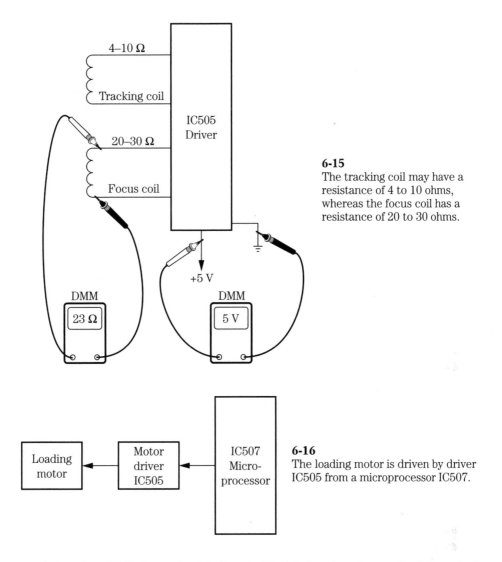

6-15
The tracking coil may have a resistance of 4 to 10 ohms, whereas the focus coil has a resistance of 20 to 30 ohms.

6-16
The loading motor is driven by driver IC505 from a microprocessor IC507.

control system IC (in the auto CD player). The CD is set on the motor drive spindle after opening the door (Fig. 6-16).

Intermittent operation of the loading tray may result from foreign material within the track area, a slipping drive belt, or a jammed mechanism. A dirty open/close switch can cause intermittent operation. Check the tray switch by shorting across its terminals. Notice if the tray rack or gears are binding. When the loading motor drives a belt to load and unload, check the belt for oil spots. Check the drive belt for worn or cracked areas. Inspect the loading pulley or plastic gear assembly for stripped or broken teeth. Check the CD changer loading tray for wires caught on the bottom of the moving tray. When measuring the voltage applied to the motor, you might find 14 V to open the tray and 24 V to close the tray and pull the CD into the player (Fig. 6-17).

A defective loading motor will not load the CD. Check for a broken or loose motor drive belt. Check all the photo interrupters and replace when the CD will not load. The

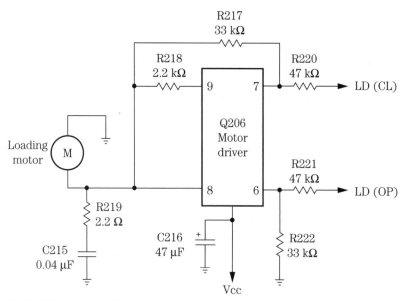

6-17   When the loading motor was tapped on the end with the screwdriver handle, the motor started to rotate in an Onkyo DX-C606 CD player.

tray would not move back and forth in a Magnavox MX3702 CD player with a leaky C04 (470-μF, 10-V) electrolytic capacitor. Take a critical supply voltage (Vcc) measurement on the loading motor driver IC.

Disc and SLED motors

The SLED and slide motors move the optical pickup assembly on rails from the inside to the outside of the CD, whereas the disc motor rotates the plastic CD. The disc motor might be called a *spindle motor,* whereas the SLED or feed motor sometimes is called a *carriage motor.* Both these motors are controlled by a system control or servo driver IC. The disc motor starts out in the center of the CD at approximately 500 rpm and moves toward the outer rim at 200 rpm (Fig. 6-18).

The slide or SLED motor brings the laser pickup within the fine tracking control range. A tracking servo signal is used to move the pickup horizontally. The tracking coil keeps the pickup assembly on the track.

Check the spindle disc motor and slide motor with voltage and continuity measurements as you would the rest of the motors in a CD player. Check the spindle motor transistors or IC if the motor tests normal. Measure the voltage applied to each motor. Most of these motors operate from a 5-, 8-, or 10-V source at the driver transistor or IC.

Clean the lens assembly with Windex when the disc motor spins slowly or will not move. Intermittent playing of the CD or slow spinning of the CD results from a dirty lens assembly. Check for open resistors in the carriage or slide motor circuits when the player plays only part of the CD. Replace the spin motor when the CD will not rotate, skips, or stops. Check the table height when replacing the disc motor.

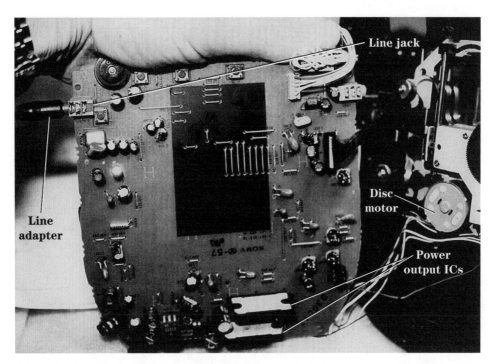

6-18 The CD sits directly on the spindle of the disc or spindle motor.

In a Pioneer PD-7010 model, the carriage or slide motor would not rotate. Very low voltage was found on the slide motor terminals. Both the 10-V sources feeding the driver IC3 were quite low, indicating a defective 110- or 210-V source or a leaky IC3. When the voltage source was removed from pins 5 and 10, the voltage returned to normal. Driver IC3 was replaced, and this restored slide motor operation (Fig. 6-19).

Intermittent skipping

Clean the lens assembly when there is intermittent skipping of a CD. A defective laser scanner assembly can cause intermittent CD operation. Check for a leaky IC in the 5-V source when the CD stops in the middle of a program. Replace a leaky drive motor IC when the motor becomes intermittent. Suspect a defective drive motor when the CD rotates and then stops. Monitor the voltage at the drive motor and notice if the voltage changes as the CD rotates. Erratic voltage applied to the drive motor from a defective transistor or IC driver circuit can be caused by a defective IC or transistor.

A worn receptacle causing unstable clamping can cause intermittent skipping. Check the table height of the spin motor, and replace the spin motor when the CD skips, stops, and then will not spin. Replace a bad worm gear for mistracking and skipping of the drive or spin motor.

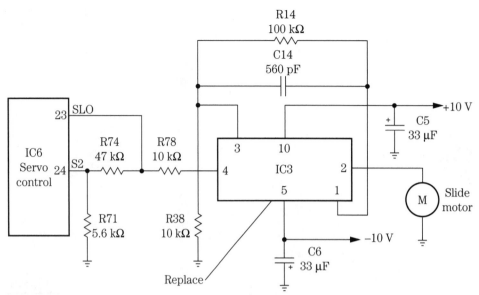

6-19 A defective slide motor driver IC3 caused no rotation in a Pioneer PD-7010 CD player.

A Fisher AD885 CD player would stop in the middle of a program and sometimes appear intermittent. A critical voltage test on the IC115 (7805) 5-V regulator revealed a very low voltage of 3.5 V. After replacing IC115, the voltage returned to normal at 5 V, and the CD player returned to normal operation.

No play, dead operation

No play, no power, and a dead CD player can be caused by many different components. A leaky or shorted diode in the low-voltage power supply can cause a transformer winding to open, resulting in no dc voltage on the large filter capacitors. Notice if the display lights up. Make sure that the voltage from the low-voltage power supply is normal. A defective on/off switch or open fuse can cause a dead CD player.

Critical voltage measurements on all transistors or IC voltage regulators can cause motors, drivers, and servo circuits to fail (Fig. 6-20). Check for a leaky or open electrolytic filter capacitor for very low or no output voltage. Make sure that the dc-dc converter power supply is turning out the correct voltage.

Check the drive motor circuits when the drive motor will not operate. Measure the voltage to the drive motor from the transistor or IC driver stage. Measure the voltage (Vcc) applied to the drive motor transistor or IC. Often negative or positive voltage is applied to these components. Likewise, check the supply voltage of the servo or control processor for no drive motor rotation. Make sure that the EFM waveform is found at the RF IC. The CD player will shut down at once when there is no EFM waveform at the RF IC. This indicates a defective RF or optical pickup assembly.

Intermittent components within the power supply can cause intermittent operations. Monitor the various sections of the power supply to determine where the intermittent

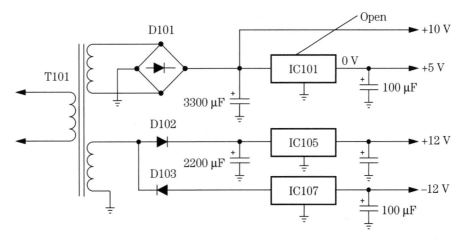

6-20 A defective open IC101 caused no motor operation in the play mode of a CD player.

component is located. Connect the voltmeter across the largest filter capacitor and notice if the voltage is constant or changes when the CD player becomes intermittent.

If the intermittent component is not in this section, proceed to the various voltage sources with the digital multimeter (DMM). Often the intermittent part is a transistor or IC voltage regulator. Monitor the voltage at the emitter terminal and output-regulated transistor voltage with the DMM. Inspect the transistor and IC regulator terminals for poorly soldered contacts. Check all filter capacitors (330 µF) in a Sony CDX120 CD player for dead, no power operation.

Noisy sounds

Hum in the sound can be caused by dried-up filter capacitors. Check each electrolytic with the ESR meter. Replace the D/A converter with background noise during the quiet play parts of the CD. Replace the RAM IC that produces a ticking sound when a CD is playing. Replace the optical pickup assembly for static heard while a CD is playing. A low-level noise can result from a defective crystal in the VCO circuits (Fig. 6-21).

Suspect IC components in the sound circuits when there is garbled audio. Distorted audio can be caused by a defective D/A converter IC. No audio can result from a faulty demodulator IC. A missing reference voltage from an IC regulator can cause intermittent static and popping noises with no audio output. Check the D/A converter IC for static in both channels. Replace small decoupling capacitors (10 to 50 µF) when there is noisy audio. Excessive noise in the audio output when the power is turned on and there is no clock signal can be caused by a defective crystal in the VCO circuits. Garbled sound in a Sony CDPC5F model was caused by defective IC501 and IC503.

The portable CD player

The portable CD player was designed for the ardent, on-the-go music lover (Fig. 6-22). A portable CD player operates from batteries or a battery pack. A portable CD

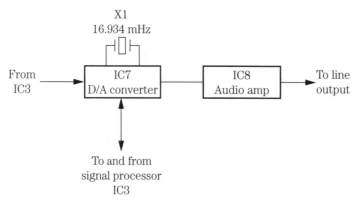

6-21 A low-level noise can be caused by a defective VCO crystal in the microprocessor.

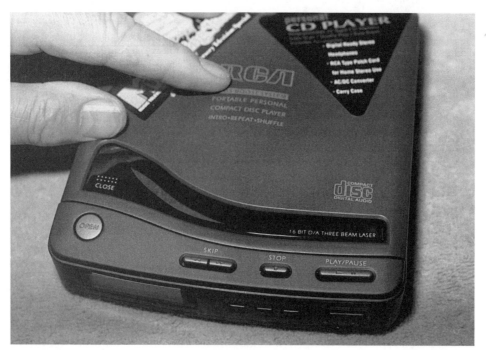

6-22 The door or lid must be closed on a portable CD player before the CD will spin.

player has circuits similar to those of a large tabletop or changer CD player. Of course, there are only two motors, slide and disc motors, because the player features top loading. The motor circuits may be controlled by one IC. You can listen through a pair of headphones with either battery or ac operation. The portable CD player has two line output jacks to play through an external amp.

All portable CD players have a top-lid interlock switch system that provides protection for the operator when the top lid is opened for loading of CDs. The lid interlock

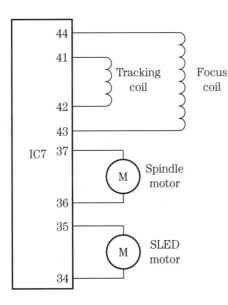

6-23
IC7 drives both tracking and focus coils and both spindle and SLED motors.

switch disables the laser signal. Remember, the optical lens assembly shines upward toward your eyes, so power to the laser diodes is removed when door is open.

The focus, tracking, and motor circuits in a Realistic CD-3370 player are controlled by IC7. The focus coils are connected to pins 43 and 44, whereas the tracking coil connects to pins 41 and 42. The focus coil test points are TP10 (F2) and TP12 (F1) (Fig. 6-23).

IC7 also provides drive for the spindle and SLED motors. The spindle motor connects to pins 36 and 37. The SLED motor terminals connect to pins 34 and 35. In this portable CD player, the coils and motors are controlled with one IC7 driver. The slide motor moves the laser assembly out on sliding rods, and the disc motor rotates the playing CD.

The portable CD player contains many surface-mounted device (SMD) components, gullwing ICs and microprocessors, and several PC boards. These small components, PC wiring, and parts make servicing portable CD players more difficult. More IC components are found in portable CD players because of their physical size. The suspected IC may control two or more different circuits. Check the supply voltage, and scope the signal in and out of the suspected IC. Take voltage and resistance measurements on each terminal pin before replacing an IC.

Boom-box CD circuits

The boom-box CD player comes in a number of large units and tabletop or portable players. Today, a CD player also may contain an AM/FM/MPX receiver, as well as CD and cassette players. Most boom-box players have top loading of the CDs without a loading motor. A rotating keeper at the top of the lid holds the CD in playing position. A boom-box player will not play until the lid is locked down. Portable CD players can be operated from batteries or the ac power line (Fig. 6-24).

Most boom-box CD player circuits are the same as those of portable CD players, except that the components are not quite jammed together as much as in a portable CD player. A boom-box CD player can have preamp circuits that switch into the regular

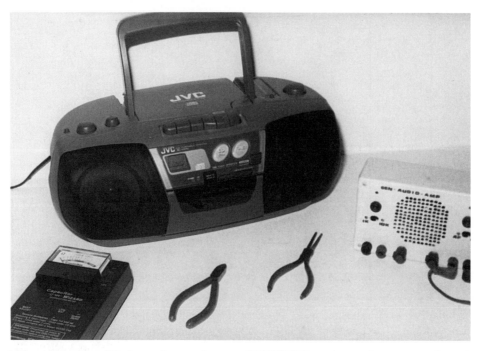

6-24 A boom-box CD player also may have an AM/FM/MPX receiver and a cassette player.

stereo cassette and radio amp circuits. You will find many SMD components within boom-box CD circuits. IC components are found throughout the CD player circuits, with only a few transistors within the mute circuits. The stereo signal from the D/A converter is switched into the input of the power output IC (Fig. 6-25).

The RF IC amp provides signals to a servo large-scale integration (LSI) as well as to the digital signal processor IC. The digital signal is sent to a D/A converter IC, with one audiofrequency (AF) amp IC serving both stereo channels. The servo IC controls one large IC driver that provides signal to the focus and tracking coils and disc and slide motor driver ICs. A large processor IC controls the LCD, signal processing, and servo LSI components. You may find 8 to 10 large ICs controlling all the circuits within a boom-box CD player.

The power supply in a large boom-box CD player might be dc-dc converter circuits, whereas a smaller boom-box player may operate from a power line cord or an adaptor power supply. Usually, the power supply transformer voltage is rectified by a bridge rectifier component or separate silicon diodes, with a sliding switch applying dc voltage to the cassette, motor, receiver, and amplifier circuits. Sometimes the power supply may have transistors and IC regulator circuits with a 2200- to 3300-μF filter capacitor (Fig. 6-26).

Similar diagrams

If possible, try to locate a similar schematic when the correct diagram is not available. A block diagram of a boom-box CD player, an auto CD player, or a portable CD player may help in isolating the problem and locating the defective part. Often universal ICs and processors are not provided for CD players. Use the original part numbers.

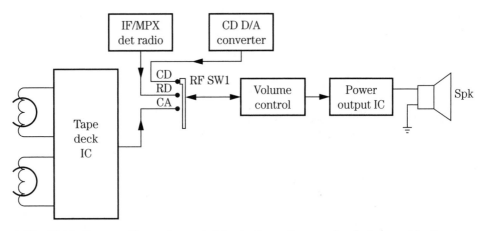

6-25 Block diagram of boom-box switching to the audio amp circuits in a combination tape deck, radio, and CD player.

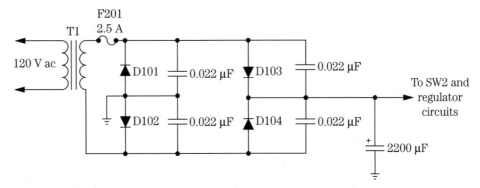

6-26 A simple ac transformer power supply provides voltage to the receiver, cassette, and CD circuits in a boom-box CD player.

The defective CD motor

The slide or feed motor moves the optical pickup assembly, whereas the disc or spindle motor rotates the CD. Suspect a defective motor when the CD will not spin or the optical pickup assembly does not move. Most of these motors operate from a 5- to 12-V source. The slide and spindle motors may operate from one motor driver IC. Often, the servo control IC provides signal to the motor driver IC, and the motors begin to operate (Fig. 6-27).

A defective CD motor may be open or intermittent. Check the dc voltage at the motor terminals when either motor is supposed to be operating. No voltage indicates a defective transistor or driver IC. Suspect a defective motor when normal voltage is present. Take a continuity test with the low-ohm scale of the DMM across the motor terminals. No measurement or a very high reading indicates that the motor should be replaced. Sometimes, by tapping the end of the motor with the handle of a screwdriver, the motor begins to spin; replace it anyway.

An intermittent motor will cause intermittent CD rotation. A sticky slide motor can cause the optical pickup to keep skipping the track. The bad spindle motor can

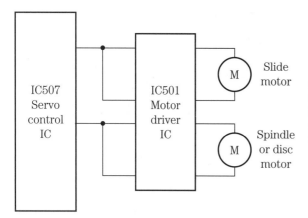

6-27
Servo control IC507 provides signal to the driver IC501 and voltage to the slide and spindle motors.

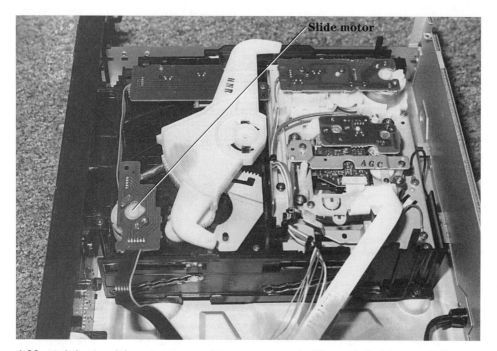

6-28 A defective slide motor can result in no movement of the optical pickup assembly.

cause the pickup to skip the track. Intermittent voltage supplied to the motors can cause the pickup to skip the track or play only one portion of the CD. Check the table height of the disc motor when the tray will not spin, skips, and stops (Fig. 6-28).

Do not overlook defective parts that can keep the disc or slide motor from operating. Reposition the SLED gear when the CD starts skipping on the first track. Clean the limit switch when the CD skips on the first track. A chattering noise may result from poor gear meshing between the worm gear and the first gear. Suspect a bad plastic worm gear when mistracking and skipping occur. Replace the power switch when the pickup intermittently loses tracking after a few minutes. Replace the reset switch if the pickup skips at the beginning of a track.

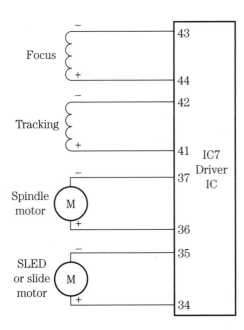

6-29
A negative and a positive voltage is applied to the coils and motors from a large driver IC7.

Clean and lubricate the guide rails when erratic skipping occurs. Wipe all foreign matter that might prevent the pickup from moving from the guide rails. Clean the lens assembly of dirt and dust that can cause the pickup to skip and jump tracks. An intermittent transistor or IC voltage regulator can cause intermittent motor operation by providing intermittent voltage to the motor driver IC circuits.

Motor drive circuits

Early CD players had transistors as motor drive components in the disc and spindle motor circuits. Today, the servo driver IC applies voltage to one large IC that drives the disc and SLED motors. The same driver IC also may drive the focus and tracking coils (Fig. 6-29).

Suspect a shorted or leaky driver IC when no or low voltage is applied to the motor terminals. Measure the supply voltage (Vcc) applied to the driver IC. A defective transistor or IC regulator may not apply the correct voltage to the driver IC. A low 1.5 to 3 V applied to the focus and tracking coils can make them move, indicating that the coil assemblies are okay. Apply a 5-V dc source to suspected motor terminals and notice if the motors take off. Transistor or IC driver motor components can appear open, leaky, intermittent, and shorted.

SMD components

Be very careful when taking voltage and resistance measurements on SMD ICs, processors, and transistors. Use a magnifying glass to get the correct terminal. Sometimes these gull-like leads are so close together that its very easy to short two terminals together with regular test probes. Either sharpen the test probe to a fine point or purchase a pair of thin probe tips with leads.

The SMD signal processor might have 80 terminals. Usually the signal processor IC can be located as the one having the most terminals. The servo signal processor might have around 50 terminals. In portable CD players, you may find the focus, tracking, spindle, and slide motor circuits on one large IC having 44 or more terminals. Of course, this processor IC can be located by its PC board wiring connected to the motors, focus, and tracking coils.

Troubleshooting D/A circuits

Check at both the left and right stereo channels for the same amount of audio with an external audio amplifier. Improper or unbalanced audio may indicate a bad D/A converter IC. Go directly to the power supply when low or no voltage is found at the D/A converter IC. Remove the solder around the voltage supply pin with a solder wick and iron, or if it's an SMD D/A component, lift up the terminal as heat is applied from the soldering iron. Make sure that the supply pin is completely loose from the voltage PC trace. Suspect a leaky D/A converter IC if the dc voltage increases to its original value; replace the defective IC (Fig. 6-30).

Do not overlook a possible mute transistor that may be shorting out the audio from either channel to ground. Replace the D/A converter IC401 for no audio output in a Sony CDP101 CD player.

Defective mute system

Sound muting is often provided to suppress noise that is produced when the power is turned on. In some players, muting is automatic when the CD stops, during acces-sory operations, and during pause mode. The sound muting system contains transis-

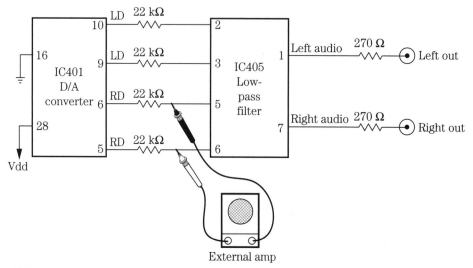

6-30 Check the right audio output from the D/A converter (IC401) for a weak, intermittent, or dead audio channel.

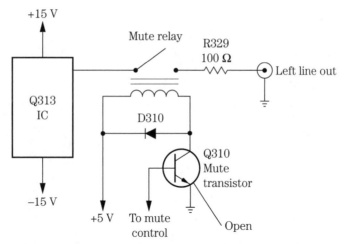

6-31 An open mute transistor (Q310) prevented audio in the left channel of an Onkyo CD player.

tors, ICs, and relays to cut out or ground out the music. Muting is also done in some line output jacks and in headphone circuits (Fig. 6-31).

A defective mute system can cause either a low or dead left or right audio output of the D/A converter. Suspect a defective mute system in the digital signal processor when both channels are weak. Check for a defective transistor when only one channel is weak or dead. Trace the audio output from the D/A converter to the line output jacks with the external audio amplifier or scope.

Disconnect the mute transistor emitter or collector terminal from the circuit with a weak or dead channel. Replace the defective mute transistor or IC when the audio returns to normal. Replace the mechanism micro IC (8-759-971-41) in a Sony CDPC70 CD player for no muting during program mode.

The CD changer

A CD changer may be found in a home entertainment system, a portable boom-box, or an auto receiver with cassette and CD changer (Fig. 6-32). Usually an auto CD changer is found in the trunk. A boom-box CD player might have a three- or five-disc changer. An auto receiver may include a six-disc changer or more. The Phillips HIFI system contains a three-disc changer, whereas the Panasonic has a 300-W five-CD minichanger system. Sony keeps the music playing with its 60-disc storage mini HIFI CD changer audio system with 100 W.

The tabletop or auto CD changer may contain five or six discs. The turntable or roulette motor rotates the tray holding the CDs and stops for the correct selection (Fig. 6-33). The roulette sensor circuits provide correct start and stop indicators for the roulette motor. The typical roulette motor may be controlled by transistors or ICs or a combination of both. A signal from pin 21 of system processor IC201 controls the transistors and ICs in the roulette motor circuits.

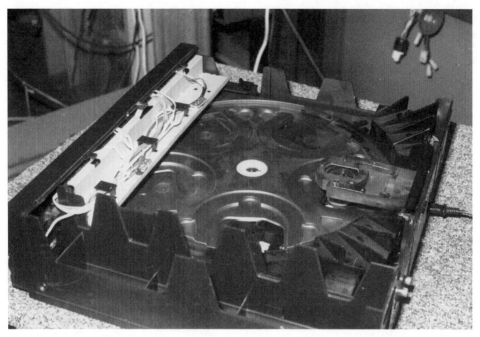

6-32 The rotary tray in a Magnavox CDC745 tabletop CD changer selects the correct CD for playing.

6-33 The roulette motor turns the tray that contains several loaded CDs in the CD changer.

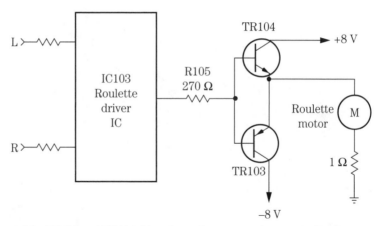

6-34 TR103 and TR104 drive the roulette motor in an early CD changer.

The roulette right and left input is applied to pins 5 and 6 of roulette motor driver IC103. IC103 provides a right and left voltage to the base of driver transistors TR103 and TR104. The reverse voltage applied to the base of each transistor can control the rotation of motor to the left or right. Voltage at the emitter terminals of TR103 and TR104 is applied directly to the roulette motor terminals through CB111 and CB103 (Fig. 6-34).

Most electronic circuits found in a CD changer are similar to those found in any large table-top CD player. The big difference is the four to six motors found in the CD changer player. A slide or SLED motor moves the optical assembly along the rails from the inside to the outside edge (Fig. 6-35). The loading motor loads the tray by moving it in and out. The roulette or select motor rotates the five or six discs in a turntable for

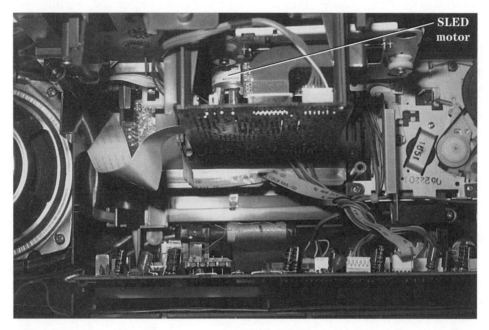

6-35 The slide motor moves the pickup assembly on rails as the disc motor rotates the CD in the playing mode.

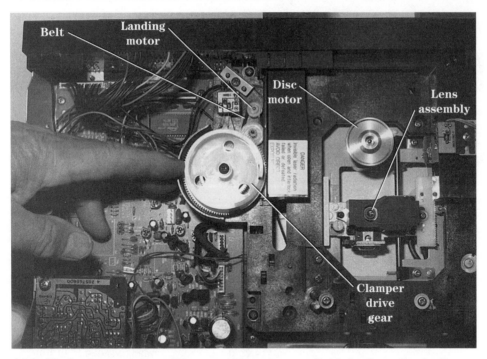

6-36 The loading motor has a different polarity voltage applied in loading and unloading.

selection. A disc or spindle motor rotates the CD at 500 to 200 rpm. The up/down motor assists in loading and playing of the CD, whereas the magazine or carousel motor rotates the turntable. In some CD changers the carousel, tray or loading, and turntable motors operate from the same microprocessor (system control) (Fig. 6-36).

You will find several different PC boards in a CD changer. A large board contains the ac power supply with voltage regulators, signal processors, and system control circuits, whereas the different motor circuits are tied to a servo control board. The servo control board has motor transistors and ICs that make the different motors operate. Removing the bottom cover may reveal the servo board and motors. The resistance of the SLED motors may range between 10 and 20 ohms; roulette motors, 15 and 18 ohms; and loading motors, 5 and 20 ohms. Check continuity or motor resistance when voltage is applied at the motor terminals and the motor does not rotate.

Servicing headphone circuits

Headphone amplifier circuits can be signal traced with an external audio amp. Often the headphone amplifier connects to the audio signal before the line output transistors. The audio is amplified by one or two separate IC amp drivers. You might find one large IC for both channels in portable CD players.

The audio is fed to a dual volume control and input circuits of the headphone driver amp. The amplified audio is capacity coupled to the earphone jack with two muting transistors in the stereo channels (Fig. 6-37).

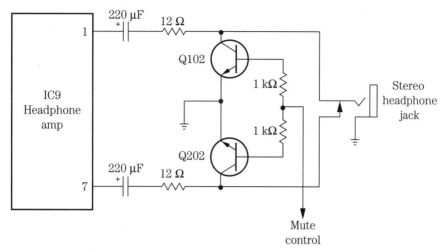

6-37 Q102 and Q202 provide muting in the headphone amp output to a stereo headphone jack.

Signal trace the audio in and out of the headphone driver IC. Suspect the driver IC or defective mute circuits when both channels are dead. With one defective audio channel, suspect a mute transistor, coupling capacitor, or driver IC. Take voltage measurements on the driver IC to determine if the IC is defective. Remove the collector terminal of the dead or weak channel's mute transistor. Locate the driver IC by tracing the PC board wiring from the jack to the first component.

Troubleshooting servo circuits

The servo processor provides power to the tracking and focus coil, spindle, and slide motors through respective IC or transistor devices. The servo IC must receive tracking error (TE), focus error (FE), and mirror data and focus okay (FOK), interface, system control, and clock data from the system control IC before the various circuits can function. The data latch, phase comparator input, VCO, spindle motor drive, and spindle motor on/off control signal from the digital processor are then routed to the input servo processor.

Improper operation of these functions may point to a defective servo control IC. Take critical voltage tests on each terminal pin. Locate the servo processor IC as the one with 40 or more pin terminals.

Check the focus and tracking coils with continuity ohmmeter tests when there is no or improper focus and tracking action. Measure the voltage on each driver transistor or IC. Take critical waveforms at each focus or driver IC.

When either the spindle or SLED motor will not function, take motor continuity tests with an ohmmeter. Measure the voltage at the motor terminals. Take critical voltage measurements on each driver IC or transistor. Scope the motor control and driver IC. Locate the driver and control IC by tracing the motor PC board wiring back to the first driver IC.

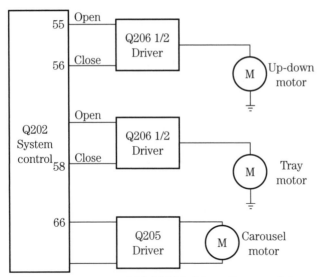

6-38   A system control IC (Q202) drives the up and down tray and carousel motors in a CD player.

Loading problems

The loading motor moves the tray out to be loaded with the CD and then returns into the player. Usually the tray is driven by a plastic gear box next to the tray assembly. A clamper or flapper assembly may operate when the tray is loading and unloading in some CD players. As the tray is closing, the clamper places a spring-loaded pressure on the CD, holding the disc in position up a disc spindle. The loading motor operates from a motor driver transistor or IC. Different polarity voltages are applied to the loading motor driver IC from a large microprocessor (Fig. 6-38).

A dirty open/close switch can cause intermittent or erratic tray operation. If the tray will not open, check all possible mechanical problems first. Visually inspect the drawer gear assembly for foreign objects. Notice if the tray rails or gears are binding. Clean the rails and gears, and then apply a light coat of lubricant to the sliding areas. Clean them with alcohol and a cloth, and lube with a light oil or grease.

Suspect a defective photo interrupter when the drawer will not load. Improper voltage to the loading motor can be caused by a shorted electrolytic capacitor in the power source. Clean the leaf switch for intermittent loading and unloading of CDs.

Do not overlook a broken or cracked loading motor belt when a CD will not load. Listen for the rotation of the loading motor. Replace the loading motor drive belt when it is loose or cracked. Replace the motor for intermittent or no loading of CDs. Sometimes a jammed gear will not let the motor rotate. The tray would not move back and forth in a Magnavox MX3702 CD player as a result of a short in the C04 (470-μF, 10-V) electrolytic capacitor.

Actual CD case histories

The CD starts and stops in the middle of the program, and this was caused by IC115 (7805), a 5-V regulator with voltage down to 3.9 V in a Fisher AD885 CD player. Improper tracking was caused by a split groove gear in a Onkyo DX530 model. Sometimes the player was inoperative and would not play the outer track, and this was caused by a bad ribbon cable on the laser head assembly to the main PC board in a Pioneer PDM501 model. No audio in the left channel of an Aiwa DXM77 CD player was caused by a defective low-pass filter (LPF101).

A background noise during the quiet parts of music was caused by a defective D/A converter (IC302) in a Sony CDP203 CD player. Inoperative functions in a Sony CFD460 player with an okay display were caused by a defective flameproof R306 (1.5-ohm) resistor. Resolder all connections on the regulator transistors for intermittent sound and clicking of the protection relay in a Pioneer SX880 CD player. Replace the 14-V zener diode when the protection relay will not turn on the speakers in a Pioneer SX750 CD player.

Critical waveforms

One of the most critical waveforms is the RF EFM from the optical pickup assembly and the RF amplifier. The weak signal from the laser assembly is amplified by transistors or an IC component. Take a quick waveform with the scope at the output terminal of the RF amp before checking any other parts on an inoperative or dead CD player. Often a defective optical pickup assembly determines if the cost is too expensive to finish servicing the rest of the CD player (Fig. 6-39).

The focus gain waveform taken right off the focus coil can indicate if the focus circuits are functioning. Although the focus waveform is not too high, a very narrow or straight line would indicate that the focus circuits are dead (Fig. 6-40). The focus waveform might fail if the focus gain control is set too small.

A tracking gain waveform across the tracking coil will determine if the tracking coil and circuits are working. Connect the scope probe across the tracking coil winding. Play a test CD. If the sound jumps when the machine is jolted or bumped, the tracking gain control might be set too close or too small. If a test CD with a small scratch is played and the audio jumps, the tracking gain may be set too large (Fig. 6-41).

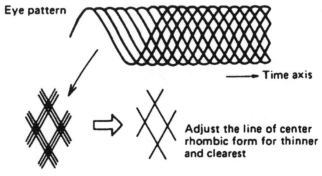

6-39 With a dead or inoperative CD player, check for an EFM waveform at first.

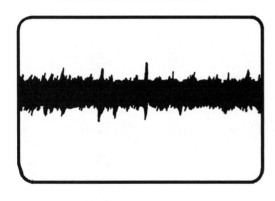

6-40
A focus waveform taken across the focus coil winding.

6-41
A tracking gain waveform taken across the tracking coil winding.

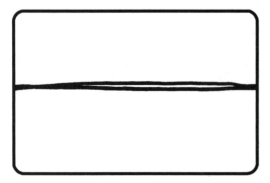

6-42
The motor waveform is kind of flat but moves when the motor rotates.

Take a waveform across the slide, carriage, or SLED motor with the scope. The motor waveform will move slightly up and down when the motor is rotating (Fig. 6-42).

To determine if the VCO is performing, locate the VCO crystal and take a scope waveform from one side of the crystal. This waveform will show that the microprocessor is functioning (Fig. 6-43).

For additional troubleshooting tips, see Table 6-1.

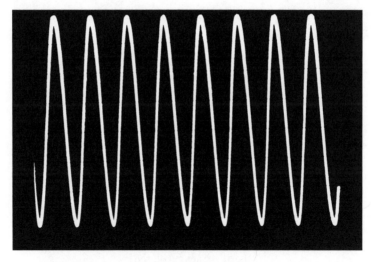

6-43   A VCO waveform taken from the crystal on the microprocessor.

Table 6-1. CD player troubleshooting.

Symptom	Remedy
Dead	Check for the open fuse.
	Check the fuse holder.
	Check for an open transistor or IC regulator.
	Check the various voltage sources.
Keeps blowing fuse	Check for a leaky transistor regulator.
	Check for a leaky audio output transistor or IC.
	Check for a leaky filter capacitor.
	Check for a defective voltage source.
	Check for a burned fuse cable and harness.
No light display	Check for defective power sources.
	Check for no negative voltage to display.
	Check for a defective display circuit.
No loading	Check loading motor operation.
	Check to see if the motor belt is off.
	Check for foreign objects in gear operation.
	Check for a jammed belt.
	Check for voltage across loading motor.
	Check binding loading mechanism.
Will not eject	Check the solenoid or plunger mechanism.
	Check the voltage at the plunger solenoid.
	Check the eject mechanism .

Table 6-1. CD player troubleshooting (*continued*).

Symptom	Remedy
Starts up and shuts down	Quickly scope for the RF or EFM waveform. Notice if the focus and tracking are searching. If there is no RF waveform, check the RF amp. Check the laser diode with a power meter. Check the servo loop.
No disc movement	Check the defective EFM. Check for a defective spindle circuit. Check the voltage on the spindle motor. Check for a defective transistor or IC driver. Check the servo or signal-processor circuits.
No disc rotation in playback	Has focus been locked? Check for a defective mechanism. Has the tracking closed? Check for a defective tracking mechanism. Check the EFM. Check the spindle or disc motor circuits.
Search operations abnormal	Check the EFM waveform. Check the voltage at the carriage or SLED drive voltage. Check the waveform at the carriage or SLED motor. Check for motor voltage at the terminals. Check the carriage or SLED motor. Suspect motor driver transistor or IC. Check the servo loop circuits.
No sound	Does the disc rotate? Has the focus been locked? Check for normal EFM. Check for sound at the output of D/A. Check the audio amp circuits. Check the headphone circuits.
Noisy sound	Check the sound circuits. Suspect a noisy output IC or transistor. Check the D/A converter circuits. Defective audio output circuits.

7
CHAPTER

Troubleshooting the TV chassis

The TV chassis has come a long way since the 6- by 10-in black-and-white table model was first introduced on the market. Today, the screen size on a direct view TV has increased to a 32- to 35-in cathode-ray tube (CRT), flat screens from 15 in on up, projection TVs of 61 to 65 in, and plasma TVs from 42 to 60 in. High-definition TVs (HDTVs) have screens that range from 43 to 65 in. Besides all the TV circuits, today's TVs may have VCR and DVD players within the same unit.

TV chassis components

Most TV chassis have a tuner; an intermediate frequency (IF) section; syncronization (sync), automatic gain, horizontal, vertical, chroma, and luma controls; high- and low-voltage power supplies; and sound circuits (Fig. 7-1). Before tearing the TV chassis apart and jumping into troubleshooting, check the symptoms. Do you have a raster? Is the picture pulled up from the bottom. Do you have sound but no picture or raster? Start with the symptom apparent on the picture tube or heard in the speaker. Use your eyes, ears, and nose to help troubleshoot the TV chassis.

No raster or high voltage may cause problems in the horizontal or high-voltage circuits. No raster, no picture, or no sound can be caused by failure of the low-voltage power supply. Hum in the sound and hum bars in the picture may result from dried-up or open filter capacitors in the power supply or a leaky voltage regulator. Only a horizontal white line indicates a defective component within the vertical circuits, and on it goes.

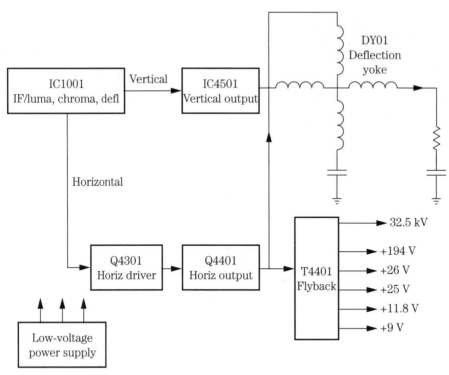

7-1 Block diagram of the vertical and horizontal deflection circuits in a TV.

You can hear a frying noise in the sound, which indicates a defective transistor, integrated circuit (IC), or ceramic bypass capacitor. A tic-tic-tic in the flyback indicates problems within the horizontal output and flyback circuits. The arc-over of a spark gap on the CRT may be caused by excessive dust and dirt in the gap area or too high a voltage. A cracking and arcing around the anode terminal of the CRT may result from excessive high voltage or poor rubber insulation of the anode button.

The most troubled sections

Most of the troubles found in a TV are in the horizontal, vertical, and low-voltage power supplies. Nothing operates in a TV with a defective power supply because the power supply provides voltage to most of the TV circuits. The most troublesome section is the horizontal circuits, which provide sweep, high-voltage, linearity, and secondary voltage circuits. Today, the horizontal circuits must function before supply voltages are developed in the flyback or horizontal output transformer circuits, providing voltage to many other circuits in the TV chassis. The TV low-voltage power supply circuits are discussed in Chap. 9.

Next most troublesome are the vertical circuits. The vertical section provides vertical sweep to the yoke and picture tube circuits. About 85 percent of the troubles found in a TV chassis are located within the horizontal and vertical circuits (Fig. 7-2).

7-2 Locations of the various stages within a color TV chassis.

Locating TV components

After determining the symptoms, isolate the defective section on the chassis. If the symptom points to the horizontal and flyback section, locate the flyback and horizontal output transistor on a heat sink. Remember, the regulator, vertical output, and horizontal output transistors may look alike. Extremely high dc voltage on the regulator and output transistors separates away from the vertical circuits. A scope waveform on the horizontal output transistor and mounted close to a driver transformer locates the horizontal output transistor.

Likewise, the vertical output transistors are found mounted on separate heat sinks. The vertical power output IC is found on a heat sink, and it is much fatter and longer compared with the line voltage regulator. Both the vertical and horizontal sweep signals are found in a large deflection IC with other circuits. Horizontal and vertical circuits are located easily on a TV chassis. Check the numbers and letters on top of the IC, and look them up in the semiconductor manual for the correct sweep IC.

Relay problems

The relay found in the input circuits of a TV may turn the set on or degauss the picture tube. An open relay solenoid can cause no power. Intermittent turn on and off can be caused by a poorly soldered connection on the solenoid winding. A defective

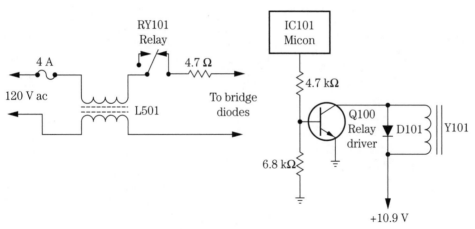

7-3 The TV is turned on by a relay controlled by IC101 micon.

relay driver transistor can cause no turn-on operation. The relay will not click when there is an open relay driver transistor or solenoid.

When the relay does not click with a dead chassis, suspect an open relay coil, bad electrolytic capacitors in the relay circuit, or a damaged relay. Suspect defective relay point contacts stuck in one position when the TV cannot be turned off. Check for a defective relay driver transistor with the diode tester of the digital multimeter (DMM) when the relay will not click and the TV will not shut off. A shorted or leaky diode across the solenoid winding prevents relay action. Remove one end of the diode from the circuit, and check it on the diode tester of the DMM.

Intermittent shutoff or intermittent turn-on can be caused by dirty or bad relay switching points. The relay driver transistor can cause intermittent shutoff operation. Check all relay connections, and resolder them when there is intermittent relay action. Check all electrolytics with the equivalent series resistance (ESR) meter when the relay begins to chatter. Low voltage from the power supply and relay chatter may be caused by a bad filter capacitor. Intermittent startup can result from a defective relay. The chassis can be intermittently dead, and this also is caused by a defective relay. When a TV intermittently goes off and comes back on by itself, this also can result from a bad relay (Fig. 7-3).

Horizontal circuits

The horizontal sweep circuits start with a horizontal oscillator and countdown or deflection circuits. The horizontal drive waveform is coupled to the driver or buffer transistor and to a horizontal driver transformer. The secondary winding of the driver transformer is coupled to the base circuit of the horizontal output transistor. The horizontal output transistor provides horizontal sweep to the yoke assembly and also develops the high voltage within the flyback. High-voltage diodes are found in the flyback windings within the integrated high-voltage transformer (IHVT) that supplies high voltage to the anode terminal of the CRT.

Horizontal circuit problems

Many different horizontal problems and interesting circuits are found in the horizontal circuits. The horizontal oscillator circuits can cause no sweep, horizontal drifting, and off-frequency symptoms. The oscillator circuits may be contained in one IC or transistor, or all horizontal circuits may be found in one large IC. The countdown circuits in the TV chassis provide both vertical and horizontal sweep circuits.

The horizontal driver transistor takes a weak drive signal and amplifies it, and the transformer couples the drive signal to the base of the horizontal output transistor. A leaky driver transistor can damage the primary winding of the driver transformer or voltage-dropping resistor, causing chassis shutdown. If the horizontal drive voltage is left off the base terminal of the horizontal output transistor too long, the output transistor also can be damaged.

The TV chassis can be dead when there is a shorted or leaky horizontal output transistor. Often the horizontal output transistor becomes shorted or leaky between the collector and emitter elements. Erratic TV startup and shutdown can result from poorly soldered connections on the horizontal driver transformer.

Horizontal bars in the picture may be caused by the horizontal oscillator being off-frequency as a result of poorly soldered deflection IC terminals or defective filter capacitors. Poor horizontal width can result from a leaky or defective output transistor, improper drive pulse, and improper voltage supplied by a line voltage regulator IC (Fig. 7-4). Horizontal foldover can be caused by leaky bypass capacitors in the pincushion circuits, a change in resistance, and poorly soldered joints on the flyback.

Horizontal tearing can be caused by poorly soldered joints on the horizontal driver transformer, zener diodes in the fail-safe circuits, high-voltage capacitors in the flyback circuits, and electrolytic and safety capacitors. The tic-tic noise can result from a bad socket connection on the horizontal transistor or emitter resistor, a badly soldered joint on the output transistor, or a badly soldered connection on the horizontal driver transistor (Fig. 7-5).

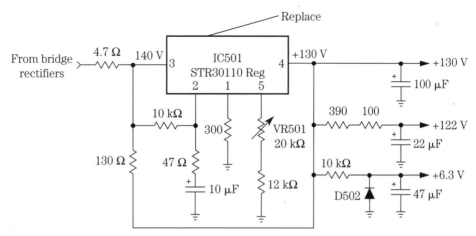

7-4 A defective line voltage regulator can cause dead, intermittent, and shutdown symptoms in a TV.

On metal
shield

Horizontal
output

Flyback

7-5 Locate the horizontal circuits by locating the flyback and horizontal output transistor on a metal heat sink.

A dead chassis can be caused by an open fuse, no relay operation, bad relay contacts, improper voltage from the power source, a defective chopper transistor or line voltage regulator, a bad yoke, poorly soldered connections on the flyback, defective electrolytics, and open fusible resistors. A dead chassis can be caused by just about any component within the horizontal or low-voltage power supply circuits, and on it goes.

Go directly to the horizontal output transistor when the line fuse keeps blowing or the chassis will not start up. Check the resistance from the collector terminal (body) of the output transistor to the chassis ground. A low resistance on the DMM indicates a leaky transistor or damper diode. In the latest TV chassis, you might find the damper diode built inside the horizontal output transistor. An open output transistor may have high dc voltage at the collector terminal with no high voltage or sweep. In some Sears and Sanyo TV chassis, the metal body of the output transistor may not be the collector terminal.

A leaky horizontal output transistor can cause chassis shutdown, open both fuses, overload the low-voltage power supply circuits, or destroy the flyback. Likewise, a leaky or arc-over flyback or output transformer can destroy the horizontal output transistor. Always check the damper diode when replacing a leaky output transistor. Improper or no horizontal drive signal can destroy the horizontal output transistor. When checking for a horizontal output waveform, hold the scope probe next to the flyback.

Always use a variable isolation transformer when servicing the horizontal circuits. After replacing the leaky output transistor and checking the horizontal circuits, apply

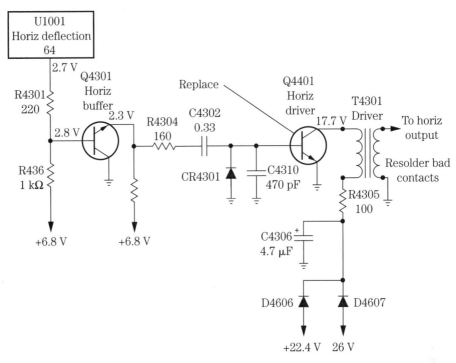

7-6 A dead RCA CTC167 chassis tries to start up several times with a defective Q4401 and poorly soldered terminals on the driver transformer T4301.

about 65 V ac from the variable transformer. Check the base of the output transistor for a drive waveform and dc voltage at the collector terminal or flyback. If the transistor appears warm, shut down the chassis and check for improper line voltage, a leaky flyback, or overloaded circuits. Keep raising the line voltage if the voltage and waveforms appear normal at the base of the output transistor (Fig. 7-6).

What keeps destroying the horizontal output?

The horizontal output transistor (HOT) can be destroyed by improper drive voltage from the horizontal circuits. An arcing or leaky yoke assembly also can destroy the HOT. A grounded flyback winding also can destroy the HOT. A defective safety capacitor can cause the HOT to fail after several minutes. Poorly soldered terminals on the driver transformer can destroy the HOT. A defective bonded picture tube can destroy the HOT after the transistor has been replaced.

Shorted or leaky bypass capacitors in the yoke circuits can damage the HOT. Resolder all terminals on the driver transformer. Check for open electrolytic bypass capacitors in the driver transformer circuits. Measure the resistance of a resistor that is in series with the B1 voltage source and primary winding of the driver transformer for correct resistance. Resolder all voltage resistor terminals on the printed circuit (PC) board (Fig. 7-7).

In an RCA CTC169 chassis, the HOT would blow as it was installed. The base waveform was fairly accurate when Q4401 was out of the socket. All components were checked in the safety circuits. C4403 was replaced because it kept destroying Q4401 (Fig. 7-8).

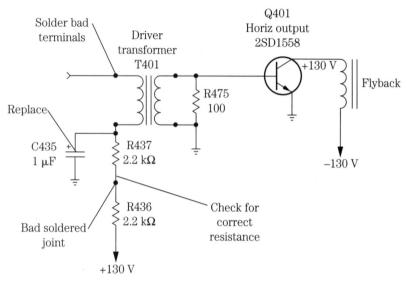

7-7 Check T401 soldered contacts, R437, R436, and C435 for repeated failure of Q401 in an Emerson TC2555D TV.

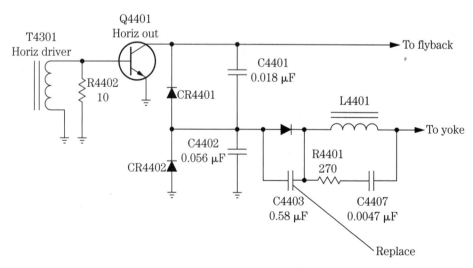

7-8 Replace the C403 (0.58-μF) capacitor in the safety circuit of the RCA CTC169 chassis that keeps destroying Q4401.

Old faithful replacement

Determine what causes failure of the HOT once it has been replaced and is damaged at once. Often the HOT becomes shorted or leaky between the collector and emitter terminals. Always resolder all driver and flyback transformer terminals. Try to replace the HOT with the original part number, if possible.

If the original is not available, look up the part number stamped on the body of the transistor in a semiconductor manual to provide the correct universal replacement. *Sams Photofacts* provides universal replacements for a horizontal transistor. For instance, in a Goldstar CMT-2612 TV, the horizontal output transistor Q402 (2SD1879) can be replaced with an NTE2331 or ECG2331 or an RCA SK10088.

Horizontal output problems

Go directly to the horizontal oscillator, countdown, or deflection IC with no wave-form at the HOT. Locate the multiple-function IC that has picture, sound, chroma, luma, and deflection circuits. Scope each terminal of the IC for a horizontal square wave. If there is no square-wave output on the base terminal driver transistor, check the supply voltage pin terminal of the IC. Measure the voltage supplied (Vcc) to the deflection IC.

Today, you may find that the dc power supply is provided by a winding and diode within the secondary windings of the flyback. This means that the horizontal circuits must function before any supply voltage can be obtained or found on the deflection IC. In the early TV chassis, the supply voltage was taken from the low-voltage power supply, and now it has been changed.

To determine if the deflection IC is okay, inject a dc voltage at this supply pin from the external power supply. At the same time, monitor the drive waveform at the deflection IC or at the base of the horizontal driver transistor. Slowly raise the dc voltage, if it is not known from a schematic, and watch the drive square wave on the scope. The dc supply voltage may vary from 7.5 to 25 V. If a square wave is found with the injected voltage, you may assume that the deflection IC and circuits are normal (Fig. 7-9). Suspect that the deflection IC is defective or components within the IC deflection circuits are bad when there is no waveform.

Horizontal voltage injection

When the oscillator sweep circuits are powered by a derived dc voltage from the horizontal output transformer, apply dc voltage to the supply terminal from an external power supply. Usually this voltage is from 7.5 to 20 V. Try a 10-, 12-, 18-, or 20-V source, and slowly bring the voltage up to get the correct waveform from the IC. You can use one or more 9-V batteries in series to reach the required voltage.

Attach the voltage source to the Vcc supply pin and common chassis ground. Solder in a piece of hookup wire, if necessary, to make a good connection and not short out other pin terminals. Pull the ac cord from the power plug for this test. Apply voltage from an external power supply, and scope the sweep output terminals. Locate the output terminal by tracing the wiring from the driver transistor base terminal.

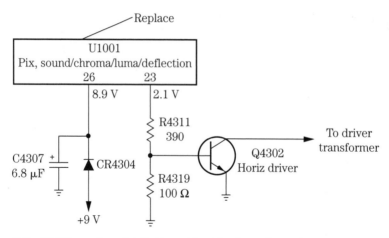

7-9 U1001 was found defective with no horizontal waveform and with an injection of 9 V at CR4304 in an RCA F35750STFM.

If the waveform is normal at the sweep IC, you can assume that the horizontal IC is normal. When a weak or no waveform is found at the countdown or oscillator IC, suspect that the IC is bad. Suspect a leaky IC if the voltage goes below half the external dc voltage. Remove and replace the defective IC.

Safety first

When the high voltage arcs over on the CRT at the button terminal with a loud cracking sound, suspect that the high voltage is way over the limit. Excessive arcing can appear inside the flyback of high-voltage diodes and may not be caused by excessive high voltage at the picture tube. Safety or hold-down capacitors within the horizontal output collector circuits to the flyback can go open or have a cracked or loose terminal. Often these safety capacitors range from 0.016 to 0.056 µF and have a voltage rating of 1.6 or 2 kV. You may find silicon diodes within the same hold-down circuits. Try to replace the safety capacitors with the same capacity and working voltage of 2 kV (Fig. 7-10).

Be very careful when measuring the high voltage with the high-voltage meter probe. Replace the safety capacitor first. A defective safety capacitor can let the TV come on momentarily with high voltage and shut down at once. Suspect the safety capacitor when the TV is only on for a few minutes. Often the safety capacitor end terminal will crack loose and then allow the high voltage to exceed the limits of the CRT. A shorted safety capacitor can cause a dead TV chassis.

A defective safety capacitor also can cause horizontal lines on the screen before the TV shuts down. The high voltage may come on momentarily, the degaussing relay then kicks on and off, and then the TV shuts down. The TV may operate for 10 to 15 minutes and then shut down; this is caused by one or more bad safety capacitors. Intermittent horizontal tearing also can be caused by a defective safety capacitor.

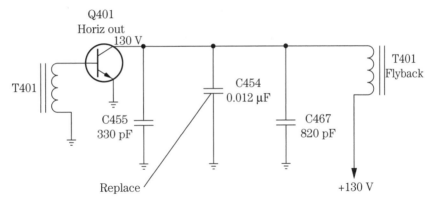

7-10 An open or loss of capacity of a safety capacitor will let the high voltage exceed the limit of the TV chassis.

Pulled in on the sides

Poor width can be caused by defective components in the horizontal, flyback, high-voltage, and low-voltage circuits. Check the high-voltage regulator circuits when there is poor width in a late-model TV chassis. The high-voltage regulator transistor, silicon-controlled rectifiers (SCRs), and zener diodes in the regulator circuits can produce insufficient width. Poorly soldered connections on the pincushion, regulator, and driver transformers can result in poor width. Open bypass or coupling capacitors in the horizontal and high-voltage circuits also can cause poor width.

Low high voltage at the anode terminal of the picture tube can result in poor width. Of course, this low high voltage may result from a defective component in the horizontal output circuit. Low drive voltage on the base of the horizontal output transistor can cause the transistor to become quite warm and produce a narrow picture. A low-voltage source applied to the horizontal output circuits can cause poor width. Check the low-voltage source for a leaky regulator transistor, zener diode, or filter capacitor.

Check the safety capacitors when the raster is shrunk on both sides of the screen. Replace the defective line voltage regulator with the raster pulled in $1/4$ in on the sides. A defective fusible resistor and voltage regulator can cause the picture to shrink on both sides and at the top and bottom. Poor width on both sides can be caused by a 5.6-ohm, 2-W resistor. Resolder all terminals on the horizontal driver transformer for intermittent shrinkage on both sides of the raster (Fig. 7-11). Poor contacts on the yoke socket can cause shrinkage on both sides of the picture. Replace D805 off of pin 2 of the STR3115 regulator for a picture that is shrunk on both sides in a Goldstar CWR-4200 TV.

Flyback problems

The flyback or horizontal output transformer supplies high voltage to the picture tube, focus voltage, and low voltage from secondary voltage sources. Separate windings on the secondary of the flyback provide different voltage sources with diodes

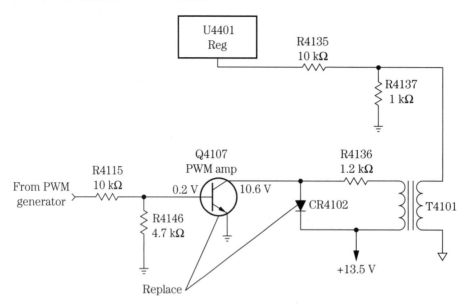

7-11   The raster was shrunk at the top, bottom, and on both sides, and it was caused by defective CR4102 and Q4107.

and filter capacitors. A defective flyback can destroy the HOT. Replace the flyback when the HOT keeps shorting out or becomes leaky after replacement. The shorted flyback can destroy the HOT and produce a chassis shutdown symptom. A defective flyback can destroy fuses and low-ohm resistors and cause a dead chassis. An arcing flyback can destroy the HOT (Fig. 7-12).

Intermittent horizontal sweep can result from poorly soldered terminals on pins 1 and 2 of the flyback. Resolder all flyback pin terminals for a dead or intermittent startup symptom. Check for a leaky HOT and poorly soldered joints on terminals 1, 2, and 10 of the flyback when a whining noise is heard from the switch-mode power supply (SMPS) circuits. Poorly soldered joints on the flyback terminals can cause the chassis to be intermittent or to go off-frequency. Suspect the flyback when the chassis tries to start up momentarily and then shuts off. An open 1.6-ohm, 2-W resistor on line with pin 9 of the flyback can cause no raster, CRT filaments not lit, audio okay, and high voltage normal.

The TV chassis may shut off after operating for 1 hour, and this is caused by a bad flyback. When the audio was okay but the chassis smokes, this also may result from a bad flyback. Suspect a defective flyback when the relay clicks on and then off, and the chassis goes dead. Immediate shutdown can result from a bad flyback. When the TV comes on briefly and then shuts off, this also can be caused by a defective flyback. Open windings between pins 2 and 3 can produce no vertical sweep. When the power switch is activated and a click followed by another click is heard and the set goes dead, this is caused by a bad horizontal output transformer.

A dead TV with a chirping noise can be caused by a defective flyback. Arcing noise followed by smoke also can result from a bad flyback. When the TV makes a popping noise and then shuts down, this is caused by a defective flyback and HOT.

7-12 A defective flyback can destroy the horizontal output transistor (HOT).

Suspect a bad flyback when the chassis will not start up. Very low high voltage of 10 kV can be caused by a bad flyback. A defective flyback may cause the chassis to come on with a tic-tic noise and then go into shutdown. Replace the flyback when noise is heard form it. A bad flyback can cause a no raster symptom with normal audio in the speakers. A bad flyback may cause a dead chassis that tries to start up with a clicking noise. A bad flyback may arc over from windings to the metal strap and core with a noisy cracking sound (Fig. 7-13).

Red-hot output transistor

A Mitsubishi CS-2656R model TV was dead with no picture, no sound, and no high voltage. Right away the HOT was checked, and there was a short between the collector and emitter terminals. Q551 was replaced with another 2SD1878 transistor. The picture came on immediately, and within minutes, the chassis shut down. Q551 was red hot (Fig. 7-14).

Before replacing Q551 once again, the terminals were resoldered on the driver transformer (T552). R552 had run very warm and was replaced. Both diodes D593 and D594 tested normal. When C556 was checked with the ESR meter, the small 1-μF electrolytic registered in the red zone. C556 was replaced.

Continuity tests on the primary and secondary windings of T552 were fairly normal. All connections on R552, C556, and T552 were resoldered. Hoping that the flyback was

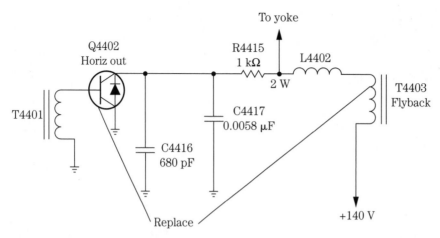

7-13 A GE CTCC146D chassis was dead with a blown fuse caused by a leaky horizontal output transistor (Q4402) and flyback.

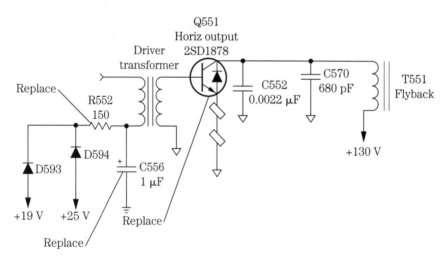

7-14 The R552 (150-ohm, 3-W) resistor, the C556 (1-μF) electrolytic, Q551, and all the driver transformer connections were resoldered after being installed for a second time, and Q551 was running red hot.

normal, another horizontal transistor was installed. The chassis came on, everything looked good, and the chassis ran for 8 hours with Q551 only running lukewarm.

Yoke problems

An open yoke winding in the horizontal sweep circuits can cause a white vertical line on the CRT. Likewise, an open vertical winding in the yoke can cause a white horizontal line on the TV screen. A bad yoke can cause the set to come on momentarily

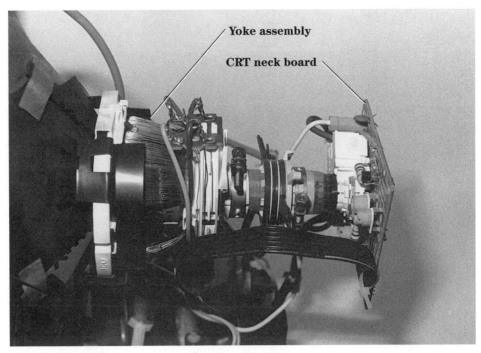

Yoke assembly

CRT neck board

7-15 The defective yoke can destroy the output transistor, arc over, have open windings, and become intermittent when it has badly soldered connections.

and then smoke and shut down. The defective yoke can blow the main or secondary fuse. A poor yoke socket connection can result in intermittent horizontal or vertical sweep. Arcing lines in the picture can be caused by a bad yoke assembly. The defective yoke can burn, smoke, and create a bad smell when the chassis starts up (Fig. 7-15).

Immediate chassis shutdown can be caused by a badly soldered yoke joint or a bad yoke assembly. The TV can come on for a minute and then smoke with no horizontal sweep when it has a bad yoke. Shrinkage from both sides of the screen can result from corroded yoke terminal contacts. You may find the chassis dead with a relay click when there is a leaky HOT and 10-ohm, 3-W resistor and a defective yoke assembly.

The arcing or leaky yoke can keep destroying the HOT. A buzzing or frying sound can be cause by arcing internally within the yoke assembly. A bad yoke may need to be replaced when a noise is heard at startup. A loud squealing noise may result from a badly soldered joint on the yoke terminal connections.

The defective yoke can cause only 4 or 5 in of vertical sweep. A bad yoke also can cause the horizontal output transistor to run red hot. Sometimes the dead chassis relay may click off and on with a bad yoke and line voltage regulator (STR30130). A badly soldered joint or trace on the yoke assembly plug can cause intermittent play or a complete loss of vertical sweep.

A JVC AV-2779S TV was dead when the relay clicked on and then off. The HOT (Q552) was found to be leaky and was replaced with a 2SD1556 replacement. After the transistor was installed, it began to run red hot and was pulling heavy current even when the power line voltage was around 82 V ac. Replacing the yoke assembly solved the problem.

Intermittent shutdown

Intermittent shutdown in the horizontal section can be caused by intermittent transistors, ICs, and driver and flyback transformers. Monitor the base terminal of the HOT with the scope, and notice when the intermittence occurs. If the waveform comes and goes, suspect that the intermittent problem is in the driver or deflection circuits. If the chassis shuts down intermittently and the waveform is normal on the HOT, you can assume that the intermittent component is in the output, flyback, or high-voltage circuit.

A bad flyback can momentarily start up the TV chassis and then shut it down. Check all electrolytics with the ESR meter (10 to 100 μF) for shutdown after the chassis operates for 15 or 20 minutes. Also check all electrolytics with the ESR meter for virtually immediate shutdown. A bad relay can shut down the chassis immediately. An arc-over on the CRT anode button can cause the set to turn on with a snapping noise and then shut down. Clean around the high-voltage anode button with alcohol and a cloth. An arcing flyback can cause the chassis to shut down.

Check for defective 2200-pF, 2-kV bypass capacitors when there is immediate shutdown. Suspect large filter capacitors (470 μF on up) when the set shuts off intermittently as it tries to start up. Leaky diodes can cause the TV to come on and shut off immediately (Fig. 7-16). The TV chassis can shut off when there are badly soldered joints on the driver transformer terminals. Badly soldered joints on the transistor regulators cam intermittently shut the chassis off and on. A badly soldered joint on the relay can cause intermittent chassis shutoff.

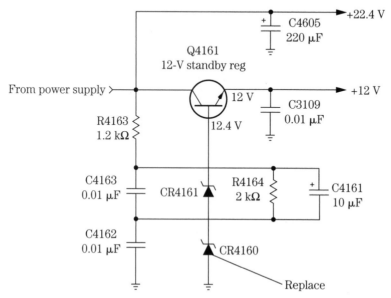

7-16 Intermittent shutdown was caused by a zener diode (CR4160) in an RCA CTC167 chassis.

Suspect bypass capacitors in the safety hold circuits (680 pF, 2 kV) when high voltage comes up followed by immediate shutdown. When the TV shuts down after it is warmed up and the high voltage increases to 33 kV, this can be caused by a bad safety capacitor. If the set comes on briefly and then shuts down, suspect a flyback with very low high voltage.

An MGA CS-3125R TV would start up intermittently or the set might come on and immediately shut down. In such a situation, check all small electrolytics with the ESR meter. Replace C940 (100 μF), C5H8 (1 μF, 50 V), C5H9 (10 μF, 50 V), D5G6, Q906, D907, and R956.

High-voltage shutdown

Excessive high voltage can make the x-ray or shutdown circuits close down the horizontal circuits and shut down the chassis. This prevents excess radiation from the picture tube and prevents damage to other components within the TV chassis. Defective shutdown circuits can cause premature high-voltage shutdown.

Monitor the high voltage with a high-voltage probe or meter at the CRT anode button (Fig. 7-17). Connect a dc meter to the horizontal fuse or flyback primary winding to check the voltage applied to the HOT. Slowly raise the variable isolation transformer line voltage while watching the high-voltage meter. Notice the voltage when the chassis shuts down, and determine if the high voltage is excessive before reaching the 120-V ac power line voltage; also note the high-voltage measurement.

7-17 Measure the high voltage at the CRT anode connection to determine if the horizontal and high-voltage circuits are functioning.

Slowly raise the variable transformer to just under the high-voltage reading that shut down the chassis. Notice if the low-voltage source is higher than normal. If so, check the low-voltage regulator circuits. When the high voltage is high compared with the line and dc voltages, suspect a defective hold-down or safety capacitor or horizontal output transformer. If the chassis shuts down before reaching the normal high voltage, suspect a defective high-voltage shutdown circuit.

Disconnect the high-voltage shutdown circuit by removing the diode or resistor terminal from the PC board wiring. The high-voltage shutdown circuit takes a pulse from a winding of the flyback and rectifies it with the voltage regulator diode to the transistor or SCR circuit that shuts down the driver or horizontal oscillator stage. Some chassis have a shutdown circuit that is fed back to the countdown oscillator IC and driver circuits, shutting down the horizontal circuits.

If the high voltage is normal and does not shut down with the shutdown circuits disconnected from the flyback circuits, repair the shutdown circuits. Always replace any terminals that were removed after the chassis has been repaired.

TV chassis shutdown

The TV chassis may shut down when there is a defective part in the horizontal or vertical circuits or the low-voltage power supply. Most shutdown problems occur in these stages. Chassis shutdown can be caused by poor terminal connections, defective transistors and ICs, poorly soldered terminals, or cracked PC board wiring. Monitor the horizontal circuits at the countdown sweep circuits, check the low voltage at the HOT, scope waveforms at the base of the output transistor, and check the high voltage at the anode of the picture tube. Notice which circuit begins to malfunction when the chassis shuts down.

A defective 10-μF, 160-V capacitor can cause intermediate shutdown. A capacitance of 100 μF at 160 V can shut the chassis off after 15 minutes. Check all electrolytics with the ESR meter. The defective flyback can start the chassis up momentarily and then shut it off. A defective relay can intermittently shut the chassis off. A defective HOT can make the chassis shut off.

The STR30135 line voltage regulator would operate 10 to 15 minutes and then shut off, although the relay stayed on. Badly soldered joints on the line voltage regulator IC or transistor may shut the set off intermittently.

Suspect a 6.8-V zener diode when the TV comes on but shuts off immediately. Badly soldered joints on small resistors can cause the TV to immediately shut off. The TV shuts off with an extremely bright raster, and when the screen control is turned down, the set may stay on, and this is caused by a low 22-ohm resistor on the CRT neck board. Resolder all joints on the driver transformer when the TV intermittently shuts off. A bad driver transistor shuts the set off after 5 minutes of operation.

A defective yoke will let the relay click, and the set tries to start up and shuts off at once. A badly soldered joint on the relay cuts the TV off at once. When the chassis appears dead when the relay clicks on and then off, this is caused by a defective HOT. When the dead chassis relay clicks on and then off, this is caused by a defective yoke assembly.

In an RCA CTC197 chassis, the set came on very briefly and then shut off; this was caused by the HOT (2SC5148), CR4115 (33-V) zener diode, R14126 (30 kilohm), R14305 (3.6 kilohm), and R13159 (15 kilohm).

High-voltage problems

Besides high-voltage arc-over, the IHVT flyback can have arcing diodes or capacitors within the molded component. The high voltage may not come up or may cause shutdown with leaky voltage diodes in the secondary winding of the flyback. Overloaded circuits in these voltage sources can be caused by leaky components within the audio, video, and vertical circuits.

Improper screen grid voltage can be caused by a burned isolation resistor or leaky diode in the derived secondary circuits. The symptoms include brightness that cannot be turned down, excessive brightness, or chassis shutdown. Disconnect each diode from the secondary circuits to determine what section is overloading the horizontal output transformer. Check each circuit after removing each diode.

Replace a bad transistor regulator when the TV has very low high voltage (19 kV). An open safety capacitor can let the high voltage go very high, and the TV shuts down at once. Check for defective electrolytics (100 to 200 V) when there is no horizontal sync, and the high voltage is raised to 33.5 kV. Suspect 4.4- to 22-μF capacitors when the set comes on and the high voltage goes above 32 kV and shuts down the chassis. Check a 100-μF, 25-V electrolytic with the ESR meter when the TV comes on with the high voltage around 10 kV and then shuts off.

Check for defective silicon diodes in the power supply regulator circuits when the picture is shrunk on both sides and the high voltage is down to only 18 kV. After replacing the HOT and the TV comes on with only 10 kV of high voltage, replace the flyback. You may have to replace the defective yoke after replacing the HOT when the high voltage is only 10 kV. A badly soldered joint on the horizontal linearity coil can cause the high voltage to go down to 10.5 kV. Replace IC400 (optoisolator) in a Magnavox 25G1-02 chassis with no raster, no audio, and a high-voltage reading of 15 kV.

The bow and arrow

When the picture is bowed inward on each side, suspect a defective pincushion circuit. In large TV screens (27 in and up), a pincushion correction circuit corrects distortion on the face of the large picture tube. You may have to look closely at a door frame or building in the picture to really notice the bowing. The pincushion circuits are defective if the outside edges of the TV raster bow. Improper adjustment of the pincushion coils may bow the picture at the top and bottom of the raster.

Check all large 1000-μF, 10-V electrolytics when the picture is curved in on both sides (pincushion problem). If the picture is pulled in on both sides like an hourglass, suspect 470-μF, 200-V electrolytic capacitors. Suspect defective bypass capacitors in the pincushion circuits for bowed pictures on both sides. Check the small resistors on the convergence board for bowed-in sides and a screen that goes out of convergence. Check the silicon diodes in the pincushion circuits when the raster is bowed on both sides and the picture goes out of convergence.

Look for open traces in the pincushion circuits for a picture that is bowed on both sides. Use the ESR meter to find breaks in the traces and PC wiring. Check Q1 and Q2 in the pincushion circuits for a curved picture on both sides with the diode tester of the DMM.

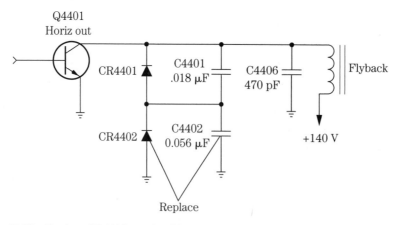

7-18   Replace CR4402 and C4402 for curved sides in the picture on an RCA F31700GG TV.

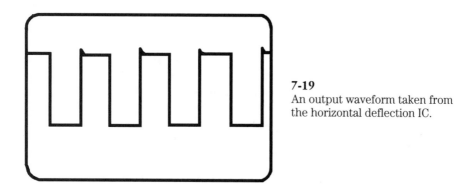

7-19
An output waveform taken from the horizontal deflection IC.

In an RCA F31700GG TV, the horizontal pincushion problem resulted in a picture that looked like an hourglass. Besides curved sides, there was static and noise in the picture. C4402 and CR4402 were replaced and solved the pincushion problem. C4402 had a changed value (Fig. 7-18).

Critical horizontal waveforms

The most important waveform in the horizontal circuits is the square wave at the horizontal deflection IC (Fig. 7-19). Next, check the horizontal waveform on the driver transistor (Fig. 7-20). A waveform taken at the base of the HOT will indicate that the horizontal circuits are normal at this point (Fig. 7-21). Take a quick waveform with the scope probe alongside of the flyback to see if the horizontal circuits are functioning (Fig. 7-22).

Vertical problems

Locate the vertical output IC on the chassis, which is tied into one of the yoke's vertical windings (Fig. 7-23). In early chassis, two vertical output transistors are found

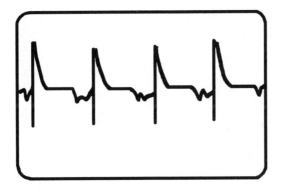

7-20
The horizontal driver transistor waveform indicating that the deflection and driver transistors are okay.

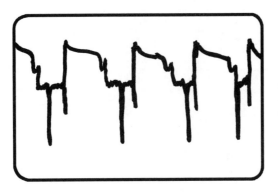

7-21
The normal waveform found on the base of a horizontal output transistor (HOT).

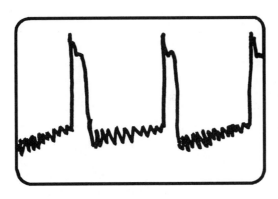

7-22
The waveform taken with the scope probe held alongside the flyback indicating that the horizontal circuits are functioning.

mounted on their own heat sinks. Most trouble found in vertical circuits results from a defective vertical deflection IC, bad silicon diodes, or defective electrolytic capacitors. Improper or no voltage from the low-voltage power supply or flyback can cause insufficient height or a horizontal white line.

Vertical fold-over and rolling problems go hand in hand, whereas retrace lines at the top of the screen can be caused by bad resistors, electrolytics, and vertical output ICs or transistors. Dried-up filter capacitors can produce vertical crawling, whereas intermittent vertical sweep can occur at just about any place in the vertical circuits.

7-23　Locate the vertical output IC on a metal heat sink.

ONLY A HORIZONTAL WHITE LINE

No vertical sweep can be indicated by a horizontal white line across the CRT. Turn the screen control up so that you can see a no-raster or vertical sweep. Always try to use the correct schematic to quickly locate a defective part. Although vertical circuits are easy to service compared with others, try to locate a schematic of another product by the same manufacturer or a similar circuit. Vertical circuits have not changed much over the years, except that now all the vertical circuits may be found in one large IC.

No vertical sweep can result from a bad deflection IC, badly soldered joints on the defection IC, open traces, low-ohm resistors, or an open yoke winding (Fig. 7-24). Resolder all pin terminals on the deflection and output ICs. Check the yoke return electrolytic with the ESR meter when there is no vertical sweep. A defective zener diode in the vertical or voltage regulator or the low-voltage power supply or voltage source can cause a no-sweep symptom.

Do not overlook a defective return electrolytic or open resistor in the leg of the deflection yoke circuits. Check all electrolytics in the vertical circuits for no vertical sweep. Suspect no voltage source when there is an open 1.5-ohm resistor at pin 5 of the flyback. Solder all pins on the flyback that can produce a horizontal white line. Replace IC301 (LA7837) when it is shorted between ground and pin 12, and also replace the FR301 (2-ohm, 1-W) fusible resistor in a Goldstar GCT-1904 TV for no vertical sweep.

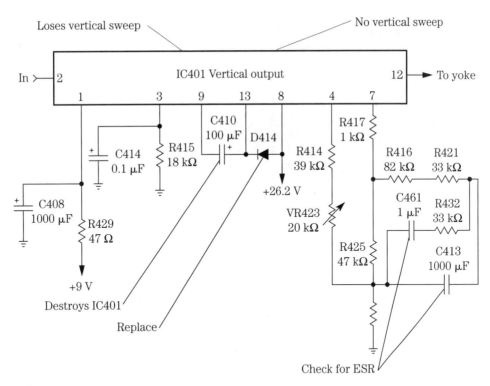

7-24 Check the critical components when the vertical section is acting up.

Not high enough

Insufficient height or only 2 to 3 in of height can be caused by a defective transistor or vertical output IC. Poorly soldered joints on the vertical IC can cause poor height. Defective electrolytic capacitors also can cause poor height. Check all electrolytics (33 to 100 μF) on the vertical voltage source for insufficient height. Only 1 in of vertical sweep can result from a defective 560-μF, 25-V electrolytic. Check all electrolytics with the ESR meter in the vertical circuits and on the height control. Sometimes the defective elctrolytic (100 μF) produces only 1 to 2 in of vertical sweep.

Check all diodes in the vertical output circuits for improper vertical sweep. Do not overlook a change in resistance in the pincushion circuits when there are vertical problems. Only 2 in in the center of the screen can be caused by a badly soldered joint or terminal of the vertical output IC (Fig. 7-25). Replace a bad vertical size control when there is only 2 in of vertical sweep.

A Sylvania 25N1 chassis could not be turned off with no vertical sweep, and this was caused by IC550 (LA7831), R445 (1 ohm), C558 (100 μF), and D551; these parts have a tendency to cause IC550 to become leaky.

7-25
Poorly soldered vertical IC connections and bad electrolytics and silicon diodes can cause various sweep problems.

ONLY A VERTICAL WHITE LINE

Usually the no-sweep problem occurs in the vertical output, vertical yoke winding, or vertical oscillator circuits. A waveform test at the vertical countdown IC or oscillator will indicate if the stage is working. Check the vertical input circuits when there is no vertical waveform.

Check the output transistors with in-circuit beta or voltage measurements. Measure the voltage supply terminal if the output IC is suspected. In vertical stage transistors, especially directly coupled circuits, scope waveforms may not be useful. With IC countdown and output circuits, the vertical waveforms can be checked at the vertical output terminal on the countdown IC, the input terminal of the vertical output IC, or the output terminal of the vertical output IC.

INTERMITTENT VERTICAL SYMPTOMS

Intermittent vertical problems are difficult to locate because they can result from many TV circuits. Monitor the vertical circuit output with the scope and voltage with the DMM at the vertical voltage source. Try to isolate the intermittent within the vertical deflection or output circuits. Attach a scope probe to the vertical oscillator or countdown circuits. If the countdown or oscillator circuits are normal, suspect the output transistors or ICs. Scope the input and output of the vertical output IC to determine if the output IC is intermittent. Take a waveform on the electrolytic coupling capacitor from the vertical circuits to the yoke winding.

The yoke coupling capacitor may be open, producing a horizontal white line and insufficient sweep, or it may be dried up, producing a bunching of white lines at the top of the raster. Always check the vertical return resistor (under 50 ohms) or capacitor found at the ground end of the yoke winding.

Intermittent and partial loss of vertical sweep can be caused by a silicon diode or electrolytic in the vertical output circuits. Badly soldered joints on the vertical output IC can cause intermittent vertical sweep. Resolder all output IC pin terminals (Fig. 7-26).

Check for badly soldered joints on both vertical output transistors. Intermittent loss of vertical sweep can result from a bad yoke connection or socket or a break in the yoke winding. An intermittent diode in the vertical output circuits can cause an inter-

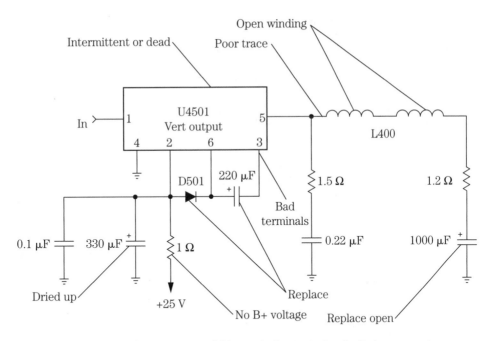

7-26 The most critical components within vertical output circuits that cause various sweep problems.

mittent vertical raster. Do not overlook bad foil or traces in the vertical circuits for intermittent vertical sweep. Intermittent loss of vertical sweep in about 5 minutes on a Hitachi CMT-2090 TV was caused by IC681 (UPC1378); one also should check the vertical coupling capacitor (C685).

CRAWLING UP THE WALL

When lines bunch together and slowly crawl up the raster, suspect the large filter capacitors or poor low-voltage regulator circuits. The raster may have dark horizontal lines along with the crawling lines. Shunt each filter capacitor with one of the same capacity or higher and with the same or higher working voltage. Shut down the chassis, and clip the capacitors across the suspected ones so as not to damage solid-state devices. Shunt the vertical output coupling capacitor when line bunching occurs.

If more than one capacitor is found inside a component and only one section is causing the problem, replace the whole capacitor component. Look for the tallest electrolytic capacitor on the chassis. Sometimes a large capacitor (650 µF or more) may be mounted on the TV PC board.

BRIGHT LINES AT THE TOP

Within the vertical transistor output circuits, retrace lines may be caused by the two output transistors, bias resistors, and diodes. Retrace lines in the power vertical IC

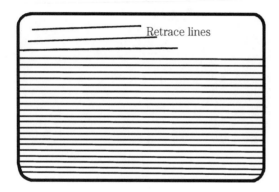

7-27
Bright retrace lines at the top of
the picture result from a defective
electrolytic or diode, a bad yoke
or IC output, poorly soldered
connections, or defective small
1- to 2.2-ohm resistors.

output can result from a defective vertical IC, electrolytic capacitor, or drive from
the deflective IC. Do not overlook a defective filter capacitor within the output IC
voltage source.

Bright horizontal lines at the top of the raster or picture and retrace lines that may
be bunched up closely together or spread apart occur at the top of the picture. Insufficient vertical sweep and retrace lines at the top can result from a bad yoke assembly. A
defective vertical output IC can cause lines at the top of the picture and some vertical
foldover. Bright retrace lines with video barely visible can be caused by a defective
video buffer transistor.

Check the 100-μF, 50-V electrolytic in vertical circuits for lines at the top of the picture. A bright picture with retrace lines can be produced by a defective 10-μF, 160-V
electrolytic capacitor (Fig. 7-27). Check all electrolytics in the vertical circuits with the
ESR meter.

A defective vertical output transistor can cause lines at the top of the picture and
some vertical foldover. Check for open small resistors from 1 to 2.2 ohms on the neck of
the picture tube board. When the TV comes on with retrace lines, TV shutdown can be
caused by excessive dust within the CRT spark gaps. A bright green picture with retrace lines can be caused by a shorted picture tube. Check the video amp terminals
when there is a very bright screen with retrace lines at the top.

A Magnavox 25B8 chassis would come on with an extremely bright raster and retrace lines at the top and then shut off. If the screen control was turned down, this allowed the set to stay on with a bright raster. Replace the R58 (22-ohm, ¼-W) resistor
on the neck of the CRT board to fix this retrace-line symptom.

Replace a defective luminance driver transistor for retrace lines. Retrace lines at the
top of the screen can result from a bad vertical output IC. Check all transistors on the CRT
neck board for loss of color and retrace lines in the picture. Open traces or PC wiring on
the CRT board can cause a bright picture and retrace lines. An open secondary winding
of the flyback between pins 8 and 10 can cause a washed-out picture with retrace lines.

No vertical deflection symptoms

The countdown or vertical deflection IC circuits are inside one large IC component
along with the horizontal deflection, color, luma, sync, and picture circuits. Simply

7-28   The location of a large IC that has luma, chroma, picture, sync, and horizontal and vertical deflection circuits.

locate one of the largest ICs on the TV chassis and take critical waveforms and voltage tests (Fig. 7-28). Locate the vertical deflection IC that goes directly to the vertical output IC. Look at the part numbers on the IC, and look them up in the semiconductor manual. A scope waveform at the vertical output terminal indicates a drive pulse. Suspect a defective IC or improper supply voltage if no drive waveform is found.

Loss of or weak vertical sweep can be caused by a defective vertical deflection IC. A leaky vertical deflection IC can cause no vertical sweep. Resolder all IC terminals for intermittent vertical sweep. Simply advance the screen control until you see a raster or a white line that indicates a bad vertical deflection IC. Make sure that the correct supply voltage is applied to the deflection IC terminals before removal.

FOLDOVER AT THE TOP

Most vertical foldover problems are caused in the vertical output IC or pincushion circuits. Vertical foldover at the top 3 in of the raster can be caused by defective 100-μF, 35-V electrolytics. Check the small 1-μF, 35-V electrolytics on the pins of the vertical output IC for only 6 in of vertical sweep and foldover at the top of the raster. Suspect a bad silicon diode when incomplete vertical sweep and foldover occur. Foldover at the top of the screen can result from a defective vertical output IC. Replace the vertical output IC when there is foldover at the top of a rolling picture (Fig. 7-29).

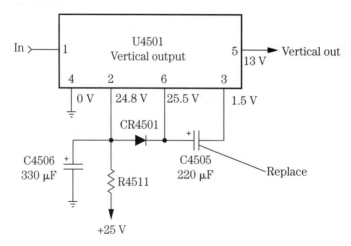

7-29 Vertical foldover in an RCA CTC177 chassis was caused by a defective 220-μF electrolytic on pin 6 of the vertical output IC (U4501).

A defective vertical output transistor can cause insufficient vertical sweep at the bottom of the screen, and vertical foldover at the top can be caused by defective electrolytics in the vertical circuits. Test all electrolyics in the vertical circuits with the ESR meter. Poor voltage regulation and a slight vertical foldover can be caused by an STR30130 voltage regulator. Suspect a defective EEPROM IC for a dead chassis that does not turn on with vertical sweep pullup from the bottom of the raster. Suspect a 1000-μF coupling electrolytic for foldover problems.

Poor vertical linearity can be caused by the vertical output IC. Replace the defective size control when there is a severely stretched picture. For severely stretched vertical sweep, check for an increase or decrease in megaohm resistors in the vertical circuits. Check all electrolytic capacitors within the vertical circuits with the ESR meter.

When the top of the picture is stretched, this can be caused by a bad 2200-μF, 25-V electrolytic on the main board. Poor vertical linearity on the top half of the picture and stretched over the remainder of the picture was caused by defective 2.2- to 22-μF electrolytic capacitors. Poor linearity on the top half of the picture and incomplete vertical sweep at the bottom was caused by a 470-μF, 25-V electrolytic. Incomplete sweep at the bottom of the picture with poor linearity was caused by a 1-μF, 50-V electrolytic. In a GE 25GT510 model, a white horizontal line was found in the center of the screen, and there was a stretched vertical linearity at the top half; this was caused by defective CR4501 (Fig. 7-30).

ROLLING UP AND DOWN

Improper vertical sync applied to the oscillator or countdown IC can cause vertical rolling. The vertical raster will not stay in one place. Sometimes vertical rolling and foldover problems are caused by the same defective part. Scope the vertical sync applied to the vertical circuits. A change in the resistance of the base or emitter circuits

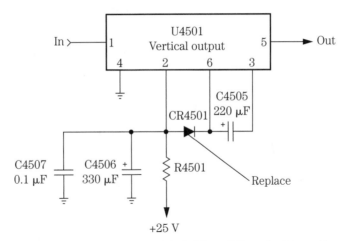

7-30 Defective CR4501 in a GE 25GT510 TV resulted in a white horizontal line and a stretched vertical linearity at the top half of the screen.

can cause improper vertical lock. Remove one end of the resistor or diode for accurate measurement.

Improper dc supply voltage to the vertical circuits can produce vertical rolling. Small 1-µF electrolytics can cause a vertically rolling picture. Partial loss of vertical sweep, plus severe vertical rolling, can be caused by a change of resistance in the vertical circuits. A vertical bounce and roll may be caused by a defective silicon diode in the vertical circuits. Replace the vertical output IC when there is a flipping vertical roll and foldover at the top on some channels. Check large 470-µF electrolytics when the vertical will not lock in until after several minutes of operation.

Suspect an STR30130 voltage regulator when there is a slight weave in the picture with stretch height. Large 1000-µF, 25-V electrolytics can cause a loss of ¼ in vertical sweep at the bottom of the picture and severe jumping after 15 minutes. Vertical jitter and incomplete vertical sweep at the bottom of the picture can be caused by a 470-µF, 250-V electrolytic. Tearing in the picture and incomplete sweep at the top can be caused by a 220-µF, 25-V electrolytic off the 16-V line from pin 7 of the flyback.

Replace the vertical IC for severe jitter. A vertical jitter when the channels are changed and especially on strong signals can be caused by a defective STR30130 voltage regulator. Replace IC421 (LA7837) and C426 (100 µF, 35 V) when the vertical output IC fails and when there is vertical jitter with a loss of vertical sweep in a JVC AV-20CM4 TV.

VERTICAL COMPONENT LOCATION

In early TV chassis, vertical output transistors were located on metal heat sinks. Today you may find one IC that contains all the vertical output circuits. Look at the numbers and letters on top of the IC for correct identification. This vertical IC also will be mounted on a small heat sink. Also check these numbers and letters out in a universal semiconductor replacement manual. For instance, an LA7835 IC is a vertical output IC that can be replaced with a universal NTE18555 IC. The semiconductor manual will show the input, output, and supply voltage terminals of the vertical output IC.

IMPORTANT VERTICAL WAVEFORMS

The waveform found at the output of the vertical deflection IC are like inverted pipes in the latest TVs (Fig. 7-31). This waveform is fed into the input of the vertical output IC. The vertical output waveform from the vertical output IC represents a kind of a sawtooth (Fig. 7-32). A feedback waveform is a regular sawtooth waveform indicating that the vertical circuits are functioning (Fig. 7-33).

System control

A typical system control IC may be controlled by one large IC. It is possible that only one control system function may be bad, whereas the rest of the functions are normal. Suspect the system control IC when several functions are not operating. Infrared circuits are fed into the system control IC to operate many system control circuits. Key input push buttons control the power, volume, display, channel up and down, and reset functions.

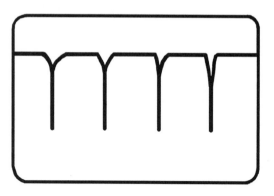

7-31
The correct waveform found at the output terminal of the deflection IC.

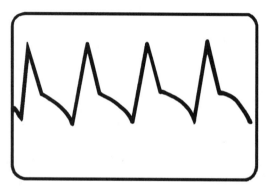

7-32
The vertical output IC waveform that is fed to the yoke windings

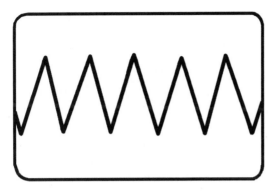

7-33
A vertical feedback sawtooth wave-form indicating that the vertical circuits are functioning.

The contrast, tint, color, brightness, sharpness, on-screen display (OSD), gain, and standby circuits are controlled by the system control IC. The vertical, automatic fine tuning (AFT), tuning, horizontal, vertical, sync, hold, and reset circuits are part of the circuits that the system IC actually controls (Fig. 7-34).

Go directly to the system control IC when any one of these circuits cannot be operated. Check the supply voltage terminal pin for the correct voltage (5 V). A dirty switch contact or diode in the switch key input circuits can cause failure of a particular function. Scope the crystal for in and out waveforms for clock data. Do not forget that many of these control features have separate transistors that help operate the various circuits after receiving the right command from the large system control IC.

Check the system, interface, and control system when one or more control functions are not operating. Check to see if both the remote and keyboard functions are working. Suspect the infrared circuits when the TV can be operated with the keyboard functions. When one function does not operate, check the signal from the control IC to a corresponding transistor or IC. A bad microprocessor or control IC may cause problems with auto programming, no screen display, no up and down volume, or a snowy condition. Check for a defective microprocessor with intermittent loss of functions such as volume up and down or channels up and down. Suspect a bad control IC when a display such as tint or closed captions is shifted to the left.

A defective control micon IC may cause vertical jitter and horizontal white lines in the picture. After warmup, a defective micon IC may lock up and not let the TV be turned off. Replace the control IC when, after warmup, the volume and channel up and down controls lock up. Replace the micon IC when there is a loss of video and green horizontal lines appear on the screen. The defective control IC may result in a dead chassis and no relay click. Replace the micon control IC when the relay begins to chatter.

After 10 minutes, the defective micon IC may cause the on-screen display to show jumbled characters. A defective control IC may intermittently let the volume go real loud, become intermittent, and lock up. The control micon IC may cause a dead chassis. A defective EEPROM that works in the control circuits can cause many different control function problems. A leaky control micon may cause no picture. Chassis shutdown can result from a bad control micon IC. Do not overlook defective components tied to the control IC that can cause many different problems.

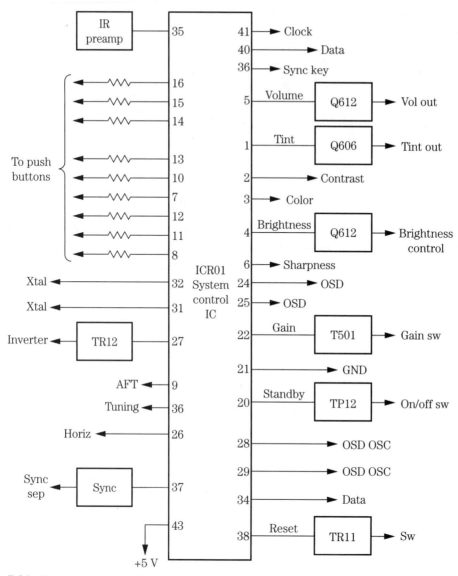

7-34 The many circuits that the system control processor controls in the present-day TV.

Tuner on the side

The old wafer tuner was found in early TVs, whereas today the varactor tuner may be located separately or in a shielded control case. The old tuner switching contacts would become dirty and had to be cleaned with a cleaning fluid. The varactor tuner has no moving parts and is tuned electronically. Today, the varactor tuner is controlled by a system control IC or a microprocessor.

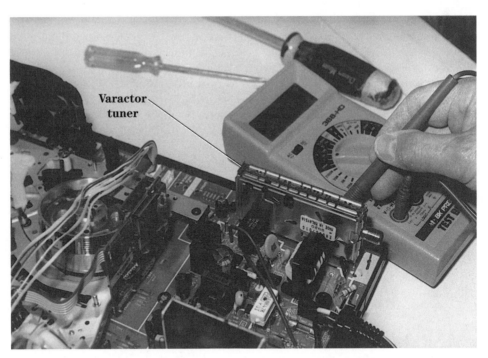

7-35 The varactor tuner is usually mounted on its side and operates with no moving parts.

The varactor tuner can be mounted on edge right on the PC board with the bottom contacts soldered into the PC board wiring. Today, some of the tuner's components are found mounted directly on the PC board under a metal shield (Fig. 7-35). Tuners under warranty should be removed and returned to the manufacturer's service center. Older tuners out of warranty can be sent to tuner repair centers.

A defective tuner can cause a snowy, weak, dim, intermittent, or flashing picture; erratic or drifting channels; or a white raster. Check for a defective tuner by plugging a tuner subber into the IF socket or connection. Voltage measurements, tuner subber injection, tuner substitution, and correct automatic gain control (AGC) voltage can help you discover a defective tuner. A defective tuner module should be removed and sent in for repair.

The RCA tuner in a CTC177 chassis is mounted on the PC board; resoldering the tuner grounds can solve a lot of different TV problems. Improper supply voltage to the mixer/OSC IC (U7301) can cause a snowy condition. An open R7401 (191-kilohm) resistor to the UHF/OSC at pin 14 can cause a snowy picture. A defective U7301 IC can cause a snowy picture on the VHF band. A badly soldered joint on pin 14 of the tuner control IC (U7401) can cause a snowy picture. A defective VHF RF amp (Q7102) is noted for a snowy picture in this tuner.

In the same tuner, open resistor R717 (100 kilohm) feeding voltage to G1 of Q7102 caused no tuner operation. Likewise, a defective VHF RF amp (Q7102) resulted in no tuner operation. No reception on channels 2 through 13 can be caused by a defective

U7301 IC (CXA1594L). A defective U7301 can cause snow on all channels, whereas the auto programming band jumps from cable to air mode. After the auto programming stops, the RCA CTC177 chassis is stuck on one channel.

Resolder all ground terminals under the tuner shield for intermittent reception on channels 7 to 13 and the UHF stations. Poor tuner ground terminals can cause intermittent and partial loss of vertical sweep. Horizontal lines flashing across the entire screen can result from shielded ground terminals in the tuner. Resolder the ground terminals for intermittent snow on channels 2 through 6. U7301 can cause snow on all tuner channels.

A very dark picture plus intermittent snow can result from ground terminals under the tuner shield. Replace U7401 when the channel display changes but the channels do not change; the picture may be snowy. The CTC177 chassis may be dead and then have a partial loss of vertical sweep when there are poor tuner grounds. Resolder each terminal on U7100 of the microprocessor for intermittent total loss of tuner action resulting in only snow and no audio.

Check for a leaky channel down switch (SW3420) when channels change by themselves. Replace Y7401, a 4-MHz crystal, for no tuner action. Intermittent display would show on the screen when there is a leaky menu button (SW3430).

No raster or a bright raster

Check the high voltage with the screen control turned up high when there is no raster. Suspect problems within the video circuits when there is normal high voltage or a defective picture tube. Sometimes when the screen is black and with the screen control advanced, you can see if the TV has video or vertical or horizontal problems on the screen. Service the high-voltage circuits if there is no raster or high voltage is seen. No raster can be caused by no heater or filament in the CRT.

A defective picture tube can cause many problems, such as poor brightness, missing colors, intermittent picture, poor focus, a single-colored raster, arcing in the gun assembly, retrace lines in the picture, negative picture, chassis shutdown, no heater or filament lit, and no raster, to name a few (Fig. 7-36). An open filament or heater can cause a no-picture or no-raster symptom. A defective CRT can have an extremely bright screen. Loose particles off the cathode element can lodge between the grids and cause an intermittent black-and-white picture. Simply tap the end of the CRT gun assembly and notice if the picture begins to flash off and on.

With a leaky picture tube, after turn-on, the raster dims and then gets extremely bright. A dim picture with no green in the raster can result from a bad green gun assembly. A heater-to-cathode short can cause the raster to change color or result in no raster at all. Replace the picture tube if there is severe arcing in the neck of the CRT.

Excessive brightness with retrace lines and vertical collapse to a thin line can be caused by a bad coil on the neck board of the picture tube. Rejuvenate the picture tube when there is a very bright green screen. Check for a resistance change on the resistors on the CRT neck board for an excessively bright picture with retrace lines. Excessive dust inside the CRT spark gaps within the picture tube socket can cause a very bright raster and then chassis shutdown. A defective video amp, luminance buffer, or reference transistor can cause a bright picture with retrace lines.

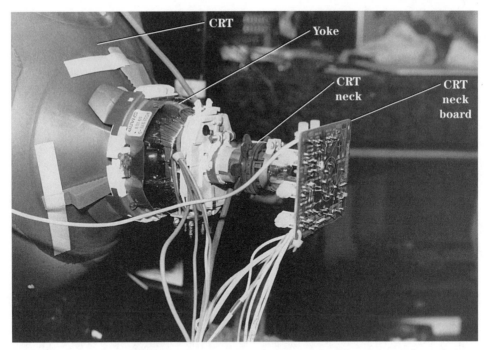

7-36 There are many CRT components mounted on the neck of the picture tube socket board.

Check the small 1- to 10-μF electrolytics on the neck of the CRT board when there is a very bright screen with retrace lines. Check for badly soldered joints and traces on the neck board of the picture tube when there is excessive brightness and retrace lines. Replace Q5106 on the neck of the picture tube board in a Zenith SS2014W TV for extreme brightness with retrace lines.

Color symptoms

Today, the color circuits are found in one large IC or microprocessor. With a normal picture and good sound, suspect the color circuits when there is poor or no color in the picture. Locate the chroma-luminance IC on the chassis. This IC contains the IF video, chroma, and deflection circuits. The chroma IC may be located on a separate video-luminance-color PC board (Fig. 7-37).

Scope the color IC for missing waveforms when there is no color. Check the color oscillator waveform (3.58 MHz) taken from the crystal pin on the IC. Check the terminals for a missing horizontal pulse from the flyback circuits. Check the three color output waveforms that feed the color output transistors on the CRT board. Suspect a defective color, picture, luma, or deflection IC for no color.

No color until the TV is on for an hour or so can result from a defective color crystal. A badly soldered joint on the color crystal can cause intermittent color. Distorted color may be caused by a degaussing problem. Weak color can be caused by improper

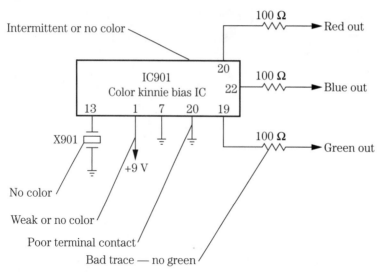

7-37   Check the following components for no, intermittent, or weak color in a TV chassis.

supply voltage, a leaky IC, leaky diodes, and open coil windings in the color circuits. In a GE 19GT313C TV, no color was caused by a defective U1001 (215542).

Look for open coils and poor component terminal connections. Check for leaky capacitors to ground. Intermittent color can be caused by leaky capacitors in the color killer circuits. Take a resistance test from each terminal pin to the chassis ground.

Green with envy

Check the color output transistors when one color is missing, and test the guns in the picture tube. Loss of one or more colors can be caused by color, sync, luma, picture, and deflection ICs. A leaky gun assembly in the picture tube can cause an all-red screen. A blue screen with retrace lines can be caused by a bad blue output transistor and poorly soldered joints at the collector resistor. A yellow tint with no blue can be caused by badly soldered joints in the blue output transistor on the CRT neck board.

Check all 33- to 100-μF, 35-V electrolytics that tie into terminals on the color IC when colors fade out. Measure all resistors on the neck of the CRT board for correct resistance when there is a greenish raster and no video. Check all diodes on the CRT neck board when there is red, green, or blue screen with retrace lines, some video visible, and normal sound. A bad line voltage regulator (STR30130) can cause the raster to shrink on both sides, as well as vertical foldover and no color, in an NEC CT-2750S TV.

Weak or intermittent color problems can occur in the color output amps. Locate the color amps on the CRT socket board. A defective green output transistor can result in a green raster, and a defective red output transistor can cause a red screen. Check the voltages on each collector terminal. Test each transistor in-circuit. Universal transistors can be used in the color output circuits.

Surface-mounted components

Besides being used in camcorders and CD players, surface-mounted devices (SMDs) are now found in TV chassis and varactor tuners. Surface-mounted transistors have three end connections, whereas resistors and capacitors have only two soldered end connections.

Surface-mounted ICs or processors may have gullwing terminals. Usually a white dot or number indicates the terminal 1. Look up the part numbers stamped on the bodies of ICs and processors to locate the correct circuits.

Resistors and capacitors may have the correct values stamped right on the body. You may find the part number corresponding with the schematic stamped alongside. Surface-mounted transistors, capacitors, and resistors are very difficult to locate without a schematic and magnifying glass.

Take voltage and resistance measurements on suspected SMD parts. Test each transistor in-circuit with a beta tester. Be very careful not to short the terminals during critical voltage measurements. Take care when removing and replacing SMD parts. Always replace an SMD part when it tests normal after removing it from the PC board wiring. Don't try to use it again. Surface-mounted resistors and capacitors can be replaced with universal SMD parts. Replace transistors, ICs, and processors with exact replacements.

The "tough dog"

A "tough dog" problem may take hours, weeks, or several months to repair. Sometimes a repair may be called tough by one technician, whereas the next technician may locate it at once. "Tough dogs" seem to come all in one week or one month. Always set such a chassis aside after working on it for an hour or so. Time can be lost if you work half a day on it. Give it up for now, and tackle it the first thing in the morning when your mind is fresh.

Ask for help if you cannot locate the defective component after several attempts. Go to the TV distributor or parts depot for help. If service personnel are not there, contact the factory. A simple telephone call and 10 minutes of your time may save you several hours of frustration. Check with factory repair depots that repair boards or the entire chassis.

On-screen display

The OSD circuits are controlled by a micro control processor to a buffer amp transistor to a driver transistor that ties into the green amp output transistor and picture tube. The system control IC determines when to output the OSD signal. In an RCA CTC167CN chassis, the micro control (U3101) provides an OSD signal to the green buffer transistor (Q2903), to a green driver transistor (Q5004), and then to the collector terminal of the green output transistor Q5002 (Fig. 7-38).

Check the OSD circuits by scoping the input terminal of the OSD, luma, and OSD black-and-white signals. Scope the waveform on the OSD green buffer and OSD green

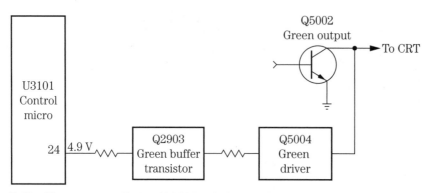

7-38   The on-screen display (OSD) block diagram found in a CTC167 RCA chassis.

driver transistors. Check for leaky diodes and transistors in the OSD circuits. A leaky diode and transistor will eliminate black around the characters. Replace the OSD green driver transistor when the display letters are black. Check for a defective EEPROM IC when there is no raster except for on-screen displays. OSD shifting to one side can be caused by a defective EEPROM. If there is no OSD, look for a leaky OSD transistor. No OSD in an RCA CTC145E chassis was the result of a leaky Q4501 transistor.

EEPROM problems

A defective EEPROM IC can cause many different problems in a TV. The EEPROM IC is found in only some TV chassis. A dead chassis can be caused by a defective EEPROM IC. No audio or video can be caused by a defective EEPROM IC. Replace the EEPROM IC when the tuner does not receive all the channels. Suspect the EEP-ROM IC when numbers are off to the right side of the screen and some are not readable. Intermittent TV channels 7 through 13 can result from a defective EEPROM. A defective EEPROM will not let the TV be shut off. When there is no vertical sweep and, once on, the TV cannot be shut off, suspect a bad EEPROM IC (Fig. 7-39).

No screen display, a very dark screen, and only a faint video were caused by a defective EEPROM. A bad EEPROM can cause a loss of vertical and horizontal sync. Suspect a defective EEPROM when the picture goes snowy, there is no video, or there is vertical collapse. The chassis can be dead with an okay-fuse symptom caused by a bad EEPROM. No menu and no channel change from the TV or remote can be caused by a defective EEPROM IC. Check the EEPROM IC when the TV will not program above channel 13. No menu or function display can result from a bad EEPROM.

A bad EEPROM can cause an intermittently dead chassis. No audio, variable audio control, and normal screen display can be caused by a defective EEPROM IC. Replace the EEPROM IC when the raster turns red or green. No color visible with only information present was caused by a defective EEPROM IC. Suspect a defective EEPROM IC when the set comes on and goes into shutdown. When a Sharp 25VT-G200 TV was first turned on, there was no horizontal or vertical sync and no color or audio; this was caused by a bad EEPROM.

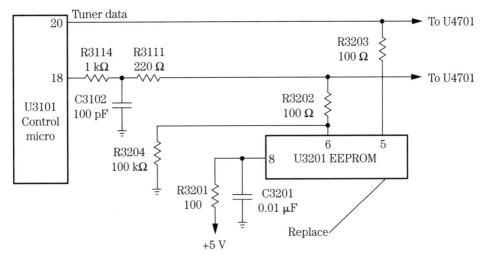

7-39 The defective EEPROM in an RCA CTC187 chassis caused the TV to stay on when it was turned off.

Audio problems

Sound problems within a TV chassis are about the same as those found in any audio amplifier. By signal tracing the audio signal with the external amp, you can locate the dead, weak, or intermittent stage. Today, TV audio systems have additional audio output jacks, fixed audio outputs, digital audio muting, and stereo sound circuits.

Low level and poor audio balance can be caused by an open or dried-up coupling capacitor. Check those 1- to 10-μF electrolytics with the ESR meter. No or distorted audio can result from a defective audio output IC. A badly soldered connection on the audio IC amp can cause intermittent sound in the left channel. Distorted audio can be caused by open filter capacitors on the voltage supply source fed to the audio circuits.

In an RCA CTC187 chassis, the stereo circuits begin with an IC1601 to the T-chip U1001. The left and right audio channels from U1001 are fed to right and left buffer, limiter, and mute transistors and then out hi-fi output jacks J1406 and J1405. For the TV speakers, the left and right outputs of U1001 are capacity-coupled (1-μF) electrolytics to the left and right input terminals of a dual audio output IC (U1901). Here the sound is amplified and capacity coupled (220 μF) to left and right 32-ohm permanent magnet (PM) speakers (Fig. 7-40). Refer to Chap. 3 for additional audio problems within TV chassis.

ACTUAL TV AUDIO PROBLEMS

Replace IC601 for no audio in a Goldstar CMT2035 TV. No sound in an Emerson EC134C TV was caused by open R354 (1.2-ohm, 2-W) resistor in the power supply. Check the C320 (220-μF) electrolytic in a Sharp 35LM36 TV when there is no audio. A bad switch transistor (Q1444) in a Sony KV20T520 TV caused no audio. Check Q603 for a leaky condition and replace an open R815 (560 ohms) off the 105-V source for no sound in a Samsung CT206S TV.

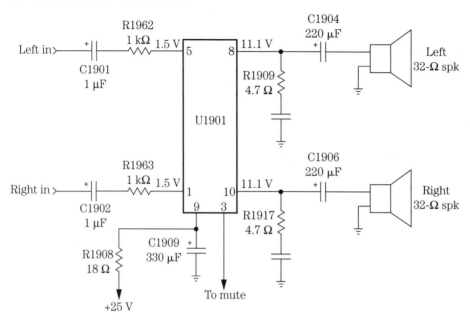

7-40 No sound was noted in the left channel and very weak audio in the right channel as a result of a leaky U1901 output IC in an RCA CTC187 chassis.

Leaky Q601 and Q602 caused distortion in a Samsung CT331 TV. Low audio and distortion in a Sanyo PC320 TV was caused by a defective C1870 (470 μF, 35 V) that kept destroying IC2850. Audio distortion in a Sharp 19MP17 TV was caused by an open R310 (100-kilohm) resistor. A defective IC201 (TA8642) caused distortion in a Goldstar CMT4842 model TV. Distorted sound in an RCA CTC140 chassis was caused by a leaky C1600 on pins 27 and 28 and a leaky IC1645 on pin terminal 12 of U1601. R301 had changed value off of IC101-4 and caused distortion at a low level of volume in a Magnavox CS2043R TV.

Check C4129 for audio buzzing in an RCA CTC175 chassis. Replace U3101 in an RCA CTC159 chassis for squealing audio. Replace the power audio output IC901 (STK563A) for no audio and a popping noise in a JVC C1950 TV. No audio, just hum was found in a Mitsubishi CS2062A TV and was caused by IC101-24 and the C113 (470-μF) capacitor. The volume would cut up and down with audio hum in an Emerson TC1372B TV and was caused by a defective IC401 (STR30130) voltage regulator in the power supply. A bad connection at the audio output 2SD401C in a Hitachi CT19A7 caused intermittent audio and a cracking sound. Replace a bad regulator in a Sony KV1951 TV for garbled audio.

Replace Q1552 (ECG123) in a Phillips E1 chassis for intermittent sound. Check IC1001 in an 19NP58 Sharp TV for intermittent audio. Resolder all connections on IC201 for intermittent audio in an NEC CT2020A TV. Replace T207 for intermittent audio in a Sony KV1943R TV. Solder connections on Q351 and Q352 for intermittent and no sound in an Emerson EC134C TV. Replace IC601 for intermittent audio in a Goldstar TS3140M TV. Check for a bad speaker in an RCA CTC145 chassis with garbled and intermittent sound.

8
CHAPTER

Servicing power supplies

Every electronic product must have some type of power supply so that the unit will function. The electronic product operates from the power line or batteries. Portable radios, cassette and CD players, and car radios operate from batteries. Large boom-box cassette players, CD players, TVs, DVD players, and VCRs operate from the ac power line. Of course, there are some radios and CD and cassette players that operate from both the power line and batteries.

You may find a dead electronic unit if improper or no voltage is found at the power supply source. Remember, the power supply is one of the most important sections of any electronic device. Go directly to the largest electrolytic capacitor, and measure the voltage across its terminals. Always check the power supply source before checking any other component (Fig. 8-1).

An ac radio, clock radio, CD player, or cassette player may have a step-down ac adapter power supply with a half-wave, full-wave, or bridge-type rectifier. Larger electronic products that operate from the power line use a power transformer with full-wave and bridge rectifiers. Early TV chassis contained a power transformer with full-wave circuits. Newer power line TVs have a bridge rectifier with a line voltage regulator. And the latest TVs and TV/VCR combos have transformers in a switch-mode power supply (SMPS) or a switching transformer circuit.

Half full

Half-wave rectifier circuits were found in many of the early radios and some TV power supplies. The half-wave rectifier delivers a half-cycle of dc voltage for every half-cycle of applied ac voltage. The half-wave rectifier is found in the transformerless TV chassis, radios, and industrial applications. You will find very large electrolytic filter capacitors in the half-wave rectifier circuits. In early radios, a selenium rectifier or one large silicon diode was found (Fig. 8-2).

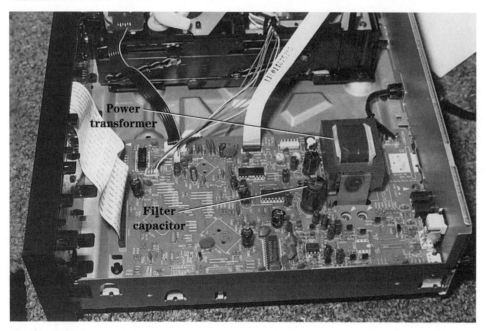

8-1   Locate the power transformer to find the power supply circuits in a Pioneer CD-5016 CD changer.

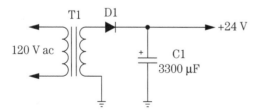

8-2
A half-wave rectifier power supply circuit contains only one silicon diode and filter capacitor.

Full-wave rectifier circuits are located today in about every electronic unit, and they operate from the power line. A full-wave rectifier circuit contains two silicon diodes. Usually, the full-wave rectifier circuit will have two silicon diodes and a center-tapped step-down transformer. Each diode rectifies an alternate half-cycle of the secondary voltage.

The ripple in the dc voltage source is equal to twice the supply frequency, and the resulting hum is much easier to filter out in the full-wave circuits. You will find smaller filter capacitors in full-wave versus half-wave circuits (Fig. 8-3).

The damaged bridge

A bridge rectifier is found in many electronic products and contains four silicon diodes. Bridge rectifiers may be made up of separate diodes, two diodes in one part, or all four diodes in one rectifier component. Bridge rectifier circuits may be found

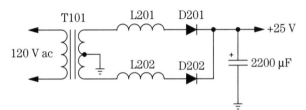

8-3   Early AM/FM/MPX receivers and cassette players contained a full-wave circuit with two silicon diodes.

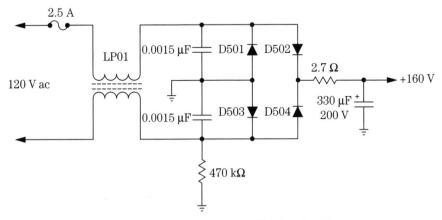

8-4   The raw ac bridge power supply found in a TV chassis without a transformer.

in a TV transformer-less power supply or in a TV with a power transformer. The negative ends (anodes) of the two diodes are grounded, whereas the two positive ends (cathodes) provide a positive dc supply voltage.

Usually the power transformer is not centered tapped with a bridge circuit, such as that found in a full-wave circuit with only two silicon diodes. The cathode end of the diode is connected to the positive output supply voltage source. You can replace a defective bridge circuit with a single silicon diode when four diodes are found in one electronic component (Fig. 8-4).

Transistor regulators

Transistor and zener diode regulators are found in clock radios, cassette players, CD players, and auto radios. You also will find several transistor regulators within TV chassis, DVD players, and VCRs. Low-voltage transistor and zener diode regulators are found in AM/FM/MPX receivers, cassette decks, FM circuits, and audio preamp and output circuits. Usually the collector terminal of the regulator transistor is tied to the highest dc voltage, and the output regulated voltage is taken from the emitter terminal. A zener diode may be found within the transistor base terminal.

Suspect an open regulator transistor when there is no or very low output voltage. An intermittent transistor can cause intermittent operation in transistor regulator cir-

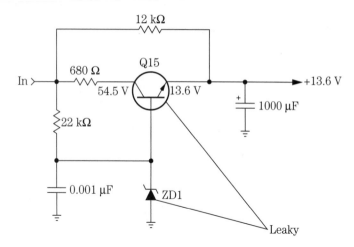

8-5 Check the regulator transistor (Q15) and zener diode (ZD1) with the diode tester of the digital multimeter (DMM).

cuits. Check for a leaky regulator transistor when the voltage output is lower than normal. Suspect a shorted or leaky diode within the base circuit of the transistor regulator when there is a shorted regulator transistor. Double-check the suspected transistor regulator, zener diode, and small-ohm resistors within transistor regulator circuits when there is no or low output voltage (Fig. 8-5).

Intermittent sound and a constant clicking of the protection relay in a Pioneer SX880 CD player was caused by a defective transistor regulator circuit in the power supply. At first, the regulator transistor was suspected of being intermittent. When the voltage probes were placed on the transistor regulator terminals, the CD player began to act up. Resoldering all connections on the regulator transistor terminals solved the intermittent CD player sound problem.

IC voltage regulators

Integrated circuit (IC) voltage regulators are found in CD player, VCR, DVD player, and TV power supplies. The IC voltage regulator also may contain several transistors and zener diodes within the same voltage sources. Often the simple IC voltage regulator consists of an input and output terminal with a common ground terminal. A leaky or shorted IC regulator may have no or very low output voltage, whereas an open IC regulator shows very little voltage on the output terminal (Fig. 8-6).

The power line voltage regulator found in a TV chassis is fairly large and has several terminal connections. Double-check the power line voltage regulator when there is a damaged or shorted horizontal output transistor. A leaky or shorted flyback can destroy the line voltage regulator in a TV chassis.

Simply measure the output voltage at the input and output terminals. If no or very low output voltage is found, suspect a defective IC regulator or a leaky component causing the reduced output voltage. You will find several transistor and zener diode voltage regulators within the secondary voltage sources of the flyback or horizontal output transformer.

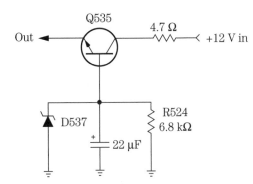

8-6
The regulator transistor (Q535) found in an auto stereo cassette player.

Critical voltage tests

An electronics technician should check the dc voltage across the largest filter capacitor first. If the voltage is low or too high, the defective component is within the power supply circuits. Check each voltage source with an accurate digital multimeter (DMM). Critical and very low voltages are found throughout electronic products. A dead power supply may have an open fuse with shorted or leaky silicon diodes, an open primary winding of the power transformer, or shorted or leaky voltage regulators. Leaky or shorted filter capacitors can blow the main power fuse. An open filter capacitor can cause a low-output dc voltage source and hum in the speakers.

A leaky transistor or IC regulator may have low output dc voltage. An open transistor or IC regulator has no output voltage. A leaky zener diode regulator can have lower output dc voltage. A transistor regulator becomes leaky or open, whereas the zener diode regulator becomes leaky and overheated with burned marks. You can easily spot a leaky or overheated zener diode in low-voltage sources. Take voltage measurements in and out of the suspected transistor or IC regulator to determine if it is open or leaky (Fig. 8-7).

Storm damage

Critical storm damage or a power line outage can destroy components within an electronic product. An open fuse, shorted or leaky diodes, and power transformer damage can be caused by a heavy lightning hit. Sometimes only a fuse is blown with no power cord damage when a lightning strike occurs nearby. A heavy lightning strike can destroy many components in electronic products found in the home, as well as any unit connected to the power line.

A direct hit on a TV antenna and power line can cause extensive damage to the power cord, on/off switch, and low-voltage diodes and can result in open foil in the tuner and several black spots on the printed circuit (PC) board wiring of a TV chassis. Besides damaged silicon diodes, large-wattage resistors may be blown open, and the line voltage IC regulator may be damaged. Sometimes a TV chassis can operate at 90 to 100 V ac with a damaged voltage regulator. Besides blown low-voltage diodes, check for

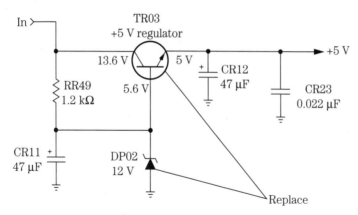

8-7 Intermittent TV startup was caused by a defective TR03 and TR05 in an RCA TX825TC power supply.

damaged zener diode and transistor voltage regulators in the power supply. Low-ohm resistors within the standby circuits of a TV chassis can be damaged by lightning or other power line outage conditions.

A high-voltage power line surge can destroy the same components in the power supply as a lightning strike. Besides diodes and regulators, a power surge may destroy ICs and transistors in and out of power supply sources. You may find open traces of wiring caused by a power line surge. Blackened and blown-apart surge components within the power line circuits can be caused by high power line voltage surges. A clicking relay with a dead chassis may result from a high power line surge that can destroy the main filter capacitor (470 to 680 μF, 200 V) within a TV chassis. A power line surge on a Magnavox 27C9 chassis caused extensive damage to D404, D405, D407, F400, Q400, R422, R430, D425, and Q402.

STORM DAMAGE, RCA CTC159

An RCA CTC159 chassis was dead and had several damaged components in its power supply circuits. No voltage was measured across the main filter capacitor (C4007). After replacing an open (2.7-ohm, 15-W) resistor (R4001), the chassis was still dead. All diodes were suspected, but they tested normal with the diode tester of the DMM. SCR4101 was tested and replaced. The CTC159 chassis still would not operate until the standby transformer (T4601) and the main filter capacitor (C4007) were replaced (Fig. 8-8).

Repairing radio power supply circuits

Most radio power supplies consist of batteries, ac power circuits, or ac adapters (Fig. 8-9). When the adapter is inserted, the batteries are disconnected from the circuit. Likewise, in large battery and ac power supply circuits, the batteries are switched in and out of the circuit.

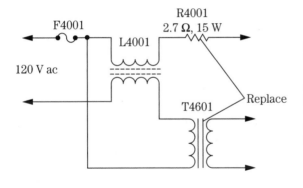

8-8
The standby transformer
(TR4601), resistor R4001 (2.7
ohms, 15 W), SCR4101, and
main filter capacitor (C4007)
were replaced in an RCA
CTC159 chassis as a result of
storm damage.

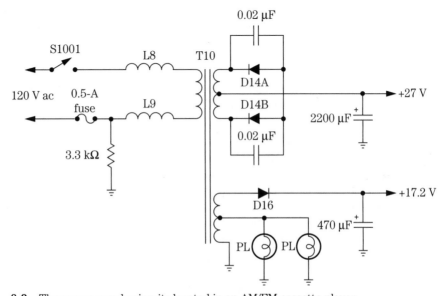

8-9   The power supply circuits located in an AM/FM cassette player.

In some models, an external battery source can be plugged into the radio, disconnecting the batteries within the radio. The ac power supply is not in use when the batteries are switched into the circuit. Poor switch and dirty jack contacts can prevent the radio from operating. Check the switch contacts and plug with the low-ohm scale of a DMM. Clean the switch and jack contacts with cleaning fluid.

Check the batteries. Weak batteries can cause weak and noisy reception or slow tape motor movement. Poor battery contact can be caused by leaky batteries and corroded terminals.

CLOCK RADIO AC POWER SUPPLIES

Most clock radio power supply circuits consist of a simple full-wave rectifier and an electrolytic filter circuit. Often a step-down transformer reduces the line voltage from 10 to 25 V ac. You may find two silicon diodes or a bridge circuit in the full-wave circuit. A low-ohm resistor and main filter capacitor are found in the dc voltage output circuits.

In larger models, diode and transistor/IC voltage regulators are found. If the radio is also battery operated (9 V), the highest supply voltage will be around 9 to 10.5 V. A 10.5-V source feeds the power IC and transistors, whereas a 6.1-V source is tied to the cassette preamp tape circuits. The lower voltage is tied to the AM/FM radio circuits.

Large table model radios with a light-emitting diode (LED) clock display have several secondary windings on the power transformer. High ac voltage (25 to 50 V) is rectified with two silicon diodes, and the resulting voltage is applied to the clock display. Another ac winding (3 to 5 V ac) goes directly to the clock display. A bridge rectifier circuit attached to another winding provides dc voltage to the tape motor and audio circuits.

Check the clock radio power transformer circuits by checking the resistance across the ac line cord plug. The resistance measurement should show the resistance of the primary winding of the step-down transformer with the power switch on. Suspect an open primary winding or power switch when there is no reading. Usually, one or two silicon diodes become leaky or shorted and place a load across the secondary winding. Since the primary winding is wound with small wire, the winding opens up.

DEAD CLOCK TIMER CIRCUIT

In a General Electric 7-4665B clock radio, no time could be seen on the electronic tube display panel, and the radio was dead. No output voltage was found on voltage regulator Q103. No ac voltage was measured on the secondary winding of T101. The primary winding of T101 was found to have no resistance and thus was open (Fig. 8-10).

TROUBLESHOOTING CLOCK POWER SUPPLIES

Check the primary winding and ac plug with the low-ohm scale of a DMM (50 to 150 ohms) if a clock is found to be dead. Often, shorted diodes and leaky filter capacitors will open the primary winding of the transformer. The transformer produces voltage when plugged into the wall outlet. The dc voltage is switched to the radio or tape player.

Check the different voltage sources by measuring the voltage across the large filter capacitor. If a voltage is not found at this point, check each silicon diode in the circuit. Measure the ac voltage across the different transformer wire leads. Check for dc voltage at the emitter terminal of the suspected voltage regulator.

In today's table, clock, and portable radios, only one power transformer is found. The transformer can be mounted on or off the large PC board. A large filter capacitor is found close to the silicon diodes. Look for larger decoupling capacitors in lower-voltage sources (220 to 470 μF). You may find the power supply circuits close to the audio output transistors.

Weak or no audio can be caused by leaky voltage regulator diodes and transistors. Open regulator transistors and burned isolation resistors can result in no sound. Don't overlook a dried-up decoupling capacitor if there is weak or distorted audio.

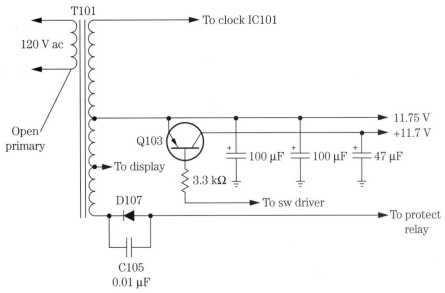

8-10 An open primary winding in a clock radio power supply circuit resulted in no radio reception.

REPAIRING THE AC ADAPTER

Check the low-voltage source of the ac adapter power supply at the metal plug. This voltage will measure a few volts higher than that marked on the adapter body. The ac adapter consists of a small power transformer, diodes, an electrolytic capacitor, a flat lead, and a male plug. If there is no voltage, measure the resistance across the male plug to determine if the primary winding is open or normal. Remove the adapter from the wall outlet when taking resistance measurements. Replace the entire adapter if the transformer winding is open (Fig. 8-11).

Often the flat rubber cord will break at the male plug or where it enters the adapter case. Lightly pull on the cord at the male end. Cut off the male plug and install a new one if the wire stretches or appears broken where it enters the plastic plug.

To check the dc cord continuity, stick a safety pin where the cord enters the ac adapter and check the resistance or continuity on the low-ohm scale of a DMM. Likewise, check both sides of the wire from the adapter to the male plug in the same manner. If there is no continuity measurement at the male plug, stick another safety pin into the insulation in the same manner as the first pin. You can check each cord for possible breakage with this method. If a break in the cord is found, cut out the broken area, solder up the two broken ends, and tape over the damaged area.

HUM IN THE SPEAKERS

Go directly to the large filter capacitor or voltage regulator for a low or excessive hum in the audio. Pull the ac plug, and discharge the large filter capacitor (1000 to

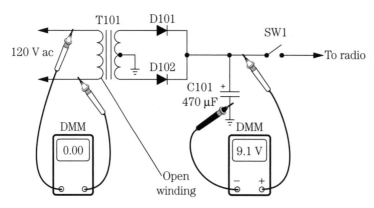

8-11 Check the primary for an open winding with an ohmmeter and the output voltage with a DMM in a dead adapter power supply.

3300 μF) with a test clip or screwdriver. Solder or tack in a good electrolytic capacitor of the same capacity or higher. Observe the correct polarity. If the hum disappears, replace the dried-up filter capacitor.

Hum can be caused by a leaky voltage regulator transistor or diode. Defective decoupling capacitors in the voltage-dropping circuits can cause a low hum in the sound. Check both the regulator transistors and the diodes in the circuit for leakage. Shunt a suspected decoupling electrolytic capacitor with one of a known value.

Motor boating or a put-put noise in the speakers can result from defective filter capacitors and diodes. A bad filter capacitor can cause a high-pitched squealing noise. A hissing sound from the power supply can be caused by small electrolytic capacitors (22 to 47 μF, 25 to 63 V). An open or dried-up electrolytic filter capacitor can cause hum bars in the sound. Check all electrolytic capacitors with an equivalent series resistance (ESR) meter.

Battery power circuits

Small electronic products may operate directly from batteries, both ac and batteries, or a battery-charging system. The batteries are switched into the power supply circuits with a power or function switch. A portable CD player may have an external jack to accept a dc adapter plug into the player and then can be operated from the ac power line (Fig. 8-12).

A weak battery can cause low or weak volume within the radio circuits and a loss of speed in a portable cassette or CD player. Batteries should be checked under load in either a separate battery tester or with a DMM. Turn on the electronic device and measure the voltage across each battery with the low-voltage scale of a DMM. Remember, only a volt or two can cause the electronic product to fail.

Clean the battery contacts with cleaning fluid, if handy. Double-check the battery and tabs for correct polarity. One battery inserted backward can make an electronic product fail to operate.

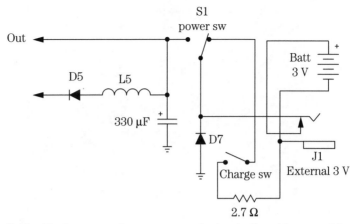

8-12 The battery and ac power supply circuits found in a portable CD player.

Locating low-voltage components

Cassette players, CD players, and ac adapters use very small power transformers. Locate the power transformer and electrolytic capacitor with the silicon diodes nearby. The transformer may be mounted off to the side or off the main PC board in cassette and CD players (Fig. 8-13). When the power transformer is mounted to one side or off the main PC board, trace the secondary leads to the power supply circuits. Check for a transistor voltage regulator and zener diode nearby in a voltage regulator circuit. The power supply circuits are found next to the power transformer and the largest electrolytic capacitor on the PC board chassis.

Servicing car radios

Some of the early car radio power supply circuits were quite simple and were fed from a 12-V car battery to a choke coil, low-ohm resistor, and the main filter capacitor. The on/off switch applied power to the radio circuits with a 5- to 10-A fuse in series with the A cable. In larger auto radios, a decoupling resistor and electrolytic filter capacitor network were found to separate the radiofrequency (RF) and intermediate frequency (IF) stages from the audio circuits. The AM and FM front-end circuits were operated at a lower dc voltage (Fig. 8-14).

In deluxe auto receivers, you may find transistor and zener diode regulator circuits. A deluxe auto receiver may operate on a 2- to 5-A fuse, on/off switch, inductance filter, and several electrolytic capacitors and low-ohm resistors to provide the different voltage sources. A function switch applies the required voltage to the AM and FM radio circuits. Another switch may turn on a cassette or CD player when a cassette or CD is inserted into the auto receiver. Several zener diodes and one or two transistor voltage regulators can provide critical regulated voltage to the front-end (AM/FM) and audio preamp circuits.

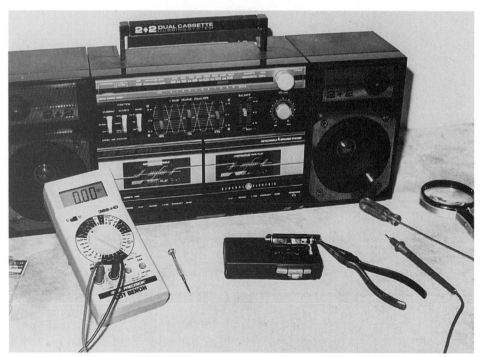

8-13 Locate the power transformer within an AM/FM/MPX cassette player to find the power supply components.

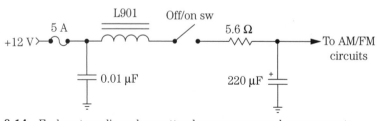

8-14 Early auto radio and cassette player power supply components.

Check the input dc voltage at the switch and at the largest filter capacitor on the chassis (Fig. 8-15). Usually the bottom cover must be removed to get at the main filter capacitor terminals. You also can check the highest dc voltage fed to the audio power output transistors or IC component. Go directly to the RF AM section power supply voltage source when the AM section is weak or dead. Likewise, check the voltage fed to the FM section when only a weak or no FM radio can be tuned in. Suspect a defective transistor or IC voltage regulator when there is no output voltage at the emitter terminal of the regulator transistor or the output terminal of the regulator IC.

DIODE TESTS

Silicon and zener diodes can be checked within the circuit if a coil or low-ohm resistor does not shunt across the suspected diode. You should have a low measurement

8-15 Locate the largest filter capacitor within a car radio and take critical voltage tests.

with the positive probe of the DMM at the anode terminal and the black probe at the collector terminal with the diode tester of the DMM. No measurement with reversed test lead probes indicates a normal diode. The positive probe of a volt-ohm-milliameter (VOM) is just the reverse of the DMM. A low-ohm measurement on the VOM with the positive probe at the collector (+) terminal of the diode indicates that the diode is normal (Fig. 8-16). Replace a defective diode with a low-ohm measurement in both directions. Remove one terminal of the diode from the PC board for a correct resistance measurement when in doubt.

MOSFET POWER SUPPLY

Metal-oxide semiconductor field-effect transistor (MOSFET) power supply circuits are found in the latest high-wattage power amplifiers to provide a higher dc voltage than the 13.8-V dc battery voltage in the automobile. Several high-powered MOSFETs operate in a parallel push-pull transformer circuit. A bridge silicon rectifier circuit is found in the secondary side of power transformer T1. Very large filter capacitors provide a smooth positive and negative output voltage to the high-powered output transistors within the amplifier circuits (Fig. 8-17).

KEEPS BLOWING THE FUSE

Suspect a defective power output transistor or IC when the fuse will not hold after replacement. Sometimes a low-amp fuse will open for no reason at all. Simply replace the

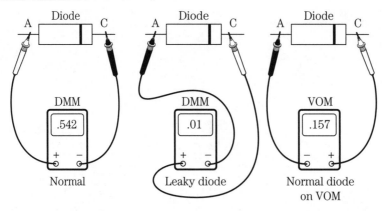

8-16 A normal or leaky diode test with the DMM and a normal diode test with the VOM.

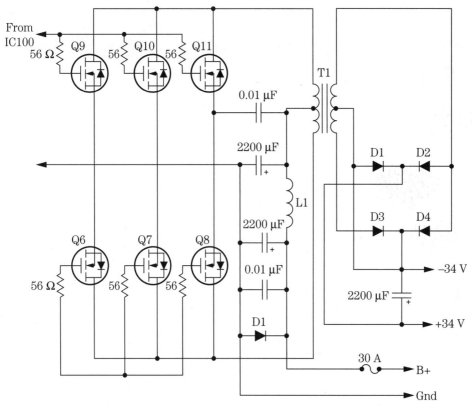

8-17 A MOSFET power supply circuit located in a 170-W auto stereo amplifier.

blown fuse. Do not replace the blown fuse with a larger-amp fuse. Check the output circuits for defective components. Test each output transistor with the diode tester of a DMM. Check the resistance across the largest electrolytic capacitor for a low reading.

Large filter capacitors and a grounded A lead can keep blowing the main fuse. If the radio harness is melted down, go directly to the transistor output stages. Test and replace the leaky or shorted transistors. Always check the fuse holder or auto fuse container for the correct size fuse.

Servicing high-powered amp circuits

Large stereo amps, from 10 to 50 W, are found in AM/FM receivers, auto radios, cassette decks, CD players, CD changers, and TV chassis. Locate the large filter capacitors, and check the voltage across their terminals. Most of these high-powered amps have a power transformer with full-wave and bridge rectifier circuits. Large power transformers are found in higher dc voltage sources that operate high-powered transistors and IC components. Check for defective transistor, IC, and zener diode regulator circuits supplying several different voltage sources (Fig. 8-18). Do not overlook an open or leaky decoupling electrolytic in a dead receiver.

Some of these large stereo units are fused, and some are not. Check for an open fuse when there is a dead chassis. Measure the dc voltage across the large electrolytic capacitors because they filter out the various voltage sources. Take a quick ESR test of

8-18 You will find several different voltage circuits with transistor and zener diode regulators within a large stereo amp receiver.

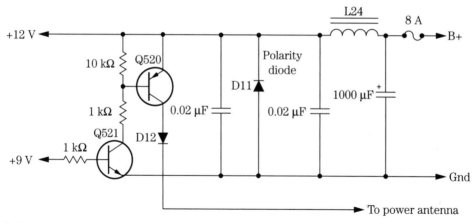

8-19 Check for leaky transistor regulators within the stereo circuits of an auto AM/FM/MPX cassette receiver.

each electrolytic in the power supply. Always discharge each electrolytic before taking any ESR tests. While in the same circuits, measure the resistance across the large filter capacitor terminals, which may indicate a shorted or leaky capacitor or a defective component in the power source. Test each silicon diode with the diode tester of a DMM. Now check each transistor voltage regulator with the diode tester of the DMM.

Sometimes an overloaded circuit can cause extensive damage to the power supply. Check the resistance across each electrolytic in the power supply for a measurement under 50 ohms. Cut into the output voltage PC trace or wiring that may be overloaded and then take another measurement. Overloaded circuits connected to the power supply are caused by high-powered output transistors and ICs and leaky transistor regulators and zener diodes (Fig. 8-19).

THERE SHE GOES AGAIN

Electrolytic capacitors have a tendency to blow their top, so to speak, with higher than normal voltage applied and gas buildup inside them (Fig. 8-20). The electrolytic can become warm and blow its top off if the polarity is reversed or it is installed backward. Small electrolytic capacitors can run too warm when a lower voltage capacitor replaces one of higher working voltage.

When replacing electrolytic capacitors, make sure that the working voltage is the same or higher. Likewise, replace defective electrolytics with ones of the same capacity or higher. Check for correct polarity before soldering. These electrolytic capacitors can blow up in your face. Excess tinfoil and insulating material may be found over the entire chassis when the top blows off.

Power transformer problems

Suspect leaky diodes, bridge rectifiers, filter capacitors, power output transistors, or ICs when a fuse blows. Go directly to the large filter capacitor and take a resistance

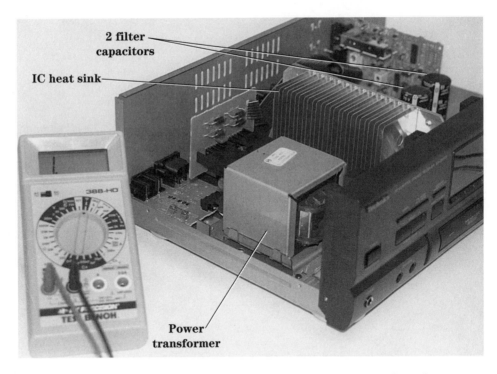

8-20 Visually inspect the electrolytic capacitors for blown or bulging sides in large stereo amp power supplies.

measurement across it. The meter hand should charge up and discharge with a very high measurement. Low-ohm readings below 1 kilohm can indicate a leaky or shorted capacitor. Here the electrolytic capacitor measures no voltage across the terminals.

Check each diode within a bridge rectifier or full-wave rectifier circuit. A low-ohm measurement is noted in one direction. If the reading is low with reversed test leads, remove the suspected diode and test it out of the circuit. Replace the whole bridge unit when one diode is found to be leaky or open.

Sometimes the power transformer can be damaged, can run warm, or can blow the main ac line fuse. If the transformer is overheating or extremely warm, disconnect all leads going to the diode circuits. Mark each color-coded wire so that it can be replaced correctly. Then power up the transformer. If it runs warm or hot without a load, replace it. Remember, these transformers are quite expensive and sometimes difficult to obtain (Fig. 8-21).

EXCESSIVE HUM

Shunt all electrolytic capacitors with a known electrolytic across the suspected filter capacitor in the power supply circuits. Clip the new electrolytic in the circuit before firing up the AM/FM receiver so as not to destroy other electronic components. Ex-

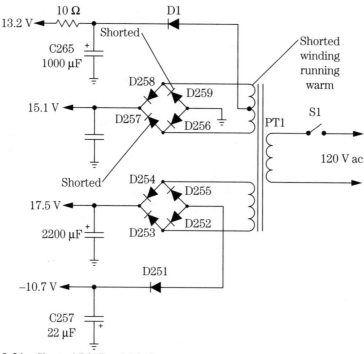

8-21 Shorted D257 and D259 caused damage to the secondary winding of a transformer (PT1) in a Sanyo GXT-300 stereo system.

cessive hum was heard in the speakers of a Sharp SC700CR receiver with no tuner action. Replacing a 470-μF filter capacitor near the heat shield solved the hum problem.

HIGH- AND LOW-VOLTAGE SOURCES

A very loud hum usually is caused by one or two filter capacitors. Check the electrolytic capacitor with the ESR meter, or shunt another capacitor across the suspected one. In both cases, discharge each electrolytic. Maintain correct polarity when installing or shunting another capacitor. The negative terminal ties to the common ground side and at ground potential. Replace all electrolytic capacitors in the power supply that have changed values, have ESR problems, and may be leaky or shorted.

A low hum can be caused by defective electrolytic capacitors and burned resistors in the decoupling stages. Locate these capacitors on a separate board with the bridge rectifiers. Weak and distorted audio can result from improper positive and/or negative voltage applied to the power output ICs or transistors. In large amps, a high positive (+40 V) and negative (−40 V) voltage may be applied to the power output semiconductors (Fig. 8-22). Leaky zener diodes and transistor voltage regulators can cause weak and distorted sound.

BAD POWER SWITCHES

Dirty or worn switch contacts within an electronic product can cause intermittence or a dead chassis. Suspect a bad off/on switch when the unit intermittently shuts off.

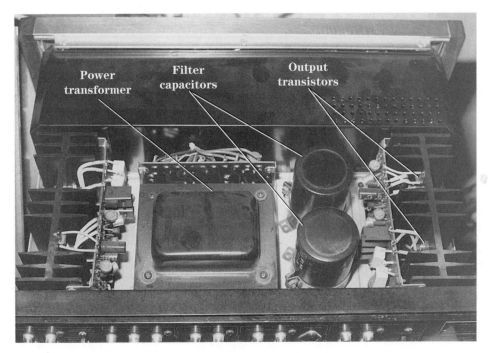

8-22 Check all electrolytics within large stereo amp power supplies with an ESR meter.

A badly soldered joint on one of the wires connected to the switch can produce intermittent operation. Sometimes the receiver turns on okay, and the next time it is dead. A dirty on/off switch also can make the relay kick off and on. Check for open foil or traces where the power switch ties into the power line. Resolder all switch terminals. Suspect a bad power switch when the product cannot be turned off and operates all the time (Fig. 8-23).

Replace an on/off switch when the switch contacts began to arc over and a loud buzzing noise is heard in the speakers. A chattering relay may be caused by a dirty or worn power switch. A dead receiver may result from an arcing power switch. Replace the on/off switch if a noisy sound is heard when the unit is first turned on.

No power shutoff

The power could not be turned off with the regular switch (S504) in a Soundesign 5959 cassette stereo amplifier. The switch tested normal. In this model, the power turnoff circuits are found in the low-voltage power supply. The same voltage was found on all three terminals of Q604. Replacing transistor Q604 with an ECG374 universal replacement solved this unusual power turnoff problem.

Repairing CD player power supplies

Small portable CD players can be operated from batteries or an ac adapter, whereas larger players have a power transformer. Some power transformers are quite small

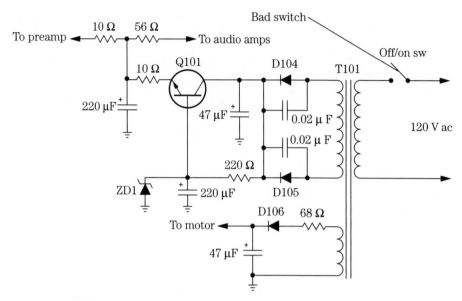

8-23 A dirty or worn power switch can cause intermittent operation and a noisy sound when turned off and on.

in size with several different ac windings. A power transformer is mounted on the chassis, whereas other power supply components are mounted on the main PC board (Fig. 8-24).

You may find three or more secondary ac windings on the transformer with full-wave, half-wave, diode, and bridge rectification. Often, only one diode is used as a half-wave rectifier to furnish higher voltage to the control and display circuits. Different voltage sources may be needed for servo, decoder, control, and display circuits. Positive and negative voltages are found with each bridge rectifier circuit.

In larger CD players, the power supply may contain several transistors and IC regulators. Dc-dc power supply circuits are found in larger CD player power supplies. Usually the dc-dc power supply provides higher separate positive and negative voltages to the CD circuits.

No loading, Goldstar CD player

The loading motor would not load a CD in a Goldstar GCD-616 CD player. Since no schematic was available and a voltage check across the motor terminals was zero, the motor terminals were traced back to a motor driver IC. No voltage was found on any terminal of the driver IC.

Since there were several different power sources within the transformer circuit, all diodes were quickly checked with the diode tester of a DMM. All diodes tested okay in circuit, and a dc voltage was found on each diode. Next, all transistor voltage regulators were tested in the circuit with the diode tester of a DMM. Q003 did not add up. No voltage was found on the output terminal of Q003. Replacing Q003 restored the 7-V supply source to the loading motor IC (Fig. 8-25).

8-24   The shorted ac power transformer within a Magnavox CD changer.

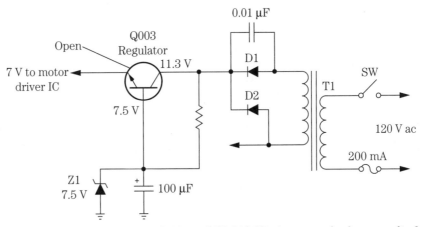

8-25   The loading motor in a Goldstar GCD-616 CD player was dead as a result of an open 7-V regulator transistor.

IC voltage regulators

Some units use a combination of transistors, zener diodes, and ICs for the different voltage sources. Look for the transistor, IC, and zener diode regulators next to the large filter and decoupling capacitors. Usually all power supply components are found in one corner bunched together, including separate diodes and bridge rectifiers.

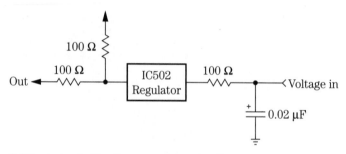

8-26 A simple IC voltage regulator circuit.

An IC regulator can become leaky and produce a lower or higher output voltage. If it is leaky, an IC regulator may run warm. If it is open, no voltage will be found at the output terminal. Check for a higher voltage at the input terminal and a lower voltage at the output terminal. If the output voltage is low after replacing the IC or transistor regulator, suspect a leaky component tied to the voltage source. When one of the positive or negative voltages is low or missing, suspect a defective IC regulator (Fig. 8-26).

Servicing color TV power supplies

Before attempting to service a TV chassis, plug it into a variable-isolation power transformer. Early TV chassis operated from a transformer, whereas the latest TVs operate directly from the power line with a power line voltage regulator circuit. Of course, today's color TV chassis may have a switch-mode power supply (SMPS) or a switching power supply that operates from a switching transformer. A variable-isolation power transformer prevents TV or test equipment damage. Besides, for a chassis that is intermittent or dead, you can raise or lower the ac line voltage to make the TV chassis act up at a lower or higher line voltage than normal (Fig. 8-27).

The typical ac power supply consists of a bridge rectifier, a filter capacitor, voltage regulators, and a fuse (Fig. 8-28). Some chassis have one diode with half-wave rectification. Look for the ac and dc line fuse and bridge rectifier circuit on the PC board main chassis. A large high-voltage electrolytic capacitor (330 to 870 µF) is found nearby. Voltage regulator transistors and zener diodes can be found around the electrolytic decoupling capacitors. In the latest TV chassis, look for a large voltage regulator IC supplying high dc voltage to the horizontal output transistor.

POWER LINE IC VOLTAGE REGULATOR PROBLEMS

Intermittent chassis shutdown can be caused by a defective line voltage regulator. Foldover at the top of the picture and a raster that is shrunk on all sides can result from a bad STR30110 IC regulator. A dead chassis and okay fuse symptom can be caused by a defective regulator. A leaky line voltage regulator may not allow the TV to turn off, or the TV may shut off at once with a leaky line voltage regulator.

A color TV may come on with a loud hum and then shut down when it has a defective high-voltage line regulator IC. The TV may shut down intermittently at turn on or may operate for several minutes or an hour and then shut down as a result of a de-

Filter capacitor

8-27 Check for a line voltage regulator and a large filter capacitor in a TV chassis to locate the power supply components.

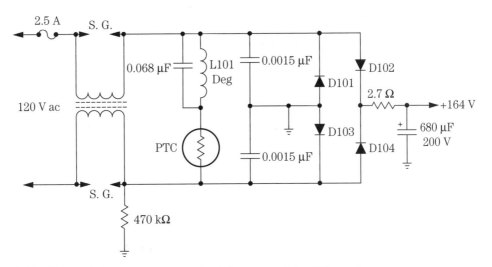

8-28 The typical ac power supply circuit in a present-day TV chassis.

fective IC regulator. Suspect a defective line voltage regulator when the TV shuts down with the power line voltage above 110 to 115 V ac (Fig. 8-29).

The line voltage regulator and main filter capacitor can cause the picture to shrink on all four sides of the screen and may cause the set to operate intermittently. Suspect a line voltage regulator when the horizontal output transistor runs warm with poor regulation

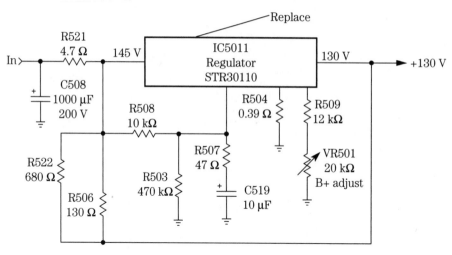

8-29 Replace a defective line voltage regulator in an Emerson TS2555SD color TV when it is dead, goes dead after a few minutes, or comes on in 10 minutes before shutdown.

and intermittent startup. Check the line voltage regulator when the TV shuts down at a normal power line voltage. Replace the line voltage regulator (STR30110) when the set comes on intermittently, the power relay clicks, and the set goes into chassis shutdown.

A dead chassis can result from a defective line voltage regulator. A pulsating picture with poor voltage regulation also can be caused by a bad line voltage regulator. Badly soldered joints on the line voltage regulator terminals (STR30125) can intermittently shut the TV off. An STR30135 line voltage regulator IC can pull the sides in and cause the vertical sweep to change. A dead TV with a blown line ac fuse can be caused by a shorted line voltage regulator IC.

Shadow-like lines floating upward through the picture can result from a defective STR30130 IC regulator. Wavy lines and poor voltage regulation can be caused by a bad STR30130 regulator. A pulsating picture, intermittent with the sides pulled in, can be caused by a defective line voltage regulator.

Hum in the audio and shutdown at high ac power line voltages can result from a bad IC regulator. Two horizontal lines floating up the picture can be caused by a defective 130-V line regulator IC. Suspect a defective line voltage regulator when the brightness is increased and the picture begins to pulsate.

TROUBLESHOOTING TYPICAL POWER SUPPLY CIRCUITS

Look at both the ac and dc fuses. Check for B+ voltage at the collector terminal (metal body) of the horizontal output transistor or across the large filter capacitor terminals. Low or no voltage at the output transistor can indicate an open dc fuse or a defective voltage regulator. If dc voltage is not found at the large filter capacitor, check for an open or leaky diode, a high-wattage isolation resistor, or poor ac wiring connections. Measure the ac voltage at the large resistor and common ground. One or more leaky diodes in the bridge circuit can cause the power line fuse to open. Be careful, this is power line voltage you are measuring.

If the main line fuse keeps blowing, suspect a leaky or shorted diode in the bridge rectifier circuit. When both fuses open, suspect a defective horizontal output transistor or voltage regulator IC. Don't overlook a burned isolation resistor in the flyback primary winding.

Heavy dark lines across the raster can indicate a defective main filter capacitor or regulator. Clip another electrolytic capacitor across the suspected one with the power off. Watch for correct voltage, capacity, and polarity before clipping in the capacitor. Replace a defective regulator when lower dc voltage and hum bars appear in the raster. These IC regulators have a tendency to open or appear leaky.

Lower output voltage from the voltage regulator may be caused by overloading in the horizontal output and flyback circuits. Remove the horizontal output transistor, and test the voltage regulator source. If it is normal, this voltage will be a little higher with the horizontal load removed. Replace the voltage regulator if the voltage is still low after removing the horizontal output transistor and the input voltage to the regulator is fairly normal.

Today, the new line voltage regulator consists of a transistor with five terminals. You can spot a line IC regulator because it has numbers stamped on the body. The last three numbers are the output voltage. The dc input voltage may vary from 145 to 175 V, depending on the filter capacitor and surrounding components. An STR30125 fixed line regulator has an output voltage of 125 V. Test the dc voltage going into the regulator and the fixed output voltage. If the right voltage is going in and improper voltage is coming out, suspect a defective regulator. Check all components that are tied to the regulator terminals.

RELAY SWITCHING PROBLEMS

Some TV chassis employ a power relay to turn on the chassis, whereas others use a relay to start the degaussing process. A defective relay can cause a dead TV chassis. An intermittent relay can cause the TV to start up and then shut down. Replace the ac relay when the set comes on okay and then shuts off after 5 minutes. When the TV shuts off intermittently and then sometimes comes back on by itself, suspect badly soldered joints of the relay on the ac power line side (Fig. 8-30).

The chassis goes dead and the relay clicks when there are bad switching points on the power relay. Badly soldered joints on the relay coil can produce a dead-chassis, okay-fuse symptom. Suspect an open relay coil with a dead TV chassis and no click of the relay. In a Zenith SMS1935 chassis, the relay did not click, and this was caused by an open relay winding.

Do not overlook a shorted diode across the relay solenoid when the relay will not even click. Bad relay contacts can cause intermittent shutoff, not shutdown, and the set may play 30 minutes or so. Check the relay driver transistor when the TV is dead and there is no click from the relay. Suspect a bad relay when the TV cannot be turned off and the set must be unplugged from the power line. The relay points may be stuck in one position. Check for a bad relay when the TV comes on when first plugged in. A Magnavox 19F1 chassis could not be turned off and had to be unplugged to turn the set off; this was caused by a defective relay.

The audio went out and the raster stayed on when a TV with a bad relay was turned off. Bad contacts on the relay caused a dead TV with a relay click and the LED channel display still on. Suspect a relay driver transistor when a TV shuts off and not a shutdown

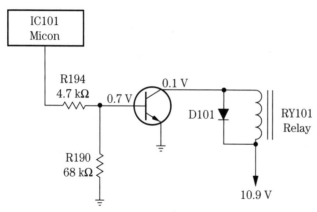

8-30   TVs with a switching relay can have many startup and shutdown service problems.

symptom. A dead TV chassis with a clocklike ticking sound that was heard if the set was left on was caused by a bad filter capacitor (33 μF, 160 V) in the relay driver circuit. Suspect a bad driver relay transistor when the TV cannot be turned off in remote operation. Replace relay driver transistor (Q108) when an Emerson TS2761 TV is dead, the fuse is okay, and there is no relay click.

Suspect a relay driver transistor or memory IC when the relay chatters constantly. A chattering relay may be caused by a defective electrolytic in the voltage supply circuit. Check all electrolytics in the relay circuits with an ESR meter. The clocklike clicking noise can be caused by large filter capacitors in the relay driver circuits. Double-check all silicon and zener diodes when the relay begins to chatter after the TV operates for 5 or 10 minutes.

Check for a badly soldered joint on a low-ohm resistor in the power supply when a TV is dead and the relay begins to make a chattering sound. Suspect electrolytic capacitors in the power supply when the picture shrinks on both sides and the relay chatters. A dead Sanyo DS13530 TV was damaged by lightning, and D002, the relay driver transistor (Q023), and an open trace on the power line to D002 were replaced and repaired.

Troubleshooting SCR switching regulator circuits

In early ac chassis, the silicon-controlled rectifier (SCR) regulator was used to control the dc voltage applied to the horizontal driver and output transistor. The ac input circuits are the same as the typical low-voltage circuits, except an SCR with driver, phase detector, and error amp provide adjustable B+ voltage. In other SCR circuits, the IC regulator with B+ adjust can be found in the derived voltage circuits of the horizontal output transformer.

Often the horizontal circuits must perform before B+ regulation at the B+ voltage source. A primary winding of the flyback is connected after the main filter capacitor and bridge rectifier in series with the SCR switching regulator. B+ can be regulated with the B+ adjust control to set the correct voltage to the horizontal output circuits. Suspect a defective voltage regulator circuit or horizontal output component when the B+ control has no effect in changing the B+ voltage.

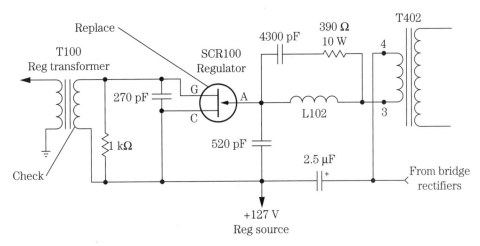

8-31   Early SCR regulator circuits found in the power supply of an RCA CTC121 chassis.

SERVICING SCR REGULATOR CIRCUITS

Check for an open ac fuse (F501). If it is open, check the diodes in the bridge rectifier circuits. Measure the dc voltage across the main filter capacitor (C505). Suspect
SCR100, T100, and associated components if there is no voltage at the output source
(Fig. 8-31). Measure the dc voltage at the anode terminal of SCR100. No voltage can
indicate an open winding of the flyback or bad connections at the PC board. Measure
the resistance of the flyback winding. An open winding or poor connections can prevent 165 V from reaching the anode terminal.

If 165 V is found at the anode terminal of SCR100 and there is no or low gate
voltage (119 V), check the IC regulator, zener diode, and associated circuits. Check
waveforms and voltages on the SCR driver, phase detector, and error amp when attached to the SCR regulator circuits. Check voltage sources from the horizontal output transformer and horizontal circuits when B+ cannot be found at the cathode
terminal of the SCR regulator. Don't overlook the IC power regulator in the horizontal flyback circuits.

DEAD CHASSIS, SCR4101

The chassis was dead in an RCA CTC159 TV, and the voltage on the horizontal output transistor was around 155 V. This meant that the raw dc voltage from the bridge
rectifiers was normal. A voltage check on SCR4101 indicated +167 V at the anode
terminal and no voltage on the collector terminal. The output voltage across C4120
(33 μF) should be around 125 V (Fig. 8-32).

A continuity test was made across the primary winding of T4101 and was normal.
CR4102 and R4115 tested good. On checking SCR4101 in the circuit, no diode test was
found across any two different terminals. SCR4101 was replaced with an ECG5424 universal replacement, and the chassis came alive. Do not overlook CR4104 in the 33-V
source for a dead-TV symptom.

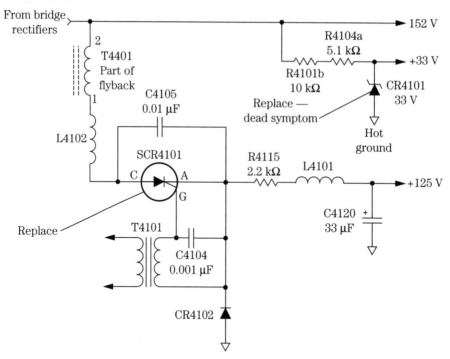

8-32 A defective SCR4101 in an RCA CTC159 chassis caused a dead TV symptom.

TV STANDBY CIRCUITS

TV standby circuits provide a voltage source at all times so that the remote control will operate. Go directly to the remote infrared (IR) circuits when the remote cannot operate or turn the TV chassis on. In some TV chassis, the remote IR pickup system and system control IC provide a standby switch with on/off switching transistors to operate the remote features (Fig. 8-33).

The dead TV chassis with an okay fuse may be the result of bad silicon diodes in the standby power supply. When a TV is turned off and the sound goes off but the picture turns black and white, this may be caused by a zener diode in the standby circuits. Check for open low-ohm resistors within the standby circuits when a TV is dead and the line fuse is open.

The standby power supply is on all the time and provides voltage to the control circuits so that the remote control can turn the TV on. Usually a small standby power supply and bridge circuits provide the required standby voltage. The small power transformer may be fused but is wired directly to the ac power line. The standby voltage provides 5 to 20 V dc to the control circuits.

Suspect the standby circuits when a TV can be turned on manually but not with the remote. Locate the standby transformer and the corresponding parts nearby (Fig. 8-34). Locate the standby transformer fuse and bridge rectifiers. Visually trace the transformer leads to the bridge rectifiers. You may find a combination transistor and zener diode regulation circuit in the output voltage source. Sometimes this transformer and

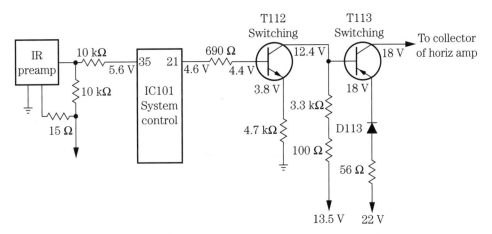

8-33 The IR standby circuit originates within the system control IC101 that controls a standby switch and on/off switch transistors.

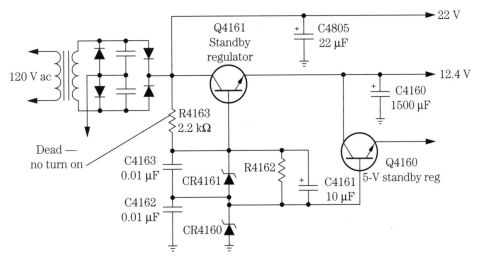

8-34 R4163 had increased from 1200 to 1720 ohms and caused a dead symptom in an RCA CTC167 chassis.

fuse are located on a separate metal chassis. The protection fuse may be in the primary or secondary winding of the transformer.

INTERMITTENT STARTUP AND SHUTDOWN SYMPTOMS

Intermittent startup and shutdown symptoms found in a TV chassis are very difficult to resolve. By monitoring the power supply voltage sources and horizontal waveforms, you can sneak up on the defective component. Monitor the raw dc power line voltage from the line voltage regulator IC.

Suspect a fusible resistor (47 ohms) at one of the pin terminals on an STR30125 regulator for a narrow picture when the TV changes channels or when the TV goes off and on intermittently. A bad flyback can cause the TV to play for days, then intermittently shut off, and maybe not come on again. Check the line voltage regulator for shutdown when the power line voltage is greater than 110 V ac. Do not overlook an intermittent transistor voltage regulator for shutdown symptoms. Check for a bad relay when the TV intermittently shuts off.

Check all electrolytic capacitors within the power supply with an ESR meter for intermittent startup. A bad IC regulator can cause a momentary startup and then shutdown. Suspect an STR30130 IC regulator when the TV shuts down after warmup and the relay clicks off and on. Check for a defective line voltage regulator IC when there is a shaky picture and the TV sometimes turns on with loud volume and intermittently shuts down. Suspect a bad system control IC or microprocessor when the TV starts up intermittently, the channels do not lock in (or sometimes do with the wrong channel number), and the sound is normal.

Small 4.7-µF, 50-V electrolytics with bad ESR problems caused a TV chassis to shut down at once. Check all electrolytics in the power supply with an ESR meter when a TV that is left on begins to chug-chug and sometimes starts up or pulses off and on. When a TV shuts down around 95 V and operates at 90 V ac, suspect a defective STR30130 IC regulator. Check for a badly soldered joint when a TV tries to start up and shuts off.

Replace the voltage transistor regulator and diode in the secondary voltage sources of a TV when the power switch is pushed and the TV appears dead and sometimes comes on later or at once. Do not overlook a defective startup transistor when a TV intermittently starts up or there is no startup. A defective EEPROM IC can cause a TV to come on momentarily, the high voltage to go dead, and then the unit not to start up.

When a TV shuts off intermittently, suspect a defective hold-down or safety capacitor in the horizontal output circuits. Check for a leaky zener diode in the power supply regulator circuits when a TV shuts off and does not shut down. Badly soldered joints on pins 1, 2, 3, and 10 of the flyback can cause a TV chassis to be intermittently dead, not turn on, and after 2 or 3 hours to start up again. A badly soldered joint on R404 and R405 (20 kilohm, 3 W) connected to T400 on the SMPS circuits caused a Ward's 26B1 TV to turn off (not shut down) intermittently and the LED to blink and then slowly fade away.

SWITCH-MODE POWER SUPPLIES

Several different types of SMPSs are found in various TV chassis. In the RCA CTC130 chassis, a variable-frequency switching power supply (VIPUR) is found, and a switch-mode power supply is found in the Sylvania C9 chassis. The ac input transformer circuits are above ground and sometimes are called a *hot ground connection.* A diamond-shaped form represents a common hot ground. The components found on the secondary winding of the transformer use a regular negative ground symbol.

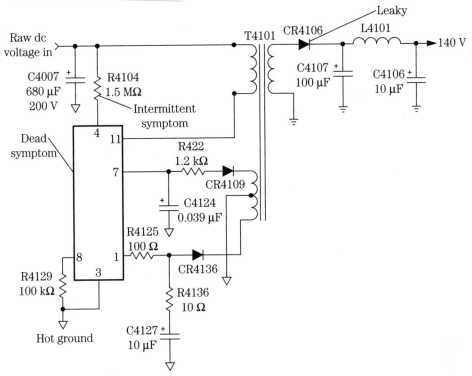

8-35 Always take critical voltage measurements from the capacitor's (C4007) common hot ground in an RCA CTC187 chassis on the hot ground side.

The raw dc voltage from the bridge rectifiers is fed to the primary winding of the power switching transformer (T4101) in the latest RCA CTC187 chassis. In other TV chassis, a heavy duty transistor works in a regulator circuit, whereas the IC component (U4101) serves as the dc regulator. All primary components on the transformer input circuits operate on the hot ground side.

The components found in the secondary of the power transformer (T4101) work in a common ground circuit (Fig. 8-35). Always use the negative terminal of an electrolytic capacitor (C4007) when taking voltage measurements in the primary winding. Likewise, use the common negative ground on the secondary side for correct output voltage sources.

A badly soldered joint on the silicon diode in an SMPS caused intermittent startup with a squealing noise at turnon. A pulsating noise was heard in the sound and was caused by a defective silicon diode in the SMPS. The TV chassis was dead, and the power light came on and stayed on even when the TV was turned off. This was caused by a shorted zener diode in the SMPS.

Check all electrolytics in the power supply circuits and flyback secondary voltage sources when the TV will operate for days, shuts off, and will not come on again. Check the electrolytic filter capacitor with an ESR meter when a TV does not start up or may start up with a thumping noise in the speaker, the sides of the picture are pulled in, and there is very low B+ voltage. Check for leaky diodes, open resistors, bad electrolytic capacitors (47 μF), and a defective output transistor in the SMPS circuits when the chassis is dead.

Check the switch-mode regulator transistor, resistors, and horizontal output transistor for a dead chassis symptom. Test the switch-mode output transistor, and look for a change in resistors or a leaky horizontal output transistor when a TV chassis is dead and the LED display goes on and then goes off after 10 seconds or so. Intermittent startup in a Zenith SE2505 TV was caused by badly soldered joints on Q3410 and C3413 (100 μF, 25 V).

SMPS, RCA CTC140 CHASSIS PROBLEMS

A dead RCA CTC140 chassis can result from a defective regulator (U4100) and Q4100 output transistor. A leaky Q4100 transistor can burn or open resistor R4111. An intermittent turnon and turnoff symptom can be caused by a defective CR4101. A shorted CR4701 can cause an intermittently dead symptom. Resolder all secondary diode terminals for dead or intermittent operation (Fig. 8-36).

TV switching circuits

The latest switching power supply circuits are found in the latest color TVs and TV/VCR combos. A raw dc voltage is fed from a bridge rectifier circuit in the primary winding of transformer T801. Filter capacitor C805 (470 μF, 200 V) filters out the ripple within the dc power supply. IC801 provides the switching action in the primary winding of T801. A hot ground is found on the components that are connected within the primary winding of T801. Use the negative terminal of C805 for all common hot ground voltage measurements (Fig. 8-37).

Separate secondary windings of transformer T801 provide several different voltages to the various TV circuits. A fixed silicon diode with electrolytic filter capacitors rectifies and filters the ac voltage in the secondary sources. Always use the common ground found at any filter electrolytic in the secondary voltage sources when taking critical output voltage measurements. If not, the voltages measured will not be accurate.

DEAD TV, PANASONIC TV/VCR COMBO

A Panasonic PV-M1327 TV/VCR combo came in with a dead chassis. No voltage was found on the silicon diode in the transformer (T1001) secondary circuits. A higher than normal dc voltage (+139.5 V) was found on the collector terminal of the switching transistor (Q1001). Q1001 was found to be open with a quick diode test with a DMM. At the same time, Q1002 tested okay. Replacing switching transistor Q1001 solved the dead TV/VCR chassis (Fig. 8-38).

TV/VCR switching circuits

The TV/VCR switching power supply circuits may contain all the TV and VCR voltage sources or just the TV circuits. A raw dc voltage from the ac bridge circuits is applied to the primary winding of the switching power transformer. Usually a switching transistor and control transistor are found in the bottom leg of the primary winding. All voltages taken from the primary side should be taken from the negative terminal of the main filter capacitor as the common hot ground (Fig. 8-39).

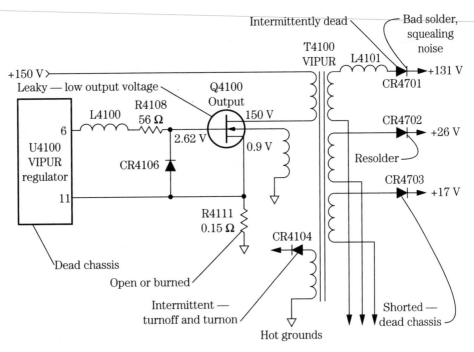

8-36 The various service problems found in an RCA CTC140 chassis SMPS.

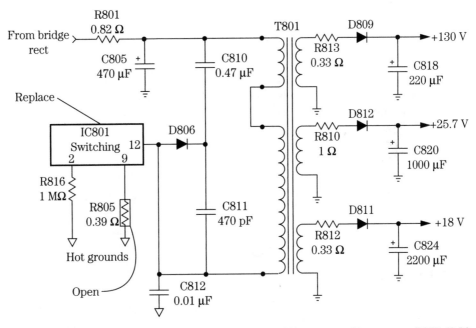

8-37 A dead symptom in a Panasonic CT-31S185/CS TV was caused by an open R805 (0.39-ohm) resistor and a defective IC801.

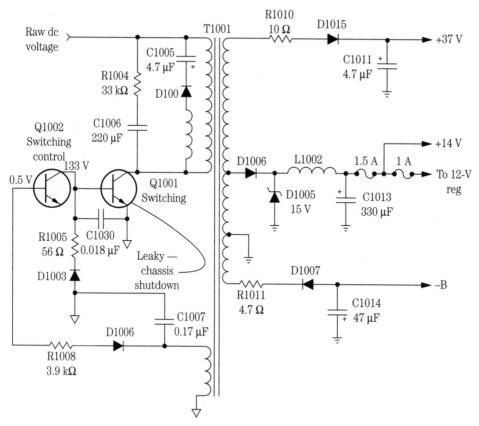

8-38 Leaky switching transistor Q1001 in a Panasonic TV/VCR caused a TV shutdown symptom.

All secondary voltages are taken from the common ground in the secondary winding of the switching transformer. The various voltage sources are rectified by silicon diodes and electrolytic capacitors. You will find zener diodes and transistor voltage regulators in the various circuits. Some TV/VCR chassis have a fuse that protects the various circuits.

TV/VCR SWITCHING PROBLEM, QUASAR

In a Quasar VV1317W model TV/VCR combo, a 1-A fuse was found blown in the secondary winding of the switching transformer (T1001). After replacement, the fuse was blown again. All silicon diodes were quickly checked with the diode tester of a DMM and appeared normal. A quick diode-transistor test on the regulator transistor (Q1051) showed a low reading of 0.17 ohm between the emitter and collector terminals (Fig. 8-40). Replacing the regulator transistor (2SC3852) with an ECG560 universal replacement solved the blown-fuse problem.

Check Table 8-1 on p. 276 for additional power supply problems and solutions.

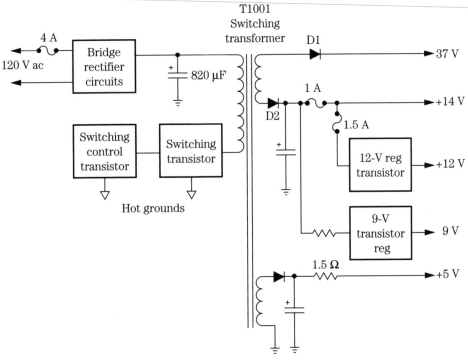

8-39 Take all the primary voltages of T1001 from a common hot ground on the main filter capacitor (820 µF, 200 V).

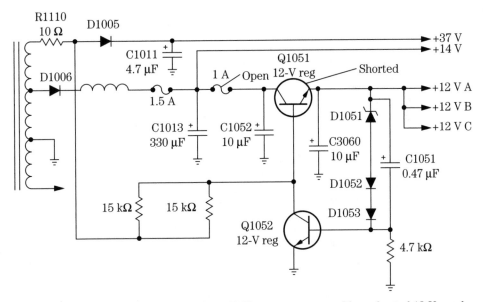

8-40 An open 1-A fuse in the secondary 12-V source was caused by a shorted 12-V regulator transistor (Q1051).

Table 8-1.   Low-voltage troubleshooting.

Symptom	Location	Repair
Dead chassis	Low-voltage power supply	Measure the dc voltage across the large electrolytic capacitor. No voltage—check fuses.
	Diodes	No voltage—check silicon diode with the diode test of the DMM.
	Bridge rectifier	Check each diode inside the bridge component. Test for leakage. Most diodes become shorted or leaky.
	Transformer	Measure the ac voltage across the transformer wire terminals. Check the primary winding for an open winding.
Loud hum in sound	Filter capacitors	Shunt the large filter capacitors. Clip the leads across the terminals with the power off.
Low hum	Decoupling capacitors	Shunt each small electrolytic capacitor in the circuit.
Intermittent voltage	Regulators	Monitor the output source with the DMM. Check for an open or intermittent regulator transistor—it often goes open. Check for an open regulator IC. Check all the wiring connections. Check the zener diodes.
Fuse keeps blowing	Diodes	Check for shorted or leaky silicon diodes. Suspect a leaky filter capacitor. Check for shorted turns in the power transformer.
Overheating transformer	Diodes	Check for one or more shorted rectifiers. Check for a leaky filter capacitor. Test for shorted turns in the transformer. Remove the transformer wire leads and see if the transformer still runs warm. Replace the defective transformer.
Flyback voltage sources:		
Dead	Horizontal circuits	The horizontal circuits must operate before any voltage sources.
No low-voltage source	Low-voltage circuits	Check for an open isolation resistor. Check for a shorted or leaky silicon diode. Shunt the electrolytic capacitor. Check the winding of the flyback to the voltage source.
Really low source	Diodes	Test the diode for openness or leakage. Check the regulated transistor. Check for an overloaded circuit in that particular voltage source.
Hum, smeary, or odd symptoms	Filter capacitors	Shunt the small electrolytic capacitors in the voltage circuits. Check for overload in that voltage source.

9
CHAPTER

Servicing stereo
sound circuits

High-powered stereo sound is now found in large AM/FM/MPX receivers, auto radios, and deluxe TVs. A tabletop high-powered receiver may be joined with a CD player to provide 100 to 500 W of entertainment. A five-shelf system may produce 50 to 100 W of power, whereas the auto receiver might have a high-powered amplifier attached with over 1000 W (Fig. 9-1). Of course, color and projection TVs do not have high-powered sound compared with other consumer electronic products. Many different stereo circuits can be connected to a high-powered surround amplifier and several speakers.

Most high-powered stereo components are laid out in a line to their respective audio channels and are fairly easy to service. When a weak or dead audio channel is found, the defective component can be located with the external amp and scope. Critical voltage measurements in the bad channel can be compared with those in the normal audio channel. The same applies to resistance and semiconductor tests. A quick test of silicon diodes and transistors can be done with the diode tests of a digital multimeter (DMM). Simply take critical signal in and out tests of the suspected integrated circuit (IC) with the external audio amp or oscilloscope. Defective components can be located and replaced without a schematic or manufacturer's diagram.

High-powered receiver circuits

Although some of the early home tabletop receivers may have 5 to 50 W of power, a deluxe home theater AM/FM/MPX stereo receiver has a 100, 300, or 500 W of power. A high-powered AM/FM/MPX receiver may have radio circuits plus CD, DVD, CD changer, cassette, or karaoke circuits. Often, lower-wattage receivers employ a higher-powered IC mounted on a heat sink, whereas high-powered units have many high-powered transistors. The AM/FM/MPX circuits are divided at the FM/MPX circuits into right and left audio channels.

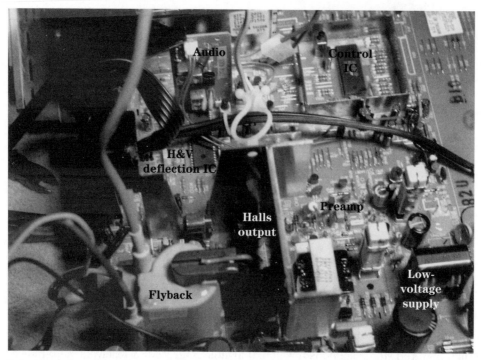

9-1 Try to locate the TV sound circuits on the latest sound IC output components.

When one stereo channel is dead, weak, or distorted, locate the power output ICs or transistors. The stereo signal may be switched manually or electronically by diodes and separate IC components to the two output stereo channels. If the stereo indicator light is not on, you may assume the FM/MPX circuits are not functioning. Check to see if all functions operate, indicating that the trouble lies in the stereo audio circuits.

Locate the defective channel by tracing the speaker connections back to the transistor output or IC component (Fig. 9-2). Most high-powered ICs and transistors are found mounted on metal heat sinks. Take critical voltage and semiconductor tests to isolate the defective component. Excessive distortion was found after warmup in a Sony STRVX22 receiver and was caused by a defective voltage regulator (STR3091).

DISTORTED RIGHT CHANNEL

The right channel in a 200-W Radio Shack 12-1967 stereo amplifier had hum and distorted audio. The right channel was traced back from the speaker terminal board to a coil that shunted a low-ohm resistor. The coil was tied to two very low ohm resistors (0.15 ohm). The bias resistors were connected to Q209 and Q212. Both transistors were checked with the diode tester of a DMM, and Q212 was shorted. Q211, which is coupled directly to the Q2 driver transistor, also was leaky. The rest of the transistors tested normal. All bias resistors were good. Replacing Q211 and Q212 cleared the distorted right channel.

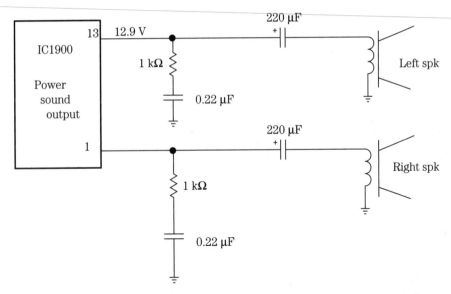

9-2 To locate the power output IC, trace the circuit from the speaker terminals or the outlet jacks to the suspected power output IC.

Auto stereo circuits

Today's auto stereo radios contain practically the same front-end AM/FM/MPX circuits as any other receiver with cassette and CD player circuits. Most of these audio circuits employ silent switching, with several silicon diodes switching in the receiver, cassette player, and CD player to the left and right audio output circuits. For higher output wattage, the auto radio might be connected to a separate 270- to 1000-W power amplifier (Fig. 9-3).

If the unit's stereo function is okay but no stereo bulb or light-emitting diode (LED) will light, suspect a defective bulb, improper bias adjustment, defective indicator light circuit, or improper voltage source. The audio signal can be signal traced right up to the stereo indicator circuits. Signal trace the audio to the input of the indicator IC. Take critical voltage tests on the IC and LEDs. Remember, FM/MPX stereo channels can function with a defective indicator circuit.

When indicator lights stay on all the time, the cause is either the indicator light circuitry or an improper threshold adjustment. Usually the indicator light or bank of LEDs is powered by a transistor or IC. Suspect a defective transistor or IC or an improper voltage source when the stereo light stays on very bright (Fig. 9-4).

EARLY AUTO STEREO CIRCUITS

Early auto radio stereo circuits consisted of two or three transistors as a driver and push-pull power output circuits. Early auto IC circuits may use separate power output ICs or a dual IC component. The stereo circuits from the cassette player and radio circuits are switched manually into the input circuits of the power output radio circuits.

9-3 Remove the printed circuit (PC) board from one end of the metal heat sink to get at the defective output transistor within a Pyle 750-W amplifier.

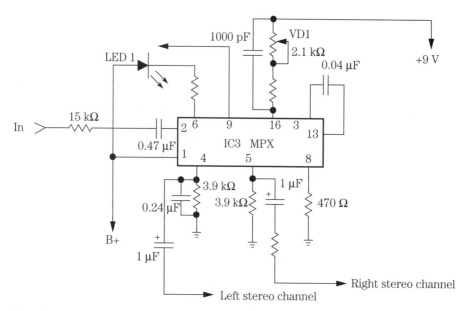

9-4 Suspect the MPX stereo IC when the stereo light does not come on or is on all the time.

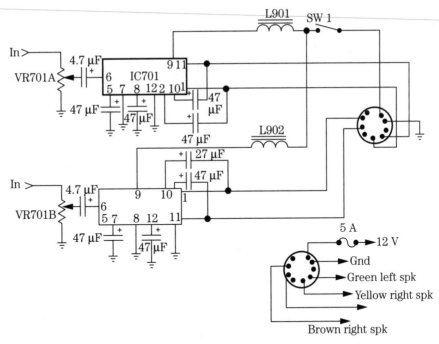

9-5 Early auto receivers had separate power output ICs in each stereo channel.

A dual volume control applied audio through 4.7- to 10-μF electrolytic coupling capacitors to each input terminal. The audio stereo outputs were connected to a socket found at the rear of the metal chassis (Fig. 9-5). After a Chevy 21APB1 car radio was on 10 minutes, the radio would pop and go dead. Replacing the intermittent output transistor solved the problem.

Auto cassette input circuits

Auto cassette player sound circuits may consist of a single IC that includes both tape head input signals. The audio signal is picked off the tape by the stereo tape heads and fed directly into a dual IC. This cassette audio is coupled to a function switch or has fixed diodes in the line to prevent radio audio from entering the tape circuits.

D101 and D102 isolate the tape head stereo sound from the other audio circuits. When a cassette is played in an auto radio/cassette player, the sound is amplified and allowed to pass through two small diodes to the input circuits of the stereo output amplifier. Likewise, when the audio stereo signal is played, the stereo sound cannot enter the tape circuits because D101 and D102 will not let it do so. Fixed diodes are used to switch the two signals, eliminating a radio/tape function switch (Fig. 9-6).

High-powered amps

The power output of a high-powered cassette receiver may range from 16 to 25 $\times$ 4 W with a peak root-mean-square (RMS) power rating of 40 to 50 W. An auto CD

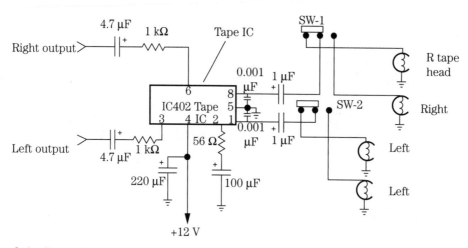

9-6 Check SW-1 and SW-2 when one channel is intermittent in the tape input circuits in an auto reverse cassette player.

player and receiver might run from 17 to 23 × 4 W with a peak RMS power output of 35 to 50 W. The auto receiver may be connected to an extremely high powered amp that has 50 × 2 to 500 × 2 W. A bridged-type amplifier may start at 300 × 1 to 1000 × 1 W or higher. The high-powered amp may drive 10- or 15-in woofers with several midrange speakers and up to seven different tweeter speakers (Fig. 9-7).

The high-powered amp may consist of high-voltage ICs or transistors. In a 250-W amplifier, five power output transistors are found in each stereo channel. All power output transistors and IC components are mounted on a large heat sink, usually the metal-fin body of the amplifier. Each power output transistor can be tested in or out of the circuit. Remove the suspected transistor for accurate leakage testing. Always replace each audio component with the exact replacement part.

REPLACING HIGH-POWERED AMP TRANSISTORS

Always try to replace defective audio output transistors with the original part numbers. These parts are easy to obtain if you are a service warranty station for a certain brand of auto receivers. The original output transistors will fit in the same mounting as those that come with the auto receiver. You know that the originals will work in the output circuit every time. Sometimes the same high-powered output transistors can be obtained from electronics mail-order firms.

If universal power output transistors are used, make sure that the part number is correct. Check them out in a semiconductor manual. Make sure that the transistor leads are the same length and are cut before mounting. When removing a defective output transistor, make sure that a piece of insulation is placed between the new transistor and the metal mounting screws and framework. Apply silicon grease on both sides of the insulator, and mount it correctly before trying to install the new transistor. Align the insulator to the correct mounting holes. Make sure that the output transistor is of the correct length and will mount on the holes of the heat sink.

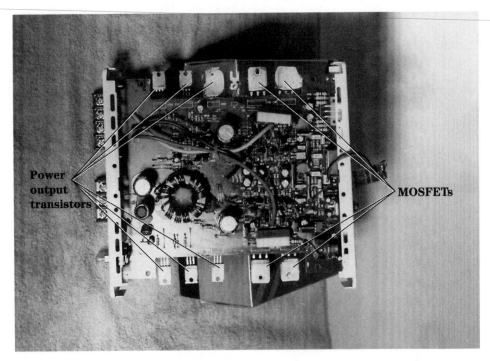

Power output transistors

MOSFETs

9-7 The high-powered amp connected to an auto radio may contain many power output transistors.

Portable cassette and radio circuits

Audio stereo circuits are found in boom boxes, cassette players, AM/FM/MPX receivers, and TVs. They are about the same, except that the higher-power-output units require more electronic components and speakers. Although a boom box may have more audio power and several speakers, the front-end circuits are similar to those of table model radios. Some large players can have up to three speakers in each channel. The function switch provides audio from the AM/FM/MPX radio, cassette player, and CD player.

The audio components are easiest to locate in boom-box stereo circuits (Fig. 9-8). You will find power transistors and IC components bolted to a separate heat sink or the metal chassis. Locate the output components on separate heat sinks with the audiofrequency (AF) audio section nearby. Wherever you find large electrolytic capacitors and power output capacitors, the audio section is close by.

Portable audio stereo circuits may consist of transistors or ICs or both. Muting transistors can be found at the line output or audio output to the volume control. A right and left meter system may be amplified by one or two transistors from the muting system. The radio, tape, or line output signals are switched into the circuit and amplified by a driver transistor or ICs. Often, a dual audio IC is found for the driver of both audio channels.

In low-powered portable stereo units, a power IC component can be found between the volume control and the speakers. A tone and balance control may be found before

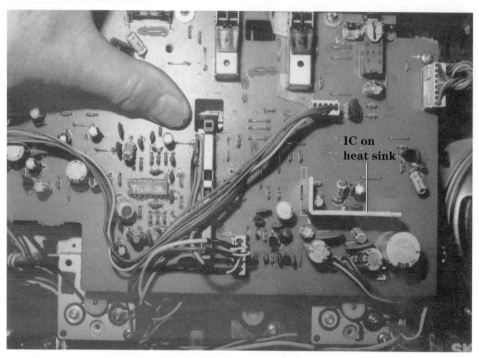

9-8 Locate the output transistors or ICs mounted on a metal heat sink in a boom-box radio/cassette player.

the volume control. The dual volume control is found with separate balance controls. Separate volume controls can be used in some audio circuits. Capacity coupling is found between the volume control and the power IC (Fig. 9-9).

The audio signal is amplified by IC3 in each channel and coupled through a 220-μF capacitor to the speakers. Most problems in the stereo channels results from a defective IC, a worn volume control, or open coupling capacitors. Weak, intermittent, or no sound can be caused by a defective electrolytic coupling capacitor. Suspect IC3 when either channel is distorted, weak, or dead. Both channels may be affected by a leaky output IC3. Sometimes only one channel is defective and the other normal. The suspected IC must be replaced in either case. Do not overlook an improper voltage supply source. Low supply voltage can result from a leaky IC3. Replace IC3 if it is running red hot.

CD stereo circuits

CD audio circuits began at the digital-to-analog (D/A) converter IC, are fed to an IC amplifier, and then are fed into line output jacks that are capacity coupled to another power amp IC and into a stereo headphone jack. The audio output circuits are switched into the regular audio stereo circuits in a boom-box cassette player, an auto receiver, and a CD player, with or without headphone reception (Fig. 9-10). By rotating or pushing the function switch to CD, the audio portion is switched to the internal audio circuits.

The audio signal can be traced from the D/A converter IC to the function switch or headphones. Use an external audio amp to signal trace each audio channel at the D/A converter. The audio signal can be traced with the scope with a CD playing. Rotate the function switch to the AM/FM/MPX radio or cassette player to determine if the CD circuits are defective and the audio circuits are normal.

Check the signal at the output of the D/A converter, input sample and hold (S/H) IC, and preamp stages through the low-pass filter network and preamp audio circuit (Fig. 9-11). When the music stops, check the IC voltage on each terminal.

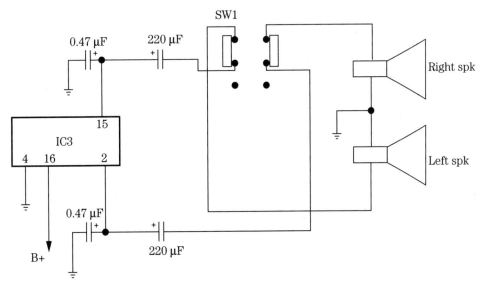

9-9   A dual output IC may be found in the stereo circuits of a boom-box cassette player.

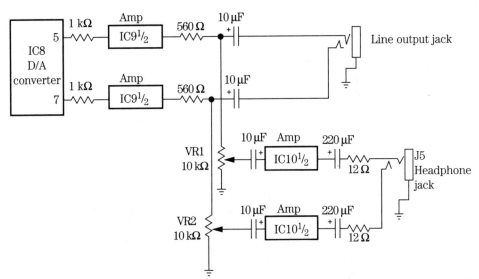

9-10   A portable CD player may contain separate IC components within the output jacks and headphone circuits.

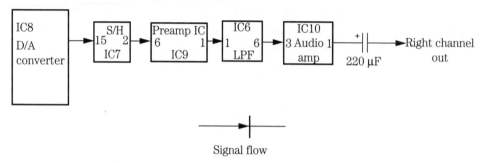

Signal flow

9-11 A block diagram showing how the signal flows in the right channel of a CD player.

Leaky audio ICs can run warm and have a low supply voltage. Remember, the CD signal is separated at the D/A converter into left and right audio channels. Muting is accomplished within the line output jacks and controlled by a control processor (Fig. 9-12).

No muting was found during the program mode in a Sony CDPC70 CD player. No control signal was found at the line output jacks. Replacing the mechanism microprocessor IC (8-759-971-410) solved the muting control problem.

Stereo cassette decks

An auto reverse cassette player might have two pickup heads and two different fly-wheels in a reverse-play audio system. When the cassette reaches the end of the tape, the player will automatically reverse and play the other track in reverse order. Often the tape head will be packed with excessive tape oxide, and sometimes one or two tape head wires are torn off during operation, with no sound from the channel in reverse order. Double-check the tape heads for packed tape oxide, and clean with alcohol and a cloth. Check each tape head for a broken wire right at the tape head terminal.

The stereo circuits found in cassette decks or players function around two tape heads with audio in the left and right channels. Two tape head windings are found in one tape head to pick up recorded audio to both channels. These magnetic tape heads are fed to identical preamp circuits that amplify the weak signal. This signal is then switched to high-power audio circuits (Fig. 9-13).

A dirty tape head can cause weak or distorted sound or a dead channel. Sometimes the head gap becomes packed with excess tape oxide, creating poor pickup to the transistor or IC preamp circuits. Look for broken head wires when one channel is intermittent or dead. Loss of high frequency can be caused by a worn tape head. Cross-talk is caused by poor adjustment of the tape head (Fig. 9-14).

Isolate the preamp circuits from the tape head to the function switch. Signal trace the audio with an external amp from the tape head winding to the base of the transistor or the input terminal of the IC. Remember, the signal directly off the tape head is very weak. The audio should be amplified after the preamp and AF circuits. With a 1-kHz test cassette, signal trace with a scope through the entire audio circuit.

A loud rushing noise in the right channel was heard in a Soundesign CD301A cassette player with the volume wide open. On examination, the red wire was found to be

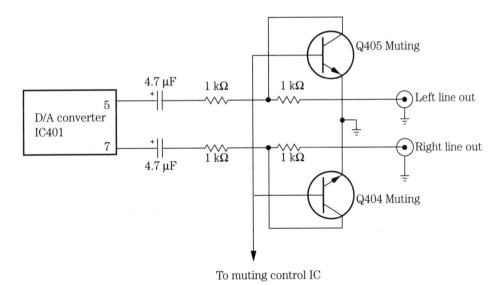

9-12 Suspect muting transistor Q404 when no sound is heard in the right line output jack.

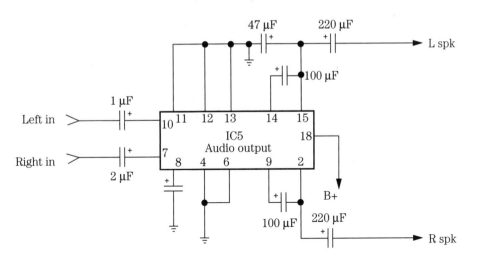

9-13 Check IC5 when the left channel is dead and the right channel is distorted in an AM/FM/MPX cassette deck.

broken off the tape head. Resoldering the red wire to the tape head solved the loud rushing noise.

Hot output transistors

Too-hot-to-touch transistors or IC components can point to a defective audio output circuit. Transistors and ICs that run red hot should be replaced. These power output

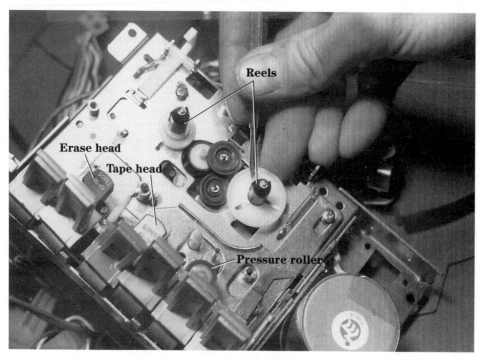

Reels

Erase head

Tape head

Pressure roller

9-14 Clean the tape heads when one channel is noisy and the other distorted.

components can run warm but not red hot. Sometimes a hot IC power output will turn a gray color and may have a piece of plastic popped out of the outside case area. When a hot output transistor is found, test the driver transistor for a leaky or open condition. Make sure that the power output emitter resistors are of the same value and not burned or open. Take critical resistance and voltage measurements, and compare them with the normal channel.

A 25 × 4-W audio amplifier may have a high-powered IC with four different sound output channels to four different speakers. Usually the amplifier feeds both the left and right front speakers and the left and right rear speakers. Here, two 1000-μF electrolytic capacitors couple the ground speaker returns to each set of speakers. A defective IC10 may result in front left channel intermittence and right front channel distortion (Fig. 9-15).

It is possible in balanced audio output circuits that no electrolytic coupling capacitors are found between the speakers and the amplifier. When one of the audio output transistors becomes red hot or appears leaky, this also will damage the speakers. The normal balanced audio output circuit has zero voltage at the speaker terminals. If a speaker is found to be mushy with a frozen cone, suspect that some type of voltage has been applied to the speaker or that too much high-powered volume has destroyed the speaker voice coil. Quickly measure the dc voltage on the speaker terminals. Do not connect another speaker until the audio amp has been repaired.

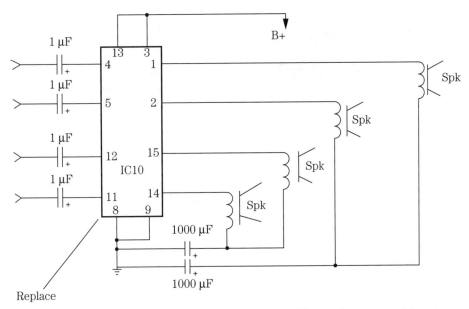

1 µF
1 µF
1 µF
1 µF

B+

13 3
4 1
5 2
12 15
11 14
8 9

IC10

Spk
Spk
Spk
Spk

1000 µF

1000 µF

Replace

9-15 The front left speaker was distorted and the rear right speaker was weak in a stereo CD and cassette player and receiver.

Dead right channel

When either the left or the right channel appears dead and the other channel is normal, signal trace the dead channel with the external audio amp and compare the same signal with the normal channel. If the dead right channel is in a cassette player, insert a cassette and trace the signal with the external amp or scope as an indicator. Slip a 1-kHz test cassette into the player, and take stage-by-stage signal checks.

Often the dead channel is caused by a leaky or open transistor or IC, bad electrolytic coupling capacitors or power resistors, or low voltage from the power supply. On large amplifiers, insert a 1-kHz audio generator signal at the input, and trace the signal with the scope. Always keep a speaker connected as a load or a 10-ohm, 10-W resistor on the speaker terminals while servicing the defective amplifier.

In high-powered audio circuits, five or more high-powered transistors are found in each stereo output channel. When a transistor goes open or becomes leaky in directly coupled circuits, the power output transistors may be damaged. A change in resistance can upset the bias on a driver transistor. The leaky bias diode can cause transistors to overheat and be destroyed. Check for burned or leaky zener diodes in the circuit that will upset the whole output amplifier. Suspect a change in resistance, a leaky diode, or an open or shorted driver transistor when the output transistors run red hot and repeatedly become damaged.

When the signal quits, take critical voltage measurements and inspect the parts found around the high-power transistors and ICs. Test the components in-circuit, and when one is found to be leaky or open, remove it and test it again. Look for shorted or

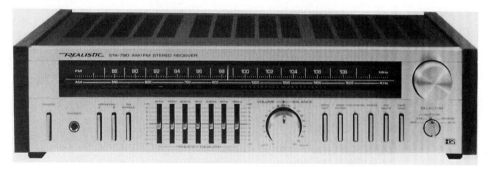

9-16 The right channel was dead in a Radio Shack AM/FM stereo receiver.

open capacitors in the IC power output circuits. Leaky or open electrolytic capacitors in power output circuits can produce a dead channel.

Look for open fuses and burned or open bias resistors in the audio circuits (Fig. 9-16). Check all the transistors in-circuit when burned resistors are found. Remove the suspected transistors, and check the bias resistors while they are out of the circuit. Replace resistors with the exact resistor found in the good channel. Look for power output transistors on heavy heat sinks.

The right channel was dead in a Sharp RT3388A cassette player; no volume units (VU) meter movement was noted. Signal tracing the right channel with a cassette inserted and then signal tracing each stage located the problem after IC4 was removed.

HOT OUTPUT IC, MEDALLION 63-030

The 3-A fuse was blown in a Medallion 63-030 deluxe auto receiver. The fuse kept blowing after replacement. A resistance measurement across main filter capacitor C321 (1000 μF) showed a leakage of 0.015 ohm. This 13.6 V dc is fed directly to the power output ICs. IC4 showed signs of overheating. Cutting loose the voltage trace to pin 1 of the IC4 (TA7205P) AF amp cleared up the shorted power supply. IC4 was replaced with an ECG1155 universal replacement, and this solved the hot output IC symptom (Fig. 9-17).

Weak left channel

Sometimes the defective component that may be causing a weak symptom is much more difficult to locate. Weak reception can be caused by dried-up coupling capacitors, open emitter bypass capacitors, bad AF and driver transistors, and a dirty tape head. One side of the tape head in a cassette player can cause a weak sound and a low hum. Leaky 4.7- to 10-μF coupling electrolytics or a low-voltage source can cause a weak audio stage. A weak left channel can result from a defective dual power output IC (Fig. 9-18).

A low rushing noise in the left channel with weak audio can be caused by a power output IC. Check all output transistors for weak sound when they are too warm to

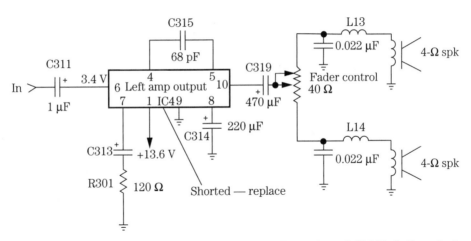

9-17 The hot power output IC4 in the right channel was shorted (0.015 ohm) at pin 1 and kept blowing the fuse.

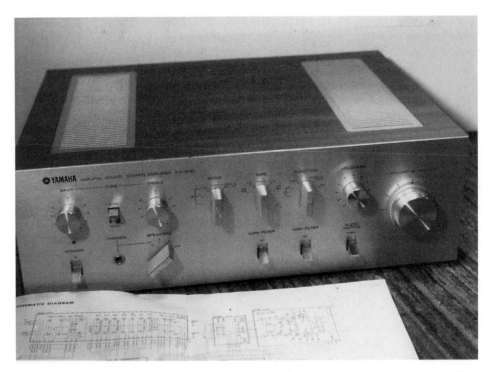

9-18 Check all electrolytic capacitors within the high-powered transistor stereo amplifier with an ESR meter.

touch. Suspect a change in emitter resistors for weak audio. Compare the normal voltage source with that feeding the weak left channel. A bad earphone jack can cause weak and intermittent sound. The weak channel can be caused by a grounded tone or volume control in the left channel. Low volume was noticed in an Onkyo TA2028 model and was caused by bad transistors Q203 and Q203 (2SC1384).

Intermittent stereo channel

Are both channels intermittent? Monitor the audio at the volume control with the external audio amp, and use a speaker tied to the output terminals of the intermittent channel. If the audio is intermittent with the external power amplifier, you know that the defective component is in the audio input circuits. If the audio is normal at the volume control and intermittent at the speaker, you know that the intermittent component is in the audio output circuits. If the sound is intermittent at the volume control, the defective part is in the audio input circuit. You can use the radio circuits or the cassette signals to troubleshoot an intermittent channel (Fig. 9-19).

If the intermittent component is in the output circuits, use a regular speaker or an external amp as a monitor. Sometimes when a test instrument is touched on one element of a transistor, the sound will return. Check the voltages on the transistors and ICs when there is intermittent sound. Applying coolant and heat to transistors and capacitors can make the part act up. Probe around small components with an insulated tool to find poorly soldered terminals or parts. Pushing up and down on sections of the PC board can cause the chassis to become intermittent. Resolder the intermittent PC board wiring section to locate a badly soldered joint. Don't overlook large blobs of solder on wiring junctions. Check for a worn or broken volume control.

Check low-value coupling capacitors (1 to 10 µF) for intermittent audio. Poor PC board connections and cracked traces can cause intermittent sound. Check for broken

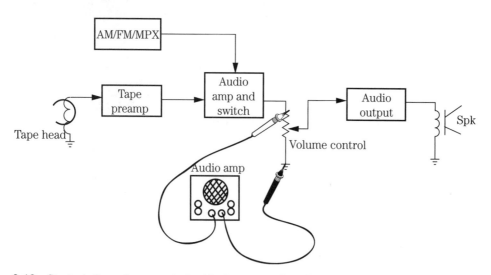

9-19 Start at the volume control with the external audio amp to isolate half the audio channel in cassette and receiver circuits.

traces and PC wiring with an equivalent series resistance (ESR) meter. Do not overlook preset volume or bias controls for intermittent audio. A dirty function switch can cause intermittent sound. A bad voice coil with broken connections to the speaker can cause intermittent audio. Double-check heat sink screws for intermittence and a crackling noise in the speakers. Replace a bad volume control when the audio cuts up and down as the control is rotated. A badly soldered connection on the volume control can result in intermittent sound. Check the speaker relay when one channel is intermittent.

Most intermittence problems within audio output circuits are the result of bad coupling capacitors, AF or driver transistors, and power output transistors or IC components. Check the driver transistors and soldered terminal connections for an intermittent audio channel. Power output transistors can become overheated and cut the sound in and out. Double-check all power output terminals for properly soldered joints.

Intermittent audio can result from a defective power output IC. Check the power output IC if it appears quite warm after operating 10 to 25 minutes. Resolder all the IC terminals. Check the sound at the input and output terminals before removing the suspected IC. Double-check the IC mounting screws for a loose mounting, which can cause intermittence and a noisy sound in the speakers. Sometimes the entire audio PC board must be resoldered for intermittent sound.

Intermittent audio can be caused by an intermittent voltage source feeding the audio circuits. When a certain stage is found to be intermittent, monitor the voltage source to that stage with a DMM. Notice when the sound cuts up and down if the voltage changes at the same time. A change in the voltage source can be caused by a defective voltage or IC regulator. Resolder all the regulator terminals. An intermittent diode or a poorly soldered diode connection in the regulator circuit can cause intermittent audio. Do not overlook the possibility of a bad protection relay, which can cause intermittent sound applied to the speaker.

INTERMITTENT LEFT CHANNEL

The left channel in a Radio Shack 200-W amplifier was intermittent after operating for an hour or so. The right channel was normal. An external audio amp was connected to each output transistor, and when the amplifier acted up, the external amp was connected to the base of the next directly coupled transistor. When the external amp was connected to the base terminal of Q110, the audio cut out. Right away the external amp was connected to the base terminal of Q108 and appeared normal. Replacing power amp Q108 cured the intermittent left channel (Fig. 9-20).

Noisy left channel

Try to locate the noisy channel by signal tracing with the external amp or scope. Start at the volume control. If the noise is heard at the volume control, proceed through the input and preamp circuits until the noisy stage is located. If you can eliminate the noise by turning down the volume control, the noise is in the input circuits. Go from base to base of each preamp or AF transistor until the noise can be heard no longer. If the noise is heard on the base terminal and not the collector terminal of a suspected transistor, you have located the noisy stage (Fig. 9-21).

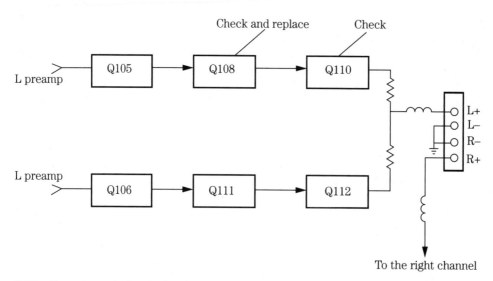

9-20 Do not overlook a leaky driver transistor when there is intermittent audio and a shorted Q110 and open Q112 output transistor.

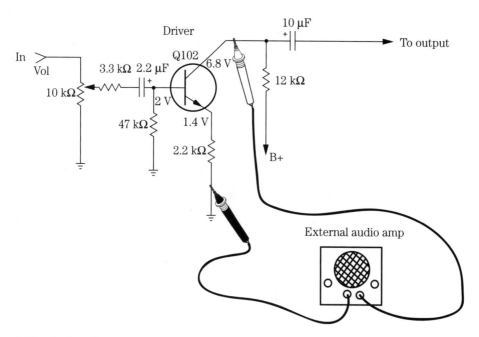

9-21 Go from base to collector terminal with the external audio amp to locate the noisy channel.

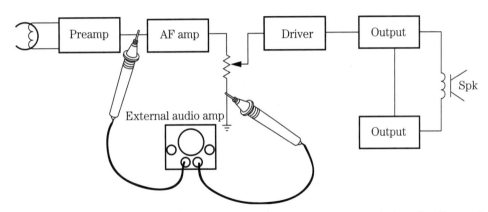

9-22 Keep the volume low on the external audio amp as you try to isolate the distorted audio stage.

Most noisy problems are caused by defective transistors, ICs, diodes, or capacitors. Replace the audio output transistors for a crackling sound after warmup. A popping noise heard in the speaker can be caused by a defective output transistor or IC. No audio and a popping noise can be caused by a hot power IC. A squealing noise may result from a defective electrolytic in the audio circuits. Tighten the heat sink screws when there is intermittent audio and a cracking sound. Static in the sound can be caused by a defective power output IC. Audio buzzing can be caused by a bad zener diode in the power source that feeds the audio channels. Resolder all connections on the power output IC for a noisy stage. Replace the defective output transistor when there is a low motor-boating noise with the volume turned down (Fig. 9-22).

Replace a 1000-μF electrolytic when there are motoring-boating and chirping noises. A motor-boating sound in the speakers can result from electrolytic coupling capacitors (1 to 10 μF) or larger filter capacitors (470 μF). Open decoupling filter capacitors can cause a chirping noise with a motor-boating sound. You may have to replace more than one filter capacitor when there is a motor-boating sound. Check all filter capacitors with an ESR meter.

An open filter capacitor can cause a squeaking noise in the speaker. A loud whistling noise can result from a bad filter capacitor. Suspect a defective filter capacitor when a loud popping noise is heard with the volume turned down. A bad filter capacitor can cause a high-pitched noise that disappears when the volume is turned up. Check the PC board for loose screws or solder, which can cause excessive hum and a buzzing noise. A bad PC board ground can result in a static-type noise in the speaker. A Yamaha R2000 receiver gave out a noisy and cracking sound. Replace a 4.7-ohm, ¼-W resistor near the 4700-μF filter capacitor causing the noisy symptom.

NOISY SOUND, ONKYO

After an Onkyo TX-V940 receiver operated for a few hours, a noisy left channel developed. Rotating the volume control clear down had no effect on the noisy audio. The noisy component was located by signal tracing the audio output circuits with an external audio amp. A noisy driver transistor (Q501) was found in the input circuits.

A bad function switch can cause a hissing sound in one channel. Replace the dual output IC when the left channel is noisy and the right channel has a firing noise after 5 minutes of operation. Replace a defective power switch when there is a constant arcing sound after the unit has been turned on. Sometimes you may find more than one defective component that can cause a noisy sound with the volume turned down, such as a defective 4.7-μF, 25-V electrolytic, silicon diode, or driver transistor.

Distorted sound

Although most distorted audio is found within the output circuits, audio distortion can occur in any audio stage. Signal trace the distortion within the stereo amplifier with an external audio amp or with a cassette and scope within the cassette deck. A low distorted channel can be caused by bad audio output transistors, burned bias resistors, and leaky bias diodes. Replace the defective driver transistor for low volume and distortion in the speaker. Always check each bias resistor for correct resistance after finding a leaky or shorted transistor. Check all low-ohm resistors in the audio circuits for a change in resistance. Suspect open or dirty variable bias controls for low distorted sound (Fig. 9-23).

Replace the audio driver transistor or IC if distortion occurs after warmup. Sometimes the driver and both output transistors must be replaced to solve a distorted audio problem. Spray coolant on the suspected AF or driver transistor to clear up or make the channel more distorted. Replace both audio output transistors for weak and distorted

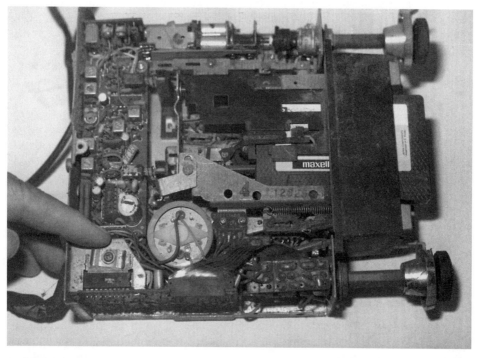

9-23 Look for a small heat sink and defective power stereo IC within an auto receiver that has distorted sound.

sound. Suspect a driver transistor when the volume cuts out or becomes weak or distorted. You may find one output transistor open and the other leaky. Quickly check all transistors in the audio circuits with the diode tester of a DMM.

Check for leaky coupling capacitors for a distorted AF or driver stage. Try to adjust the bias control when the left channel becomes distorted at high volume. Check for a change in or open large-load resistors within the driver or AF audio circuits. Remove one end of the suspected resistor, and take an accurate measurement. Do not overlook a dirty or worn function switch for distorted or garbled sound. A distorted left channel might be caused by a defective audio output transformer.

Excessive distortion always occurs in the audio output circuits. Test each power output transistor and IC for extreme distortion. An open coupling capacitor tied to the volume control can cause weak and distorted audio. Go directly to the high-powered output ICs for extreme distortion. Replace the suspected output IC when distortion occurs after warmup. A defective power output IC can cause extreme distortion. Replace the power output IC when extreme distortion is found along with a popping sound.

Replace both output ICs when extreme distortion is found in both channels. Look for something in common with both channels when distortion is found in the left and right output speakers, such as a dual power output IC, improper voltage source, or defective power supply. In high-powered amplifiers, if a positive or negative voltage is missing, extreme distortion results. Replace the defective regulator transistor when distortion occurs after several hours of operation. A weak and distorted channel can be caused by defective IC outputs and a missing negative voltage. A bad voltage regulator can cause garbled audio in the audio stages. Suspect a defective driver IC and voltage regulator when there is distorted audio after warmup.

Check for leaky diodes and dried-up or open electrolytic capacitors within the voltage sources that feed the high-powered transistors. An open 470-μF electrolytic on the 14-V line can cause distortion. Check all electrolytic capacitors within the power supply for distortion with an ESR meter (Fig. 9-24). When the audio output IC keeps shorting out with extreme distortion, replace all electrolytics in that voltage source. Replace R1904, IC1900, and U1001 for extreme distortion in an RCA CTC159 TV chassis.

High-powered signal path

Often, power output transistors within the power output circuits are directly coupled to one another or with a resistor in between. Notice that the signal path is found at the base of Q101 and that one-half of the Darlington transistor amplifies the audio signal and directly couples to the base terminal of Q103 (Fig. 9-25). The output audio from the collector of Q103 is fed through a resistor to the base of Q105. Also, the audio signal is fed to the emitter of Q107, with the collector fed through a resistor to the base of Q106.

Q105, Q106, Q108, and Q109 are in a series push-pull amplifier circuit. Q105 and Q106 operate in a push-pull audio circuit and feed the output signal from the emitter terminals to the base of Q108 and Q109. Again, Q108 and Q109 are in push-pull operation. The high-powered audio output signal is taken from the emitter resistors of both Q108 and Q109 through a coil-resistor network and speaker relay switch.

9-24   Check all electrolytics in the output audio circuits with an ESR meter.

Check the audio signal at the base of each transistor for audio gain with an external audio amp or scope probe. Notice the gain or loss of each stage. The signal at each base can be compared with the same circuit of the normal channel. A loss of signal or distortion can be signal traced to a defective transistor circuit. When signal or distortion is noted on the base of a transistor, check the preceding circuits. You may find an open driver or leaky AF transistor producing weak and distorted sound.

Checking defective stereo circuits with an external amp

An external audio amp is a very useful test instrument when monitoring audio intermittent, weak sound, noisy circuits, or a dead stereo channel. Always keep the volume on the external amp as low as possible. Plug the defective amplifier into a variable-isolation power transformer. You can go from stage to stage to find the defective component. Then compare the signal at the defective stage with the normal stereo channel. Clip the amplifier ground lead to the common ground on the defective amplifier, and start signal tracing at the volume control. If the noise or distortion can be controlled with the volume control of the defective amplifier, signal trace the audio amp output circuits with the external audio amp.

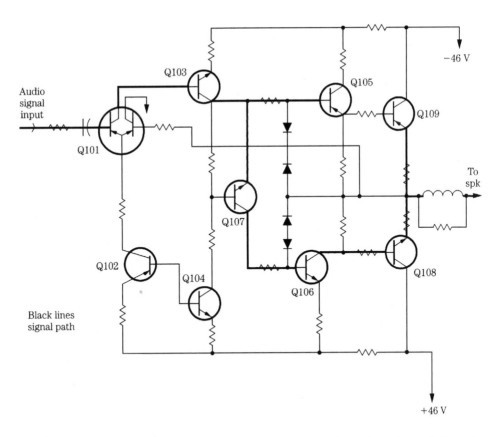

9-25 Follow the audio output signal from base to base of each transistor through the high-powered audio stereo channel.

Hum in the sound

A loud hum in both speakers can be caused by open filter capacitors or broken foil around PC board leads. Check the large filter capacitors if they appear leaky. Suspect a decoupling capacitor when a hum appears after a few minutes of operation. Shunt these capacitors with the power off so as not to destroy solid-state components. Turn the power on, and listen for a hum in either channel.

Hum in one channel can be caused by a leaky audio output IC. Open electrolytics in the voltage source of the power output IC can cause extreme hum in the audio. Suspect a defective power output IC when there is hum in the speaker when the power switch is turned off.

An open volume control can cause intermittent audio and a low hum in the speakers. Low hum can be caused by poorly soldered PC boards. Check for open low-ohm resistors and leaky diodes when there is hum in the speakers. Check for an open resistor tied to the volume control when there is a variable hum noise. Low-level hum can be caused by a leaky diode. A bad relay can cause hum and an intermittent clicking noise

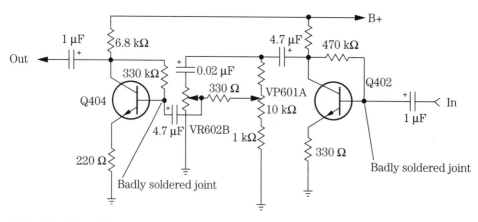

9-26 Resolder all terminals on the driver transistor for a dead channel and hum in the speaker.

in the audio. Red hot resistors in the power output circuits can cause hum in the sound. Replace resistors R501, R502, R503, and R317 on pins 10 and 12 of IC301 for hum in the sound of a Fisher CA226 receiver amplifier.

Replace a 220-µF, 10-V and a 10-µF, 160-V electrolytic in the voltage source for hum in the sound. Low hum can be caused by a defective electrolytic in the voltage sources. A slight hum can occur with a lower voltage source caused by a large filter capacitor (6800 µF). Replace large filter capacitors when there is hum and a buzzing noise in the sound.

A dead channel with hum in the speaker can result from a badly soldered joint on a driver transistor (Fig. 9-26). A leaky driver transistor can cause a loud hum in the audio. Replace both output transistors when there is low hum in the left channel. Suspect defective bias resistors in the emitter circuit of the output transistor when hum occurs after the amplifier is on for 5 or 10 minutes.

TV sound circuits

In early stereo and low-priced TV audio circuits, output transistors were found in each stereo channel. The AF driver transistor (Q101) is directly coupled to Q104 and through a low-ohm resistor (680 ohms) to the base of Q102. The sound is directly coupled from the collector of Q102 to the base terminal of Q103. Q103 and Q104 operate in a simple push-pull audio circuit. A low +13-V source feeds the driver and output circuits of Q103 and Q104 (Fig. 9-27).

A simple power output IC may contain a dual audio output stage. The whole audio circuit may be found in one large IC component. Check the dual output IC when both stages are affected with distortion, intermittent audio, noise, or a dead stereo channel and the other channel is okay. Often, when a dual audio output IC becomes defective, both channels are distorted, weak, or intermittent. Stereo sound within smaller TV screens can employ a dual or single output IC for each channel. Replacing the dual output IC can solve problems within both output circuits (Fig. 9-28).

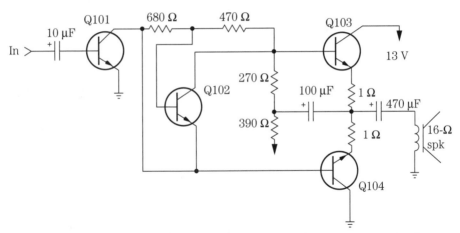

9-27 Early low-priced TVs used small transistors within the stereo output circuits.

Power output IC

9-28 The audio circuits within a low-priced TV/VCR combo have a dual output IC.

Deluxe TV stereo channels may include a stereo decoder IC, an audio switch, a buffer transistor, a mode switch selector IC, a TVB, audio preamps, and separate high-powered output ICs. The stereo audio is developed within the stereo decoder IC and is capacity coupled to an IC audio switch (U1410). A separate buffer transistor is found between the audio switch IC and is capacity coupled to the mode selector switch

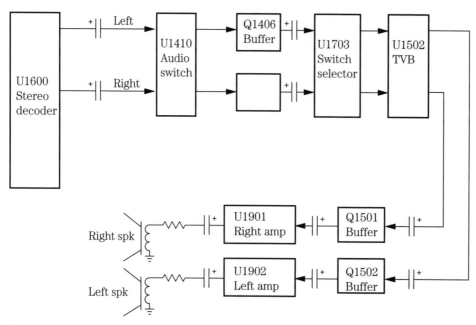

9-29   Block diagram of present-day deluxe TV stereo audio output circuits.

(U1903). Again, two buffer transistors are capacity coupled to the separate power output ICs from U1502. An 8- or 16-ohm permanent magnet (PM) speaker is capacity coupled to the left (U1902) and right (U1901) power IC amplifiers (Fig. 9-29).

A defective stereo decoder (U1710) can cause a dead channel or dead audio in both channels. Check the supply voltage (11.5 V) at pin 1 of U1710 in an RCA CTC167 stereo decoder circuit. Low voltages may indicate that U1710 has become leaky, reducing the supply voltage. The supply voltage may be low from the power source, causing weak and distorted sound. Simply remove terminal 1 from the PC wiring trace with a solder wick and iron. Flick terminal 1 with a small screwdriver or pencil to make sure that the pin terminal is loose. Now measure the supply voltage at the removed pin trace terminal. Suspect a leaky U1710 when the normal voltage returns or the voltage is a volt or two higher.

Go directly to the voltage source in the power supply if the voltage is still extremely low (Fig. 9-30). A defective U1710 has been known to cause a dead right channel, whereas a defective left Summer channel transistor (Q1703) has resulted in a weak channel. Do not overlook a surface-mounted solid component that has a poorly soldered end connection that ties the voltage circuits together at the IC component.

TV IC power output components

The latest stereo power output circuits in TV chassis are usually found with IC parts. A dual-purpose power output IC is found in lower-priced models, whereas more powerful and separate ICs are found in the latest TV sound circuits. The audio input signal is capacity coupled to the input terminal of each separate power output IC. Likewise, the output terminal is capacity coupled to an 8- or 16-ohm PM speaker. In

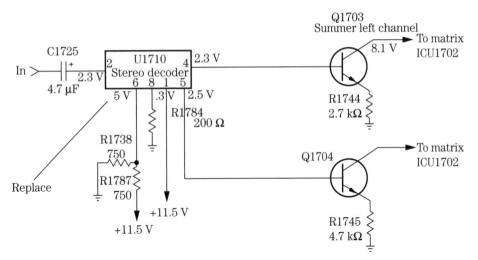

9-30 Leaky stereo decoder (U1710) in an RCA CTC167 chassis caused a weak stereo output signal.

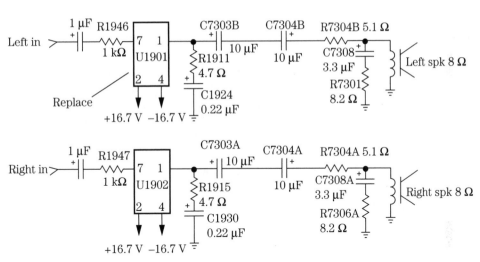

9-31 Notice that a positive and negative voltage are applied to the output U1901 and U1902 ICs in the latest stereo channels.

some high-powered IC circuits, separate positive and negative supply voltages are found (Fig. 9-31).

Signal trace the weak, intermittent, or distorted signal into and out of the power IC to locate the defective channel. Then signal trace the audio signal into each input terminal. When no schematic is available, apply the signal tracer indicator probe to each terminal until the right one is found. The output terminals can be located by tracing from the defective speaker connection. Now compare the same terminals with the normal audio output IC. Another method is to inject the audio signal into the normal audio output IC to find the input and output terminals and then to check for the same amount of audio at the defective IC.

A defective power output IC can run warm. A defective output IC might cause a noisy and crackling sound. No audio may result from a badly soldered connection or a broken trace wiring. Both audio speakers can be dead when there is a bad dual output IC. Sometimes prodding around on the noisy output IC can make the IC act up or quit. Replace the defective IC when a frying noise occurs after 5 or 10 minutes of operation.

Suspect poorly soldered joints on the power output IC terminals when the audio cuts out in both channels. The defective power IC can be okay for several minutes, and then shuts off; then only a hum is heard. A popping left channel with distorted sound can be caused by a bad output IC. You may have to replace both power output ICs when distortion is found in both channels.

The dead or distorted audio channel can be caused by missing a positive or negative voltage on the power IC. Check for a defective voltage regulator IC when the audio becomes distorted after warmup. Check for a leaky voltage regulator transistor when improper voltage is fed to the power output IC. An open large filter capacitor can cause distortion in the voltage source and in the power IC circuits.

NO SOUND, RCA CTC187 CHASSIS

A dead and distorted audio section within an RCA CTC187 TV chassis can be caused by several different audio components. The normal audio was traced, and a defective U1601 IC and R1601 (470-ohm) resistor and an open R1618 (33-kilohm) resistor were found in the 12-V source of the power supply. In another RCA CTC187 chassis, U1601 and Q1001 were found shorted with an open R1627 (200-ohm) resistor in the audio output circuits. Also check for a leaky C1617 (4.7 µF) within the same chassis (Fig. 9-32).

Speaker relay problems

A bad relay can cause the sound to cut in and out at low volume. Suspect a bad speaker relay when the volume cuts in and out at any level of audio. A static noise in

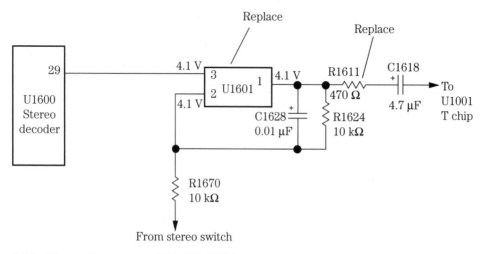

9-32 Distorted sound in an RCA CTC187 TV stereo chassis was caused by a defective U1601 and R1611 (470 ohms).

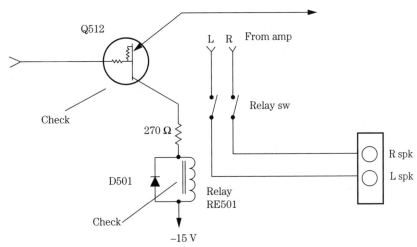

9-33 Check the solenoid driver transistor and solenoid switch for intermittent sound in the speaker.

the speaker with one channel cutting out can be caused by a bad speaker relay. Check the speaker relay when intermittent audio is found in both speakers. Replace the speaker relay when only one channel is intermittent. Check for bad connections on the relay when the sound cuts in and out. The speaker relay may not energize with a bad voltage regulator IC or transistor (Fig. 9-33).

Audio protection circuits

When the protection relay will not click on, check for a defective protection IC. A bad zener diode in the voltage source can cause the protection relay to not turn on the speakers. Check for a badly soldered joint on the protection IC when the protection circuit comes on after a few minutes of operation. The speaker may not be activated with a shorted protection IC. Suspect a bad protection IC when the relay will not click on or off. A no-sound symptom can result from a leaky protection IC. Resolder all connections of a voltage regulator transistor or IC with a clicking protection relay with intermittent audio. Suspect a bad protection IC when there is no audio in either speaker. A bad protection IC will not let the relay click on (Fig. 9-34 on p. 306).

Large speaker problems

Large audio speakers can become defective as a result of a warped or loose cone. A blatting sound from the speaker may result from a loose voice coil diaphragm. Sharp objects poked into a speaker cone can make a vibrating sound. A frozen speaker cone can be damaged by too much applied volume. A blown speaker voice coil can be caused by too much volume applied to the speakers. Check the speaker by pushing up and down with your fingers to see if the cone rides against the magnet or is frozen and will not move (Fig. 9-35 on p. 306).

For more amplifier problems and solutions, check Table 9-1 on p. 307.

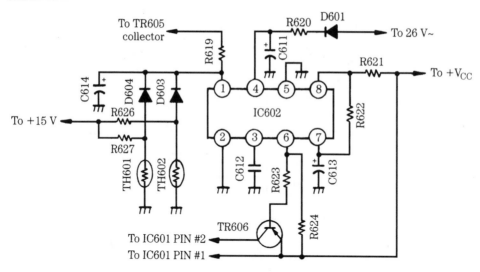

9-34 A bad speaker protection IC (IC602) in the protection circuits can cause no or intermittent audio.

9-35 Push up and down on the large speaker cone to see if the voice coil is dragging or frozen.

Table 9-1. Servicing stereo sound circuits.

Symptom	Location	Repair
Dead— no sound	Volume control	Signal trace the audio at the volume control to determine if the problem is in the output circuits.
	Output circuits	Measure the voltage across the main filter capacitor. Check the voltage on the output transistor or IC. If the fuse is blown, suspect a leaky output transistor or IC. Signal trace the volume control to the base of each transistor with a scope or external amp.
Intermittent sound	Input or output	Monitor intermittent audio at the volume control to determine if the problem is in the input or output stage. Monitor at the speakers and on the base of the second AF or driver transistor. Monitor at the input and output of the suspected stereo output IC. Look for an intermittent coupling capacitor and burned bias resistors, diodes, and transistors.
Weak sound	Input or output	If the cassette player sound is weak, check the tape heads. Clean with alcohol and a cloth. Check the weak channel against the normal stereo channel to see what stage the problem is in. Go from one side of the coupling electrolytic capacitor to the other for weak audio. Check the signal from base to base of each transistor. Check the input signal and output of the suspected IC. Weak sound is caused by open coupling capacitors, bias resistors, diodes, and transistors.
Distorted sound	Output stage	Clean the tape heads. Check the input and output of the power IC for distortion with an external amp. Check each transistor for openness or leakage with in-circuit tests. Remove the transistors and diodes, and check them out of circuit. Check the audio output bias resistors for correct values. Check the bias diodes for leakage. Make sure the voltage source is proper and both the positive and negative sources are equal. Check the distortion at a given point in the circuit and compare it to the normal channel. If both channels are distorted, suspect a common IC or improper supply voltage. Distortion is caused by open transistors, ICs, bias resistors, and bias diodes, and improper voltage. Remember, in directly coupled circuits, an open or leaky transistor down the line can affect all transistors in the stereo lineup.
Hum in sound	Filter capacitors	Shunt each filter capacitor with a good one. Shut down the chassis to shunt the capacitor. Discharge each capacitor after shunting each one in circuit.
	Low-level hum	Check and shunt each decoupling capacitor. Check for an increase in resistance at the base terminal. Check the input cables and wires.
Motor boating in sound	Audio output	Replace the audio output or ICs.
	Low-voltage power supply	Shunt all electrolytic capacitors in the power supply.
Output transistors run red hot	Output stage	Check for open or leaky driver transistors. Check the bias diodes and resistors. Check for excessively high voltage source. Check for either a positive or negative voltage source that is missing.

10
CHAPTER

Troubleshooting
AM/FM/MPX stages

Early AM and FM radios were made up of transistors, variable tuning capacitors, intermediate frequency (IF) transformers, and AM and FM oscillator coils, all mounted on a printed circuit (PC) board. The coils were tuned by variable tuning capacitors, with a dial string moving the dial pointer. The AM/FM front-end circuits were switched in and out with a slide or rotary switch. Miniature IF coils or transformers were found in such early AM/FM stages (Fig. 10-1).

By removing the back of the case or cover, you can easily locate each section. The ferrite antenna rod serves as the antenna coil, whereas the FM coils are made up of larger wire and soldered directly to the PC board. Usually these radiofrequency (RF) and antenna coils are located near the variable tuning capacitor. Both AM and FM IF coils or transformers are shielded and found in a row in front of the variable tuning capacitor. The AM coils are wound with many turns of fine wire, whereas the FM coils form a bare solid wire.

The AM and FM bands may switch in supply voltage to separate AM and FM circuits, whereas in deluxe radios a band-switching rotary switch may be found. You can locate each transistor stage to the coils and shielded IF transformers. The audio output stages are found near the volume control leads and can be traced back from the speaker terminals to locate the output transistors.

The front-end circuits

Use either a comparable schematic or a block diagram when the actual schematic is not available (Fig. 10-2). Early front-end circuits consist of an AM RF coil, a converter, an IF detector, and audio stages. The FM front-end circuits may consist of transistors

10-1   Miniature IF coils are found in portable AM/FM radios.

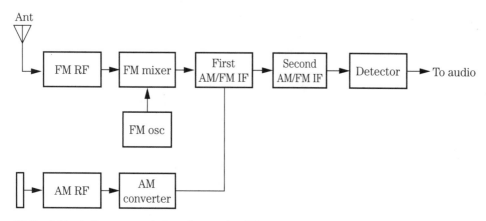

10-2   A block diagram can help to locate the different receiver stages when no schematic is available.

in the FM RF coil, an FM oscillator, an FM mixer, first and second IF transformers, a discriminator coil, and audio circuits. A separate tuning capacitor with fewer plates is used to tune the higher FM frequencies, whereas the AM stages are tuned with many aluminum plates. Both AM and FM variable capacitor sections are rotated with a dial cord and dial indicator (Fig. 10-3).

10-3   A tuning knob rotates the dial cord that turns a variable tuning capacitor in an AM/FM/ MPX radio.

Locate the separate AM and FM front-end sections. The FM coils are made of solid wire and are mounted on the PC board. Locate the AM ferrite antenna rod with coil, and trace the wires to the tuning section and RF AM transistor. The FM/MPX stereo section should be mounted close to the FM stages. Early AM/FM radios had IF transformers sticking up from PC board. The audio section may have power transistors or integrated circuits (ICs) on heat sinks, whereas the low-voltage power supply can be located by the power transformer and large filter capacitors. A quick voltage test across the filter capacitor can indicate if the power supply is working.

Note the receiver symptoms, and apply them to the components located on the chassis. If the AM reception is weak, intermittent, or has no reception, go directly to the front-end circuits. Locate the antenna ferrite coil and tuning mechanism. In early chassis, two or three ganged tuning capacitors tuned the RF and oscillator stages. Today, varactor diodes tune the RF and oscillator sections. In digital tuned receivers, a tuning controller IC provides control and tuning data that are applied to the RF, mixer, and oscillator circuits (Fig. 10-4).

Common problems

Look for circuits common to both the AM and FM sections when stations are not received. Check the low-voltage power supply and regulator source feeding the AM and FM circuits. A leaky regulator transistor or zener diode can produce a dead AM

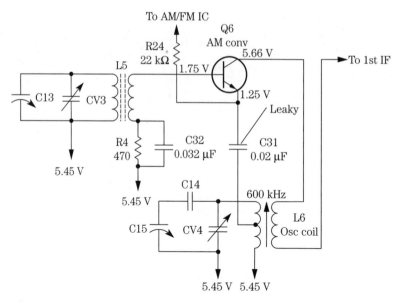

10-4 A leaky capacitor C31 (0.02 µF) caused a no-station symptom in the AM converter stage of a Radio Shack clock radio.

section and a normal FM section. Critical voltage measurements on the AM or FM transistors and IC components can help to locate a defective low-voltage source.

Check the IF stages or combination FM IF and AM converter circuits. Sometimes the first FM IF transistor serves as the AM converter. Notice if both the AM and FM IF stages are common to one another (Fig. 10-5). For instance, an open second IF transformer connection can destroy both AM and FM reception. A dirty AM/FM slide or function switch can cause dead AM or FM reception.

A big change

Today, AM/FM/MPX receivers have many new changes since tube and transistor radios. Instead of a variable tuning capacitor, varactor diodes tune the AM and FM coils. A central phase-locked-loop (PLL) processor provides the varactor diode tuning voltage. The IF transformers have been replaced with ceramic filter devices that are cut for the desired IF frequency of both AM and FM bands.

Instead of transistors, the AM and IF front-end section may consist of IC components throughout the receiver circuits. Silent switching of the various receiver sections is done with silicon diodes. Dual IC power audio outputs are found in both stereo channels. In fact, a front-end IC may contain all the RF, oscillator IF, and preaudio amp circuits, whereas one large power output IC may contain all the audio circuits (Fig. 10-6).

No or poor AM reception

The AM section may be dead as a result of any faulty transistor or IC component found in the front end, and FM reception may be normal. Dead AM reception can be

10-5 IF transformers were common to both AM and FM circuits in early AM/FM radios.

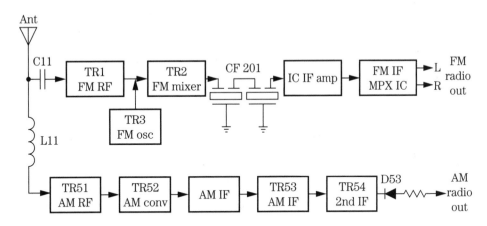

10-6 Block diagram of an AM/FM receiver with separate AM and FM stages.

caused by a leaky or open AM transistor or IC component. Check each transistor with in-circuit tests. If in doubt, remove the transistor from the circuit. You will find that the suspected transistor is open or leaky with in-circuit tests but normal when removed. Replace the transistor if in doubt. Sometimes the AM/FM front-end IC is defective in AM and normal in FM circuits (Fig. 10-7).

Clean the function switch assembly with cleaning spray when the AM or FM is erratic or intermittent. Spray cleaner into the switch contacts, and work the switch back

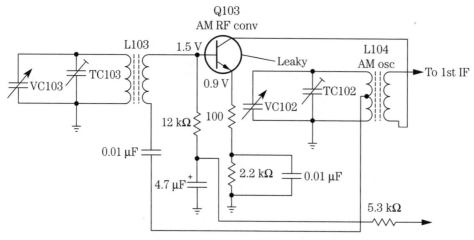

10-7   A leaky or open Q103 results in poor or no AM reception.

and forth to clean the contacts. Replace a broken function switch with an original part number, if it is available.

Suspect a defective AM RF transistor when a local station can be heard with a voltage probe on the collector terminal. Test the transistor in-circuit for open or leaky conditions. Critical voltage tests will indicate if the transistor is open or leaky. Check the antenna coil for open or broken leads when there is no AM.

Actual AM problems

Radios that use a ferrite antenna coil can produce a weak radio symptom when the ferrite core is cracked or broken. Some of these ferrite cores are round, and others are flat. Sometimes the core will break inside the coil winding, resulting in weak AM reception.

Suspect a defective front-end IC for no AM reception (Fig. 10-8). Replace the front-end IC component if there is no AM reception. Check the RF transistor for weak or no AM reception. Replace the AM IF transistor or transformer if there is no AM reception. Resolder the RF and converter transistor terminal connections for no AM signal. Replace the RF transistor even if it tests normal because it can break down under load. Check for improperly soldered contacts on the AM oscillator coil when there is no AM reception. Suspect a leaky RF AM transistor when you can get only one local station. Inspect all leads off the ferrite antenna coil when there is no AM reception. A poor board connection of the detector diode can result in no AM signal. Suspect a broken wire or trace to the AM tuner assembly when there is no AM reception. Do not overlook bad board traces around the AM section when there is poor or no AM reception.

A dead AM section can result from no or improper supply voltage to the AM circuits. Check the supply voltage to the AM RF coil, converter, and IF transistors or IC components. Suspect a leaky or shorted decoupling capacitor or transistor voltage regulator when there is no or improper supply voltage. Replace the leaky IC regulator when only a hum is heard in the AM or FM band.

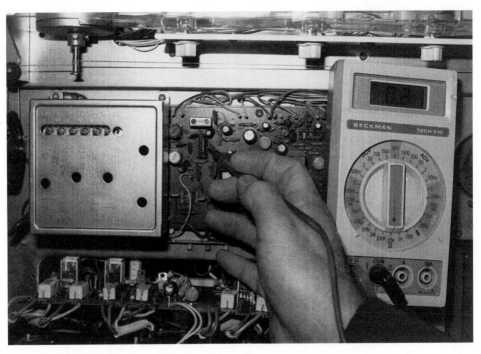

10-8   Check the AM front-end circuits when there is no AM reception.

Intermittent AM reception

Intermittent or weak AM reception also can be caused by a defective zener diode in the power supply. Weak AM reception also can result from a leaky transistor regulator in the AM supply voltage source. A weak AM symptom can be caused by a bad AM converter transistor with only 0.2 V when it should be 0.6 V between the emitter and base terminals (Fig. 10-9).

An open primary winding of the AM IF transformer can cause a rushing noise on the AM band. A dirty AM/FM switch can cause intermittent or no AM reception. Suspect a bad oscillator padder capacitor for intermittent AM reception. Intermittent AM reception can be caused by a bad or open IF transformer. Simply move the IF shielded transformer and notice if the AM signal goes in and out. When only one station can be tuned in at the lower end of the dial, suspect a broken base capacitor within the AM IF transformer. Check for a broken metal slug in the oscillator coil when some stations are weak or drift off channel.

Determine if both the AM and FM stages are intermittent or just the AM circuits. A cracked or broken antenna core can weaken AM reception. The crack or break may be inside the coil winding, so take a close look (Fig. 10-10). Some cores can be replaced. The exact replacement may be available from the manufacturer.

The small Litz wire used in antenna coils is easily torn loose or broken off, and poorly soldered connections can produce intermittent AM reception. Check the windings for continuity. Use the lowest scale on the ohmmeter, and check each winding. You should get a direct short or very low measurement.

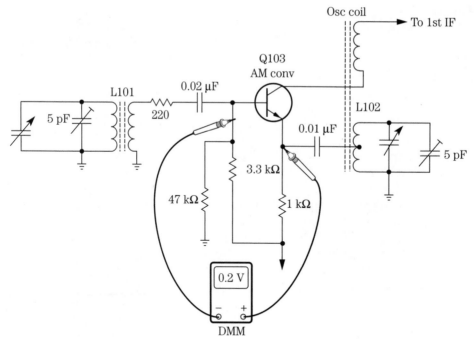

10-9 A leaky transistor can be identified by improper 0.2-V bias voltage when it should be 0.6 V on an NPN transistor.

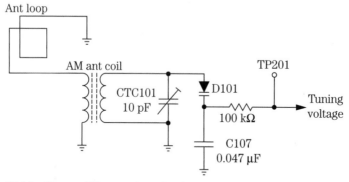

10-10 For no AM reception, check for a broken coil wire or connection on the antenna coil.

Push up and down on the PC board within the AM section. Probe around on small coils, capacitors, and resistors with an insulated tool. Sometimes by moving a small part you can turn up a poorly soldered connection or terminal. An intermittent terminal lead or junction within the transistor can be located in this manner.

The trouble may be between the AM converter stage and the antenna coil. Grasp the body of the antenna coil to see if the stations become louder. If they do, there is continuity between the converter and at least part of the antenna. Look for a break or poorly soldered connection under a layer of coil dope or wax.

Weak or intermittent AM reception can be caused by a defective RF transistor. You will find a separate RF transistor in the better receivers, but many receivers use a single transistor AM converter that functions as both the RF and oscillator stages. When a unit has only two variable tuning capacitor sections or varactor diodes, you can be fairly sure that the unit uses a converter stage with no separate RF. If several stations can be tuned across the dial, you can assume that the converter transistor is normal.

No AM reception

No AM could be tuned in on a Radio Shack 13-1267 receiver. Critical voltages were taken on IC101. Zero voltage was found on terminal 23, and the rest of the voltages were way off. The tuning voltage on D102 at test point TP201 was normal. Tracing the secondary winding voltage back to TP203 indicated very little voltage. Resistor R125 (22 kilohm) showed signs of overheating and was cracked on the bottom side. Replacing leaky LA1266 and R125 solved the problem (Fig. 10-11).

No AM, normal FM

Suspect the AM front-end section when there is normal FM but no AM reception because both bands join together in the IF section. Poor contacts or broken antenna coil leads can cause a loss of AM reception. Check for a leaky or open RF AM or converter transistor when there is intermittent or no AM reception. The RF transistor can break down under load and still test good. Clean the AM/FM band switch for no or intermittent AM reception. Check the first AM IF stage for open or broken inter-

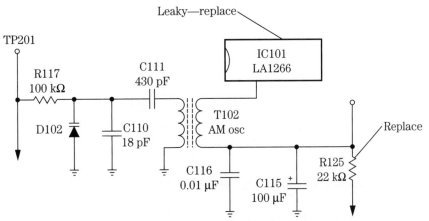

10-11 Replace leaky IC101 and resistor R125 (22 kilohms) for no oscillator voltage in the AM oscillator circuit of a Radio Shack 13-1267 stereo receiver.

nal windings. Resolder all PC board contacts when there is intermittent AM and normal FM. Replace IC3 on the tuner PC board when there is no AM reception in an Akai ATM77 receiver.

No FM

Suspect a defective FM stage when the AM reception is normal. A leaky or open FM RF FET transistor or IC can cause no FM reception. Sometimes, by placing the meter probe on the base terminal for a voltage measurement, a local station may be tuned in, indicating a leaky or open RF FM transistor. Take critical voltage tests on the RF transistor. An open FM RF transistor will have zero voltage on the emitter terminal and a higher dc voltage on the collector terminal than normal. A leaky RF FM transistor may have a low voltage on the collector terminal and no bias voltage between the base and emitter terminals. Check all transistors within the FM front-end circuits with the diode tester of the digital multimeter (DMM). Sometimes spraying the transistor with coolant can make it act up or begin to operate (Fig. 10-12).

A loud rushing noise on FM can indicate a defective mixer or oscillator transistor or IC. No FM reception can be caused by improper voltage from the low-voltage circuits. Check for leaky or open regulator transistors and zener diodes. Sometimes a dirty switch contact supplying voltage to the FM circuit can prevent FM reception.

A defective IC voltage regulator can cause no FM reception and only a hissing noise. Check for a bad zener diode on the power supply board when there is no FM reception. A

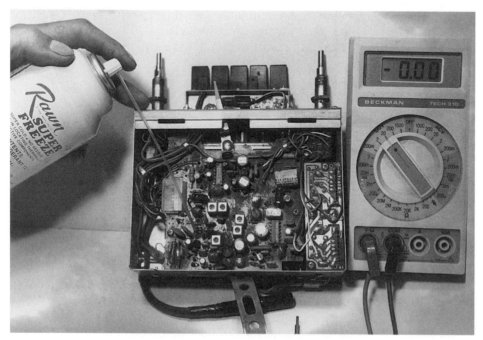

10-12 Spray coolant on an intermittent transistor to make it act up or return to normal.

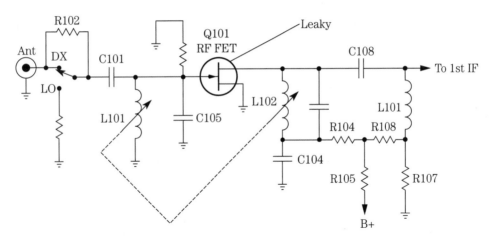

10-13　No AM reception with a leaky RF FET transistor (Q101).

red hot low-ohm resistor within the voltage supply source may be caused by a defective 47-μF electrolytic. Replace a leaky voltage regulator transistor when there is no FM reception. Replace defective bypass capacitors in a tuner section with low supply voltage.

Poor soldering on the emitter terminal of the FM RF transistor can cause no FM reception. A bad padder capacitor in the oscillator circuits can cause no FM reception. Realignment of the padder capacitor on the FM antenna coil can result in no FM stations on the lower end of the dial. No FM signal can result from zero voltage found on a leaky or shorted FM FET RF transistor that might lower the supply voltage source.

Replace the first FM IF transistor when only a local FM station can be heard. Check for a defective IC ceramic filter network when there is no FM reception. Check for an open FM coil within the tuner when there is no FM reception (Fig. 10-13). No FM reception was found in a Marantz SX5 receiver with normal AM, and it was caused by a defective C124 off of Q107-2.

No FM, AM normal

A Soundsign 5154 receiver came in with normal AM reception and no FM reception. When both front-end signals were not found at the first or second IF transformer, the FM RF coil, FM oscillator, and FM mixer transistors were checked. Try to locate the bare-wire FM RF coil and FM oscillator coil on the PC board. The RF FM transistor is located between the FM antenna terminals with a bare FM RF coil in the base circuit of the FET transistor. When the positive probe of the voltmeter touched the base terminal, a local FM station could be heard. The open SK41 FM transistor was replaced with an RCA SK9164 universal replacement.

Dirty AM/FM switch

Suspect a dirty or worn AM/FM switch when either section is dead or intermittent. The dirty contacts can cause no AM or FM reception. A dirty AM/FM switch will not

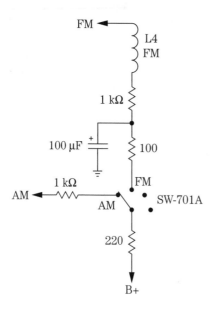

10-14
Clean the AM/FM switch (SW-701-A) when there is intermittent or no FM or AM reception.

apply a dc supply voltage to one or both bands. Replace the worn switch with a toggle or slide switch found just about anywhere. Spray cleaning fluid down inside the switch area, and move the switch back and forth to clean up the contacts (Fig. 10-14). Replace resistor R154 (1.5 kilohms) for slow switching from AM to FM in a Mitsubishi DAR11 receiver.

Weak FM reception

When only one FM station can be heard with two AM stations, replace the FM IF and AM RF transistors. Distorted and weak FM reception can result from a leaky FM/MPX IC. No voltage on the collector terminal of the FM IF transistor can be caused by a badly soldered IF connection and will result in very weak FM reception. A defective or dirty local-distance switch can result in a weak FM signal. Check for an increase in resistance of the resistors in the FM circuits when there is weak FM reception.

Replace a leaky voltage regulator transistor with low supply voltage to the RF section when it causes a weak FM signal. Weak or no FM reception with a hissing noise can be caused by a defective IC voltage regulator.

Check for a bad FM padder capacitor when the FM starts fading out with a lot of noise. Realignment of the padder capacitor may be needed on the antenna coil when there are no FM stations on the lower end of the dial. Replace the FM oscillator transistor when the stations drift off channel. Adjust the FM discriminator coil when the FM reception drifts off. Readjust the alignment of the FM detector coil when the receiver will not lock on FM reception. Resolder a bad connection and tighten the mounting screws on the PC board when an FM station drifts off with only a hum in the audio in a Randix KR4140 receiver.

If the AM section is normal but the FM is weak, prime suspects are the FM RF section and the FM antenna. The FM RF transistor is typically mounted on the PC board next to self-supporting coils. If a chassis layout is handy, use it. If not, signal trace the wiring from the FM RF variable capacitor section, varicap, or varactor diodes to the RF transistor.

The FM RF transistor can be checked with in-circuit transistor or diode tests with a DMM, which will indicate if the transistor is open or has a beta reading. Don't adjust any FM screw trimmers, variable capacitors, or FM RF transformers until proper repairs are made.

Really weak FM reception

Suspect the AM/FM tuner front-end section when the FM reception is weak. Measure the supply voltage at the supply pin terminal (Vcc). Extremely low voltage in a Realistic receiver was caused by a very low power supply source (Fig. 10-15). Remove pin 20 from the supply voltage and notice if the voltage increases. If not, repair the low-voltage supply source.

No AM or FM reception

Look for defective components that are common to both the AM and FM bands. The power supply, IF transistors, IC components, and RF transformers and ceramic filters

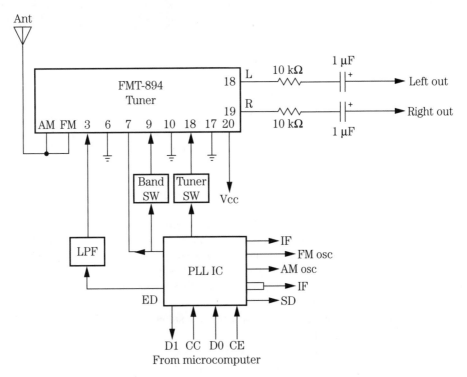

10-15 Weak FM reception was caused by improper supply voltage at pin 20 of an AM/FM tuner assembly.

10-16 Suspect the front-end section when the cassette or CD player operates in the audio section.

are common to the AM and FM sections, so if both sections are dead, chances are that the cause is one of the four. Rotate the function switch to the tape or phono position to check the audio stages. If they work, the problem is located in the IF stages (Fig. 10-16).

Take voltage measurements on each IF transistor and IC. Compare the voltage with the supply source. Improper readings can indicate a defective transistor or IC. Test all transistors in-circuit, or remove suspected ones for testing. Normal voltage readings indicate that the IF transistors and power supply are functioning but will not always identify a defective IF transformer.

A defective IF transformer can be located with signal injection or continuity tests. Use the ohmmeter to check each winding. The AM IF transformer will measure a few ohms, and the FM IF transformer will measure less than 1 ohm. Erratic or intermittent AM and FM reception can be caused by a poorly soldered connection to the transformers or a broken IF coil winding. Ceramic IF filter networks have no resistance measurements and seldom cause problems.

You can locate a defective IF transformer by signal tracing. Turn the function switch to the AM or FM position. Inject a 455-kHz signal to the base of each consecutive IF stage, starting at the last IF stage and working toward the converter stage. Likewise, inject the FM IF stages in the same manner with a 10.7-MHz signal. When no tone is heard, you have located the defective stage. Signal from a white noise generator can signal trace the RF and IF stages.

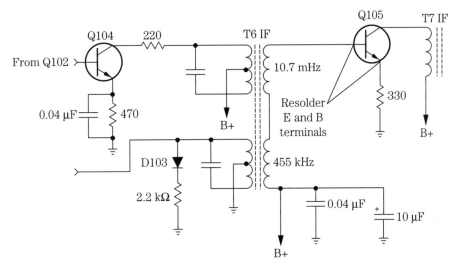

10-17   Resolder poor connections on Q105 for both intermittent AM and FM reception.

Replace a bad regulator IC in the power supply when there is no AM or FM reception. A shorted zener diode in the power supply can cause no AM or FM reception. Check for a defective decoupling electrolytic in the voltage source when there is no AM or FM signal. A leaky IC regulator can cause a loud hum on the AM and FM band when the volume of the receiver is turned up.

Clean the AM/FM selector switch when both bands are intermittent or dead. A bad front-end AM/FM IC can cause no AM or FM reception. Replace both the AM converter transistor and the RF FM transistor when there is no AM or FM reception. Resolder the antenna input connection when there is no or weak AM and FM reception. Replace the common AM/FM second IF transformer when there is no AM or FM reception. Resolder the AM/FM first IF transistor emitter terminal when there is no AM/FM reception (Fig. 10-17). Replace Q202 and Q203 in the low-pass-filter circuit near IC201 in a Pioneer SX6 receiver when there is no AM or FM reception.

NO AM OR FM, XTAL XA-800

No AM or FM reception was noted in an Xtal XA-800 car radio. A quick voltage test on the AM/FM front-end circuits showed no dc supply voltage. On checking the power supply, several components were found burned. D104 had become shorted and caused resistor R125 (180 ohms) to become charred. Replacing R125 and D104 restored the 7.75 V to the AM and FM circuits (Fig. 10-18).

Intermittent AM and FM reception

Determine if the audio section is working in both stereo channels by clicking the center terminal of the volume control with a screwdriver blade. If hum is heard in both stereo channels, you can assume that the audio section is normal. Try out the

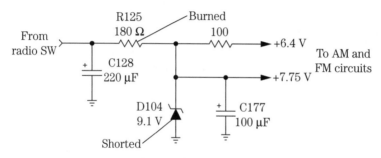

10-18 Shorted D104 and burned resistor R125 (180 ohms) caused no AM or FM in a large Xtal XA-800 car radio.

cassette or CD player if there is one in a combined AM/FM receiver. When both AM and FM sections are intermittent, you may think that the trouble lies in the front-end circuits that are common to both bands, such as the IF or ceramic filter stages. Locate the FM IF transformers or ceramic filters, which are usually wired in the series circuit (Fig. 10-19).

First, try to move the IF coils or transformers around with your fingers. You might have a badly soldered connection. Move the IF transistor or IC with an insulated tool or pencil. Sometimes a transistor lead will have a poor internal connection or poorly soldered terminal connection. Since IF transistors and ICs frequently cause intermittent conditions, spray each one with three coats of coolant. Let each coat disappear before applying the next one. Take your time with each transistor or IC before going to the next.

Monitor the low-voltage source feeding the AM and FM IF stages. Intermittent voltage can produce intermittent reception. Clean the AM/FM slide and function switches.

If the intermittent trouble still exists and it seems to be around the IF transformer, remove it from the board. You may be able to repair the IF transformer by removing the insides. Pull back the tabs or indented area, and pull out the coil from the outside shield. Check each winding for continuity. Place the coil under a magnifying light, and solder the suspected terminal. These transformers should be replaced with factory originals. After repair, tap and twist the transformer slightly to check the success of the repair.

AM up, FM down

Suspect an FM front-end circuit when the AM stations are normal and the FM is dead, weak, or intermittent. The defective component must be in the FM RF coil, FM oscillator, or FM mixer circuits, ahead of the common IF circuits for both AM and FM reception. Check for a leaky or open FM RF transistor or IC when no local stations can be tuned in. A badly soldered connection on the FM RF transistor or IC can cause a dead, weak, or intermittent condition. Check for open coupling capacitors within the FM front-end section with an equivalent series resistance (ESR) meter (Fig. 10-20).

A badly soldered connection on the first IF transistor can cause a very weak FM signal. A weak FM station can be caused with a dirty local distance switch. Intermittent FM

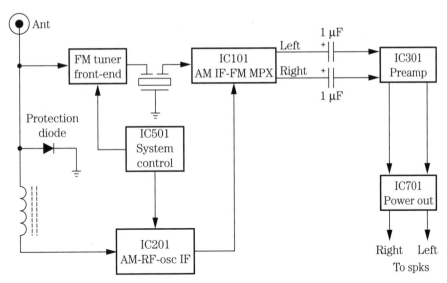

10-19 Look for common components, such as faulty IF coils, a bad ceramic filter, or improper or intermittent supply voltage, for an intermittent AM/FM symptom.

10-20 Check for bad traces and electrolytic capacitors in the front-end circuits with an ESR meter.

reception can be caused by a badly soldered IF transformer connection. Replace the FM oscillator transistor when there is only a loud rushing sound. Resolder all connections on the FM band for intermittent FM reception. Excessive static can result from a noisy FM IF IC.

No FM reception may result from a low or no voltage applied to the FM RF coil or FM oscillator transistors. Check for a broken or open voltage source resistor with low voltage applied to the FM circuits. A bad voltage regulator transistor or IC can cause a no FM symptom. Only a hissing noise on the FM band can be caused by a leaky IC regulator. Check for a leaky IC regulator for hum on the FM band.

Components that cause intermittent AM/FM

Double-check all combination AM/FM components such as the front-end IC, which might include all circuits and feeds directly to the FM multiplex and detector AM circuits (Fig. 10-21). Check for a defective ceramic filter with the crystal checker if one is handy. Test the continuity of each primary and secondary side of the filter for poor connections. Check for intermittent AM/FM IF transistors. Replace the IF transistors when there is noisy reception.

Intermittent AM and FM reception may be caused by a leaky or open voltage regulator transistor and zener diode in the voltage supply circuits. An intermittent transistor or IC voltage regulator can affect both the AM and FM bands. Replace open or dried-up filter capacitors within the same voltage supply source when there is no AM/FM reception, when the AM is okay and the FM is distorted, and when there is no FM with weak and intermittent AM reception. Check for a defective electrolytic in the power source when both AM and FM reception are weak and distorted.

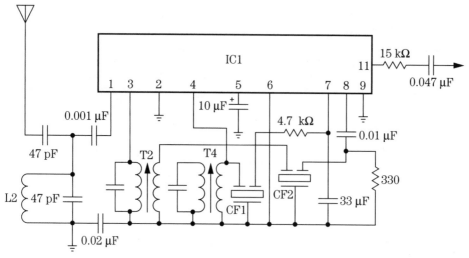

10-21   Take critical voltage and resistance tests on IC1 when there is no AM or FM reception.

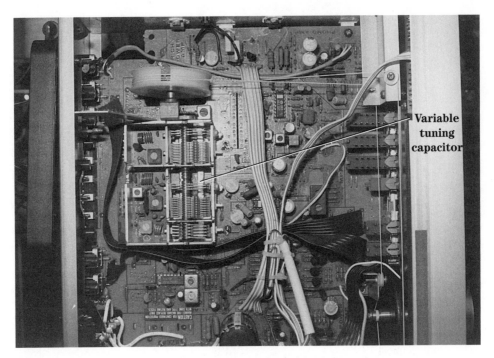

Variable
tuning
capacitor

10-22 Warped plates in the variable tuning capacitor can short out the various stations within the AM and FM bands.

No tuner action and intermittent and excessive static with intermittent shutdown were caused by a bad ground with loose screws on the main PC board of a Sony STRAV201 AM/FM/MPX receiver. Intermittent reception in early radio receivers can be caused by the metal plates meshing in both sections of a variable capacitor tuner when the tuning control is turned (Fig. 10-22).

Defective IF circuits

A badly soldered connection inside the IF transformer can cause intermittent AM and FM reception when both IF coils are in a series circuit (Fig. 10-23). Dead AM reception can result from an open IF transformer winding. Small bypass capacitors inside the IF transformer can cause a dead or intermittent symptom. The IF circuit can be dead or intermittent as a result of a leaky or intermittent IF transistor or IC component. Poorly soldered terminal connections on the IF transistors or ICs can produce a dead or intermittent symptom. Resolder all connections on the IF board when there is intermittent AM and FM reception. Intermittent supply voltage feeding the AM or FM IF circuits can cause an intermittent AM/FM symptom.

Although ceramic filter networks found in AM and FM IF circuits cause very few service problems, sometimes you will find a defective ceramic filter component. Take a continuity measurement with the ohmmeter on both sides of the ceramic filter. No

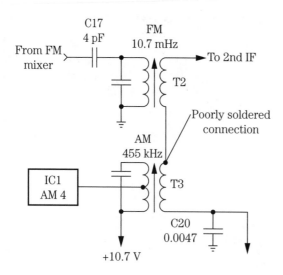

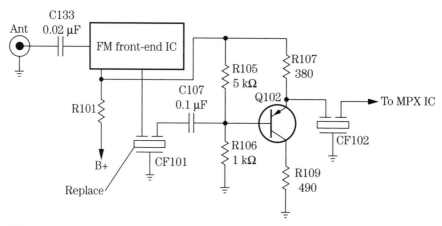

10-23
A bad first IF connection within an AM/FM IF circuit can result in no AM or FM reception.

10-24 A bad ceramic filter (CF101) in a Realistic 12-2103 in-dash AM/FM car radio resulted in no FM reception.

resistance should be measured. If a resistance below 1 kilohm is found, remove one terminal on that side of the ceramic filter and take another test. Sometimes a low-ohm component such as a transistor, resistor, or diode may be wired in parallel with the filter and can cause a low resistance measurement. Remove the ceramic filter from the PC board, and take another measurement. Both sides of the ceramic filter can be checked with a crystal checker (Fig. 10-24).

DEAD GE IF SYMPTOM

No AM or FM reception was noticed in a GE 7-4665B clock radio. Most of the voltages checked out on the RF and converter circuits. Very little voltage was found in the second IF stage and was traced through T2 and T3 IF circuits. A continuity check was made on all IF transformers, and a badly soldered connection was found at the top of T3 of the 455-kHz IF transformer. Resoldering T3 solved the problem.

Weak AM and FM reception

Sometimes you may find a defective front-end IC that causes weak reception in both AM and FM bands. Check the IF and power supply circuits when there is weak AM and FM reception. An open or leaky IF transistor or IC can cause a weak symptom in both bands. Suspect poor IF transformer connections causing weak and intermittent reception. Take critical voltage tests on each IF transistor or IC. Check all IF transistors with in-circuit tests with the diode tester of a DMM.

Low supply voltage from the power source can cause a weak symptom in the AM/FM bands. Take a critical voltage measurement on the suspected transistor or IC. Trace the low voltage source back to the power supply with critical voltage tests. Check for a leaky or open transistor or IC voltage regulator in the voltage supply source. Sometimes there is a leaky voltage transistor or zener diode within the base circuit of the regulator transistor. Check for burned resistors or a change in low-ohm resistors within the voltage regulator circuits. Make sure that the resistor is defective by removing one end and taking another reading. A change in a decoupling electrolytic can cause a low supply voltage (Fig. 10-25).

WEAK AM/FM RECEPTION, SHARP

The voltage supply source to both AM and FM circuits of a Sharp SC210 receiver was quite low at 0.58 V. Since the supply voltage on the voltage regulator IC (IC202) was around +19.5 V, the output voltage on pin 14 was only 0.58 V dc. The voltage regulator (ZD201) was suspected of leakage because the collector terminal was between R314 (100 ohms) and R257 (47 ohms). ZD201 tested normal on the diode tester of a DMM. Replacing leaky IC202 with an ECG1243 universal IC solved the weak AM/FM reception (Fig. 10-26).

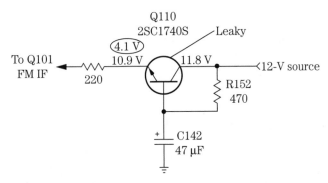

10-25 A leaky regulator transistor (Q110) provided weak FM reception in a Pioneer VSX-452 stereo receiver.

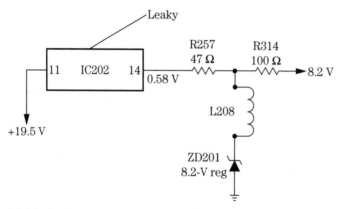

10-26 Replacing a leaky voltage regulator (IC202) in a Sharp SC210 receiver solved the weak AM/FM reception.

Rushing noise in FM, AM OK

Check the FM RF transistor or IC when there is a rushing noise on the FM band with normal AM. The open FM oscillator transistor or IC can cause a rushing noise on the FM band with the volume turned up and no stations tuned in. A noisy RF or oscillator transistor can cause a hissing noise in the FM band. Replace the FM oscillator transistor when there is a low rushing noise. Microphonic reception can result from a defective FM RF transistor. Excessive static can result from a defective FM IC component. Realignment of an FM IF and MPX IC transformer can cause a noisy FM channel. Check C412 (6.68 μF, 35 V) for a leaky condition when noise and a loud roar are heard on the high-end stations on the FM band in a Sony STRVX22 receiver.

Transistor in-circuit tests

You can quickly check all transistors in the circuit with the diode tester of a DMM or transistor tester. The quickest method is to use the diode tester of the DMM because you can complete the job in minutes. Check for open, leaky, and high junction resistance tests. A high-resistance test will be a lot higher in resistance than with a reverse test between two transistor terminals. Sometimes the transistor will test normal within the circuit and break down under load (Fig. 10-27).

At other times a transistor may test open with an in-circuit test and when removed test normal. Make sure that no low-ohm components, such as coils, diodes, and low-ohm resistors, are tied across any two transistor terminals. FM transistors should be tested with in-circuit tests because they are difficult to remove and replace.

Another method is to take a bias voltage test between the emitter and base terminals to make sure that the transistor is good. Measure the low voltage between the emitter and base terminals. An NPN transistor should measure 0.6 V, whereas a PNP germanium transistor should measure 0.3 V. Usually the transistor is normal if it has a correct bias voltage measurement. Measure the emitter voltage to ground and the

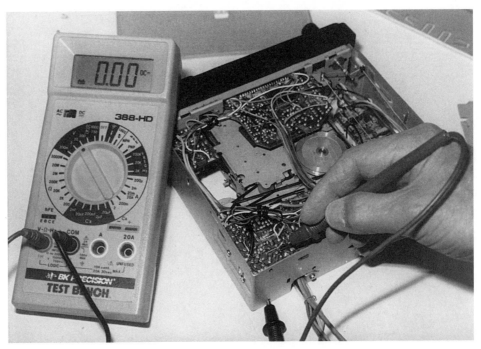

10-27 Checking suspected transistors with the diode tester of a DMM in an AM/FM receiver.

base voltage to common ground, and the difference in voltage should equal the correct bias voltage.

After locating a defective FM transistor, check the location and lead length and how the transistor is mounted. Obtain a new universal replacement, and cut the leads the same length as the old ones. Mount the transistor back in the same spot. Redress leads around the transistor. Test the new transistor in the circuit. Sometimes when replacing the FM mixer and oscillator transistors in older receivers, FM alignment must be performed. With the latest receivers, especially with ceramic filter IF stages, alignment is not needed.

Replacing AM/FM RF-IF IC components

Make sure that the suspected front-end component is leaky or open. Check the supply voltage (Vcc) that is applied to the IC part. Take critical resistance measurements from each IC pin terminal to common ground. Make sure that the low or shorted measurement does not tie directly to common ground with a wire trace. Take signal in and out tests on all audio IC components to locate the defective IC.

Remove the defective IC by applying solder wick material on each side of a row of IC terminals. Place the hot soldering iron on the mesh material, and suck up the excess solder. Move each IC terminal with a pen or screw driver blade so that the terminal is free from the wiring trace. Be very careful not to pull up the PC wiring with too much heat from the soldering iron. Double-check each terminal after the IC has been removed (Fig. 10-28).

10-28   Check all IC trace terminals after the IC has been removed.

Replace the defective IC with the original part number if available. Universal IC replacements also work well in AM/FM front-end circuits. Look up the part number stamped on top of the IC in a semiconductor manual. Make sure that pin terminal 1 is marked on the PC board. Likewise, mount terminal 1 of the new replacement in the correct position. Solder all IC terminals with a 30-W soldering iron. A battery-operated soldering iron is ideal when soldering up IC terminals.

FM discriminator and MPX circuits

Early FM discriminator circuits consisted of an FM IF transistor, discriminator coil, and two diodes in a full-wave circuit (Fig. 10-29). A left and right audio channel were found after each detector diode. Later, a multiplex IC component was found at the beginning of the right and left stereo audio channels. Sometimes, just touching up the alignment of the discriminator or multiplex can cure a loud hissing noise.

A shorted or leaky diode can cause distortion and hum within the audio channels. A red hot MPX IC can cause only a local FM station to be tuned in. Weak and distorted FM can result from a bad FM/MPX IC. Extremely distorted FM music can be caused by a leaky MPX IC. Replace the MPX IC when there is no FM reception. Check the FM IC in the tuner, and readjust the MPX for normal FM stereo. When the FM quits after 10 to 20 minutes, solder around the adjustment control in the matrix circuits. A weak FM/MPX audio can be caused by a defective IF transformer, RF-IF tracking, or an FM/MPX IC or a poor FM antenna circuit.

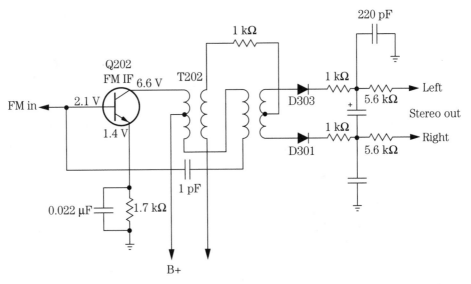

10-29 Improper adjustment of the FM discriminator coil in an early transistor AM/FM/MPX radio caused distorted FM reception.

Replace the FM detector coil when there is no or intermittent FM reception. Suspect that the FM discriminator coil is out of adjustment when the FM station begins to drift off channel. Check D26 for a shutdown of the MPX IC when the FM drifts off in a Randix KR950B receiver.

Distorted FM tuning

When each FM station was tuned in, distortion was found in a JC Penney 853-3218 four-channel receiver. T3 and T4 on the discriminator coils were adjusted to no avail. D5 and D6 checked normal on the diode tester of a DMM. When the capacitors were tested in the output, C46 (4.7 μF) showed less capacity, and ESR problems were found on the ESR meter. The tuning meter responded and distortion was eliminated after replacing C46 in the discriminator circuits.

No stereo light

Try to adjust the SVR control in the MPX IC circuits when the stereo light is not lit or will not go off. In early FM circuits, a panel light was used, and in today's FM/MPX circuits, a light-emitting diode (LED) component is found. Check all low-capacity electrolytic (10 to 47 μF) in the MPX PC board when the stereo light is out (Fig. 10-30). Replace the MPX IC when the stereo light cannot be adjusted and the light stays on all the time. Suspect low-capacity electrolytics when there is normal FM or AM audio and the stereo light will not light up. Check for an open regulator transistor (Q803) when there is no signal strength meter or no FM tuning meter movement in a Marantz 2253 receiver.

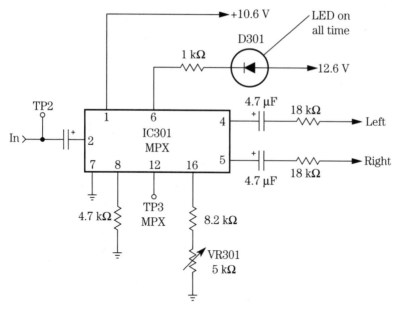

10-30 Check the low-value bypass capacitor, resistors, and IC301 MPX when D301 stays on all the time.

The signal and recording meters may have analog meter movement or a bank of LEDs. The audio signal can be traced right up to the meter or LED circuits with an external amp. These indicator LEDs can be tested with the diode tester of a DMM. If one becomes dead or weak, all have to be replaced in one large component. Determine if signal is applied to the indicator circuits before checking the LED indicator circuits. No LED display in a Sony STR53555 receiver was caused by an open lamp on the LED panel.

Drifting FM

The FM stations would drift off channel after a Sanyo FTC26 auto radio operated for 1 hour. The stereo light would go out at the same time. All voltages were checked on the MPX IC (IC301). D301 and resistor R308 (560 ohms) were replaced on pin 10 with no voltage indicated. The FM coil (T201) also was touched up to solve the drifting FM signal.

Liquid crystal display (LCD)

The latest AM/FM receivers have a liquid crystal display (LCD) mounted right on the front panel. The LCD indicates the station that is tuned in, numbers, and the various receiver operations. Usually the LCD is driven by an IC from a data-clock crystal control microprocessor. Suspect a driver IC when the display appears dead (Fig. 10-31). Check all voltages on the display and driver IC. Scope the data crystal for a waveform.

10-31 Take critical voltage measurements on the LCD driver IC for no display.

A very dim display can be caused by a leaky IC. Check for poorly soldered joints on the panel display board for intermittent display functions. For no functions and no display, check for badly soldered connections to resistors and silicon diodes in the power supply. Replace C908 for no digital display in a Technics SA410 receiver.

The varactor-diode tuner

The varactor diode changes capacitance when a dc voltage is applied to it. By varying the voltage on the varactor diode, the diode changes capacity and tunes both the AM and FM coils in the receiver. The varactor diode is found in both AM and FM front-end circuits. In early radios, a variable capacitor tuned the various bands, and now the varactor diode takes its place (Fig. 10-32).

Besides the FM varactor tuner, you may find varactor diodes in the RF coil, mixer, and AM oscillator circuits. The same signal voltage is supplied by the controller. In some chassis there is a test point to measure the controlled voltage. The voltage applied to the varactor diode is a different voltage for each station.

Determine if the varactor diode is operating by taking a voltage measurement fed to the diode. Rotate the AM dial and notice if the voltage applied to the diode changes as the knob is rotated. If so, the tuning controller IC is functioning. Suspect a leaky varactor diode when the voltage does not change across it. Check the different voltages found on the receiver's AM RF coil and converter varactor diodes. If one of the diodes'

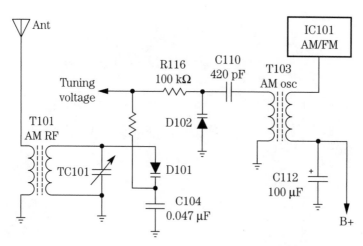

10-32 The tuner controller varies the voltage on the varactor diodes to tune the AM/FM bands.

voltage does not change, check the diode for shorted or leaky conditions. The varactor diode can be checked with the diode tester of a DMM. Remove the varactor diode's ground lead for a correct reading in only one direction. Compare the diode resistance reading with the other varactor diodes in the radio.

Suspect a bad digital tuning controller IC when there is no change of voltage across each varactor diode. Double-check all voltages measured on each diode. If the voltage does not change at all, make sure that the controller IC or corresponding circuits are defective. Measure the supply voltage to the tuning controller IC. A low or improper supply voltage (Vcc) may indicate a defective controller IC or power voltage source.

Varactor tuner controller

The varactor tuner controller IC provides the correct voltage to each varactor diode within the AM/FM RF coils, AM converter, FM oscillator, and mixer circuits. The digital tuning system provides control over the supply voltage fed to the front end (AM/FM tuning), fluorescent display, signal indicator, and switching in some digitally controlled receivers. A microcomputer chip provides digital tuning control and remote data to several stages of a digitally controlled system. The synthesizer processor (IC) provides controlled voltages to the AM/FM RF and IF circuits (Fig. 10-33).

Check the supply voltage at the FM tuner. This voltage will vary as the tuning knob is rotated. Suspect a faulty transistor regulator or improper voltage at the processor chip when there is no voltage at the FM tuner. Take critical voltage tests on IC101, TR103, and TR102. If voltage is found at the front-end tuner, suspect a defective tuning system. Check each stage in the same manner as those found in conventional front-end tuning circuits. Take a scope test of the crystal-controlled phase-locked-loop (PLL) circuit of IC101.

Very little tuning voltage was found on the AM RF coil, AM oscillator, and FM circuits in an integrated Radio Shack 13-1267 component receiver. The variable tuning

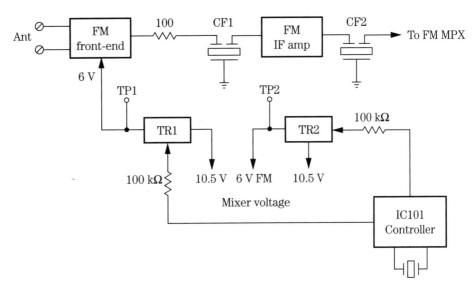

10-33 Check TP1 and TP2 for a variable voltage applied to the varactor diodes in the FM RF coil, FM oscillator, and FM mixer stages.

(VT) voltage source was traced back to AM diodes D101 and D102. This voltage is fed through a 100-kilohm resistor to test point TP201 (VT). The VT voltage is derived from PLL IC103, which provides tuning voltage to the AM and FM tuning circuits. A B+ voltage check at driver transistor terminals on Q105 was normal, except very little voltage was derived for the VT voltage source. Replacing transistor Q105 (2SK583) solved the tuning voltage problem.

Overload protection

Some deluxe receivers have a shutdown circuit when a power transformer or power supply becomes overheated. A new positor component is inserted in series with a circuit that shuts down the receiver circuits when these parts are running too hot. The positor part is mounted directly on the transformer and power output transistor heat sink to protect the power supply and amplifier components. When the temperature of the positor changes, the other circuits are shut down (Fig. 10-34).

When the temperature of positor TH601 (the one installed with a heat sink) and power transformer positor TH602 rises abnormally, the resistance of the positor becomes larger, and the pin 1 potential of IC602 will increase and cause transistor TR606 to turn off. The result is that pin 2 of IC611 is now cut off from the power supply voltage (Vcc) line, and the power transistors are protected.

Check D603 and D604 with the diode tester of a DMM when output transistors have overheated and have been damaged. Make sure that positors are mounted against the heat sink. Take a resistance measurement of each positor and compare with a new component. Replace IC602 if all other components are normal. Measure correct +14.5-V feeding positors and diodes.

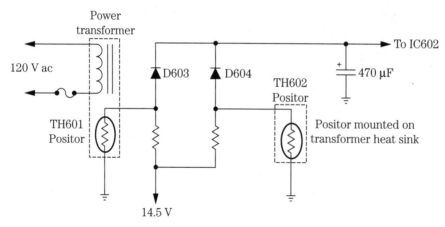

10-34 A positor component is found on the power transformer and output heat sink to shut down the receiver with overheated parts.

Power output transistor protection

A transistor within the output circuits of a high-powered amplifier protects the output transistors when abnormally high current flows through them. Transistor TR604 protects TR602 and TR603 from overheating. TR604 shuts down if there is excess input drive or when the overload impedance connected across the output is too low (Fig. 10-35). If the current increase is excessive, the voltage across R610a, b will turn on TR604. Then the pin 1 potential of IC602 will increase and cause transistor TR604 to turn off. Now the result at pin 2 is that IC601 is cut off from the power supply voltage line.

To suppress power turn-on noise, a time delay is provided by time-constant R622 and C313. This time delay is set to activate transistor TR604 to connect the +Vcc line and pin 2 of IC601 after enough time has elapsed for the tone-control amplifier and the preamplifier to reach a stable operating condition (Fig. 10-36 on p. 340).

Take critical voltage tests on each transistor and IC. Test each transistor in-circuit. Check the supply voltage on IC601 and IC602. Measure each bias resistor for a change in resistance after other tests are made to service the special circuits in the latest receivers.

Remote control circuits

The remote that comes with a large deluxe receiver may control the on and off buttons and volume up and down. The remote also may control the tuner bands to tune the different stations up and down and have a separate button for muting the sound circuits. Besides controlling the receiver circuits, the same remote can control the TV and VCR functions. The remote may be powered by two AAA batteries (Fig. 10-37 on p. 340).

Remote volume control

In many of the latest surround sound receivers, the remote control tunes in the various radio stations on the AM and FM bands. The volume of the receiver can be rotated

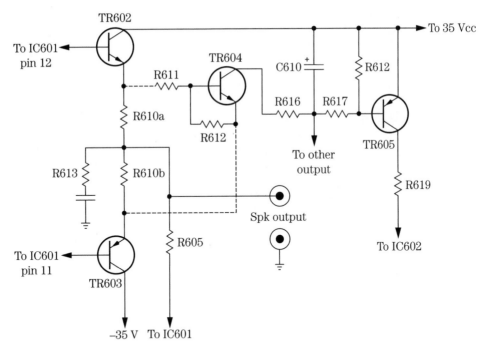

10-35   TR604 and TR605 protect the power output transistors from overheating as a result of excessive current or leaky conditions.

up or down by the remote. The up and down volume control is rotated by a dc motor and motor driver IC. The volume control driver IC provides up and down voltage to the variable dc motor. The up and down signal is fed into pins 1 and 9, whereas the motor out is fed to pins 2 and 7 of IC381 (Fig. 10-38 on p. 341).

For further troubleshooting, see Table 10-1 on pp. 341–343.

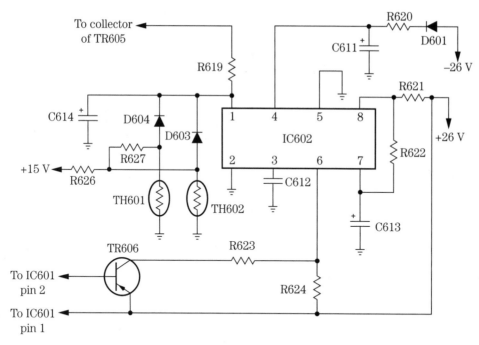

10-36 Positors TH601 and TH602 are tied to pin 1 of IC602 to shut down the receiver when components are overheated.

10-37 A simple infrared remote control can operate the tuning, up and down volume of the volume control, and off and on operations in deluxe and surround receivers.

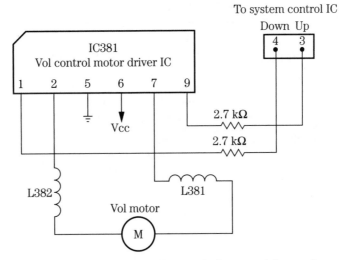

10-38 The system control IC controls the up and down volume of the receiver with a driver IC381 and dc motor tied to the volume control.

Table 10-1. AM/FM/MPX troubleshooting.

Symptom	Cause and remedy
Receiver inoperative (no LED or light)	Replace faulty cord
	Defective power switch
	Blown main fuse
	Bad power transformer
Fuse blown with power on	Defective power transformer
	Shorted silicon diode
	Shorted main filter capacitor
	Shorted output transistors
Power indicator light but no sound from both channels	Defective dual power output IC
	Defective power output transistor
	Defective speaker relay
	Defective ac relay
	Bad speaker fuse
Speaker inoperative	Speaker relay or dirty contacts
	Bad or open speaker
One channel does not work	Blown speaker fuse
	Defective output transistor
	Defective output IC
	Bad copper traces in the output
	Shorted or open speaker
	Replace burned bias resistors
	Defective coupling capacitors

Table 10-1. AM/FM/MPX troubleshooting (*continued*).

Symptom	Cause and remedy
One channel dead when signal test applied to volume control	Defective driver or AF transistor Defective driver IC Replace open or burned bias resistor Replace open or dried-up electrolytics Defective balance control
No headphone reception	Bad headphone jack Open headphone resistor Dirty headphone jack contacts
No cassette operation	Check and clean tape heads Check for torn headphone connection Defective transistor or IC preamp
No phono operation	Bad phono pickup Torn crystal connection Broken wires on crystal pickup Bad phono plug
No FM	Defective front end Defective IC front end Dirty AM/FM switch Defective voltage regulator transistor Defective MPX IC and circuits Defective IF FM circuits Defective ceramic filters
Poor or no multiplex	Improper adjustment of FM discriminator Improper adjustment of VR control Replace discriminator transistors Replace MPX IC
Stereo indicator does not light	Defective lamp or LED indicator Improper VR adjustment Improper MPX adjustment Defective resistors in MPX circuits Improper supply voltage
FM mute inoperative	Defective FM mono/FM switch Check for leaky mute transistor Check for no mute control IC
No AM	Check AM front end Defective AM IC Defective AM/FM switch Defective AM varactor diodes Defective controller circuits Damaged AM RF broken coil

Table 10-1.   AM/FM/MPX troubleshooting (*continued*).

Symptom	Cause and remedy
Bass no effect	Defective bass control
	Bad connections on bass control
	Shorted internal bass control
	Open or dried-up coupling capacitors
Treble control no effect	Defective treble control
	Torn-off connection on treble control
	Internal shorted control
	Open or dried-up coupling capacitors
No meter or display	Defective LCD
	Improper supply voltage
	Bad display IC driver
	Open voltage regulator transistor
	Defective crystal
Remote control inoperative	Replace battery
	Dropped or broken PC wiring
	Test with portable radio

11
CHAPTER

VCR and
TV/VCR combo repairs

Today, VCRs come in many different electronic products. You may find a VCR by itself or within a CD/VCR combo, a DVD/VCR combo, a DVD/VCR recorder, a DVD/VCR home theater, a CD/DVD/VCR combo, or a TV/VCR combo. Recently, VCRs have been placed within 9-, 13-, and 19-in TV chassis. Now a VCR can be found in a 27-in flat screen TV with a built-in DVD player (Fig. 11-1).

Although VCR repair can be quite difficult without a schematic, there are a lot of mechanical and electronic problems that can be solved without one. Mechanical problems can be seen, felt, and heard, whereas electronic symptoms may require added service time. Note the electronic problem, and isolate and locate the defective component. Look for defective integrated circuits (ICs) and transistors that can produce a number of electronic problems. IC and transistor parts cause most service problems in VCRs.

Locate the ICs and transistors that are tied to the various parts. Leaky and overheated ICs can be located by touch and with critical voltage measurements. Case histories may help to solve other VCR problems. Record each case history for future reference. Remember, several VCR brands may be the same inside.

What are the symptoms?

You must know the symptoms of the VCR before you can look for defective components. How does the VCR operate? A dead or intermittent speed problem may be the result of a bad drive motor. Poor VCR loading may be caused by a jammed cassette assembly or a bad loading motor circuit. When the VCR eats tape or spills tape out, check for a bad clutch assembly, erratic takeup reel, or a bad friction gear, and on it goes. Check the symptom, and apply it to the correct section within the VCR.

11-1   The VCR chassis is found mounted ahead of and on the same printed circuit (PC) board as the TV in a 13-in Panasonic TV/VCR combo.

You can compare the symptom with a block diagram or another VCR case history when a schematic or service literature is not available for a particular VCR model. Compare the VCR problem with those of other VCR models and schematics. Several symptoms can be caused by a common defective component within both play and record modes. You may find that the VCR chassis before you is the same as a schematic in the file cabinet.

SLOW LOADING, SYLVANIA

The cassette loading was slow, and then the machine would eject in a slow manner and would shut off before the tape was ejected in a Sylvania VC4510 VCR. Since the capstan motor provided loading, voltage checks on the motor driver IC (IC2005) and main capstan coil drive IC (IC2004) were taken. The voltage was down to 7.1 V and should be around 14 V. Tracing the 14-V source back to the power supply and to D1006 revealed that the circuit was still low in voltage. D1015 tested normal. C1012 (330 μF) and C1013 (330 μF) were tested on an equivalent series resistance (ESR) meter and were replaced to solve the problem (Fig. 11-2).

VCR cassette problems

The VCR cassette itself may be cracked or broken and will not load properly. Sticky substances on the tape can cause it to become jammed or spill out. The tape may record in a loose fashion and, when played, may eject or pull out, with excess tape from the cassette. Dry or worn wheel tabs within the cassette can cause slow and

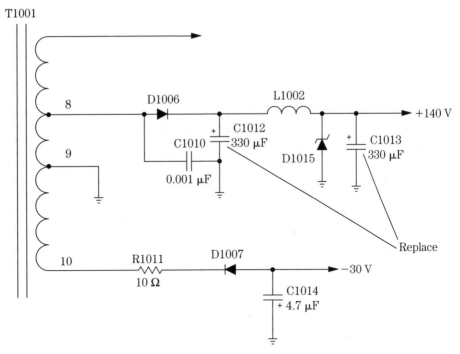

11-2 Replace the C1012 (330-μF) electrolytic for slow loading in a Sylvania VCR4510.

erratic speeds. A bad cassette holder may cause the cassette to jam or prevent loading of the tape. Try another cassette to determine if the cassette is defective. Unraveled tape inside the VCR may have to be cut loose to remove the cassette.

Head cleanup

The video head surface should be cleaned with a chamois skin stick. Clean the tape head horizontally with a cleaning stick or tab. Do not move the stick up and down because you can damage the head. Throw away the dirty cleaning stick after cleanup. Each time the VCR appears on the service bench, clean the tape head and tape path components. The tape head can be cleaned with isopropyl alcohol and a cotton swab (Fig. 11-3).

Dirty tape heads can cause a loss or deterioration of the video signal. Video dropout and audio distortion can result from a clogged tape head. Some manufacturers have included automatic head cleaning systems in their machines. Clean the tape heads, guides, rubber pinch rollers, and threading mechanism of oxide. These surfaces can be cleaned with wet, dry, or magnetic systems.

Audio head cleanup

Clean the audio control head in the same way. Move the cotton swab back and forth on the head. Wipe off all tape guides and rollers with a swab dipped in alcohol to clean the tape running path (Fig. 11-4).

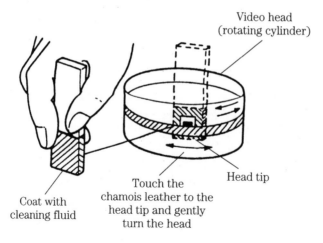

11-3   Clean the tape head by moving the chamois stick horizontally across the rotating cylinder.

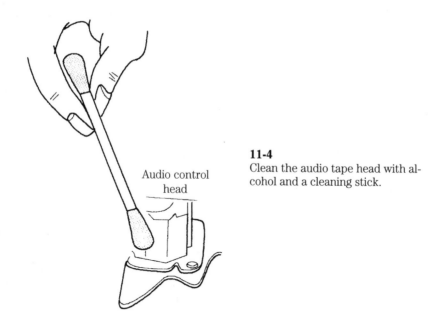

11-4
Clean the audio tape head with alcohol and a cleaning stick.

Be very careful not to damage the upper drum and other tape running parts. Clean the head horizontally, not vertically. Wait until the head area is dry before operating the unit to prevent tape damage. A dirty audio tape head may cause distorted, intermittent, or no sound.

Loading motor problems

Go directly to the loading motor and housing bracket assembly when the cassette will not load. The defective loading motor may be open or become erratic in rotation.

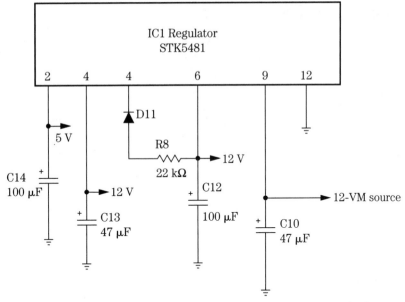

11-5 Replace regulator IC1 for an improper 12-V source that feeds the cassette motor in a Zenith VC-1820 VCR.

Bad brushes in the dc motor can make the motor not run, resulting in a no-loading symptom. The defective loading motor may be erratic in rotation. Slow tape loading can be caused by a bad loading motor. Intermittent tape loading also can result from a defective loading motor.

Measure the voltage across the loading motor terminals. Suspect a loading motor IC when no voltage is found at the motor terminals. Check for defective silicon diodes in the loading motor driver IC voltage source. Check for small 3.7- to 3.9-ohm resistors in the voltage source when the motor will not rotate. A bad 5.1-V zener diode in the power voltage source can prevent tape loading (Fig. 11-5).

The defective loading motor may spin the head for a few minutes and then shut off. Check the loading motor and mode switch when the VCR begins to load and then shuts off. A defective loading motor or binding load assembly can cause a slow-loading symptom. Replace the pulley and loading motor belt when slippage occurs and the cassette will not load. Check for a bad bracket assembly when the VCR tries to load and ejects the tape. Check the cassette load or in switch when the tape will not load.

Besides a defective loading motor, a defective loading cam can result in a VCR that will not load a tape. Check for a broken motor belt or a jammed gear when the loading motor will not load the tape. A bad front-loading gear may prevent loading of a cassette. Replace the rotary switch when there are loading problems. Inoperative cassette loading can be caused by a bad mode switch. Replace the eject gear and side plate when the cassette will not load or eject. In a JVC HRO152U VCR, the tape would not load even with a new belt, and this was caused by hardened lubricant on the cam assembly.

LOADS, THEN SHUTS DOWN

Suspect a bad loading motor when the tape loads slowly and the machine quits in the middle of loading the tape. Check for a defective loading motor when the head spins for a few seconds, and then the unit shuts off. A bad cassette down switch can cause inoperative loading, immediate tape ejection, and play if the tape is loaded manually. Look for a bad cassette holder when the cassette jams on loading.

Check for a bad electrolytic capacitor in the voltage source of the loading motor and driver IC when the VCR shuts down after loading. Suspect a servo IC when the tape loads and the unit shuts off. Monitor the supply voltage when the tape tries to load and the unit shuts off; this is caused by a defective transistor or IC regulator. The bad loading motor IC can cause the tape loading to lock up and the unit's power to go off. The VCR shuts off when loading is caused by a capstan B+ control transistor. Replace IC2001 in an Emerson VT0950N VCR when the unit shuts down after loading.

A bad mode switch can shut the VCR off when it is in the loading mode. When the loading motor is rotating, squeals, and then shuts off, this can be caused by a bad mode switch. Look for a bad end sensor when the cassette will not load or accept tape. Suspect a defective loading cam gear when the unit locks up during the loading process. Check for a bad sensor when the tape loads and right away unloads.

STARTS TO LOAD TAPE, SHUTS DOWN

The tape starts to load around the drum and then unloads in a JVC HR-D170U VCR, and the unit shuts down. A quick voltage check was taken on the drum motor drive IC1. The supply voltage should be around 5.1 V and was near zero. The 5-V source is taken from an 11.6-V source. R11 and R10 checked normal in the voltage source. The 5.1-V zener diode tested normal in the circuit. C12 and C16 checked normal with the ESR meter. IC1 was replaced with another AN6671K IC component (Fig. 11-6), thus clearing the problem.

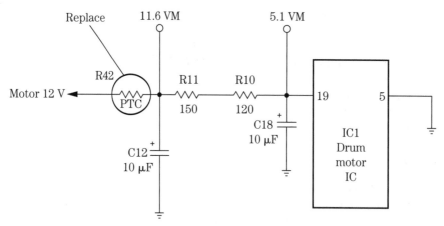

11-6 Defective drum motor regulator IC1 was replaced in a JVC HR-D170U VCR that tries to start up and then shuts down.

Power-up problems

Suspect a bad voltage regulator when the unit powers up and then in 3 or 5 seconds has no functions at all. Check for a bad electrolytic filter capacitor in the power supply for a bad power-up symptom. Check for a defective 12-V zener diode when the unit powers up and then shuts off. Check the main filter capacitor (470 μF) for no power-up when the unit is cold and the LED blinks. A shorted or leaky electrolytic on the switch 5-V line can cause the unit to power-up at once and then shut down. Check all electrolytics with the ESR meter.

The dead chassis with no power-up symptom can be caused by an open fuse in the power supply. A defective transistor regulator in the 5-V source can produce a dead symptom. Check for a defective transistor switch circuit in the power supply for a dead and no power-up symptom. A defective 12-V regulator in the power supply can cause no power-up, a display light, and no cassette loading. A dead unit with no power-up can result from a shorted or open line voltage regulator IC. Shorted or leaky diodes in the power supply can cause a no power-up symptom. Suspect low voltage to the main microprocessor or system control IC for no power-up. Replace an open power transistor regulator when power shuts off in 2 seconds.

Check for a bad end sensor when the unit powers up, tries to load the tape, and sometimes operates. Replace the end sensors if the unit powers up and then shuts off. A bad 4.19-MHz crystal can cause power-up inoperative with flashes in the display and power supply voltages normal. If the VCR power-up supply reels rotate for 10 seconds and the loading motor runs all the time, this can be caused by a bad cam switch assembly. A bad mode switch can cause the supply reel to rotate counterclockwise with the guides fully loaded and then to shut off. Replace the cassette up and down switch when the unit powers up, turns to load, and then shuts off. A bad LED on the mode printed circuit (PC) board can cause a power-up and an attempt to load the empty tray. Replace resistor R2 (680 kilohm) off of Q1001 in a Magnavox VR8520 VCR when the unit powers up, the cylinder spins, and the unit loads with a squealing noise.

NO POWER-UP, RCA

An RCA VR250 VCR would not power up. Critical voltage measurements were made in the power supply and on the main system control microprocessor (IC601). The supply voltage source was 1.1 V on pin 32 (5.3 V). At first, the 5.3-V source was suspected of being low at the power supply source. The 5.3-V source is fed to the data pin (Vdd) of IC601. The 5.3-V source is derived through D602 from a 6-V source in the power supply. A supply of 6 V was found on the anode side of D13 and 1.1 V on the collector side. D602 checked normal in the circuit. When pin terminal 32 was removed from the PC wiring on pin 32, the voltage returned to around 6 V. Replacing the shorted system control IC601 solved the problem (Fig. 11-7).

Speed problems

A cracked or a loose capstan belt can cause a slow speed problem. A stretched, loose, or shiny and smooth loading belt can produce slow speeds. Suspect a

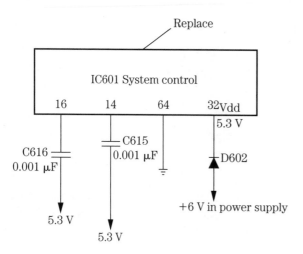

11-7
No power-up in an RCA VR250 VCR was caused by a defective system control IC601.

defective capstan motor for slow speeds. A dry flywheel bearing with hardened grease can cause slow speeds. Clean out the capstan bearings and relube. Suspect a defective brake arm for slow speeds. A dirty brake pad can cause a slow speed problem.

Check for a bad IC regulator feeding the capstan motor circuits for slow speeds. A bad transistor regulator in the 5.8-V source can cause the capstan motor to run slow. A low-voltage IC or transistor feeding the capstan supply voltage also can cause slow speeds. Check all electrolytics within the power supply source for improper capstan motor voltage. Slow drum speed can result from a bad drum driver IC. Look for an open resistor or a change in resistance in the small-ohm resistors (2.2 ohms) in the power source. A bad servo IC can make the capstan motor run slow.

A speed that is too fast also can be caused by a defective capstan or drum motor. Check for poorly soldered joints on the motor terminals. Suspect a loose magnet inside the capstan motor for fast speeds. Check for poorly soldered connections on the capstan drive IC. Replace a defective capstan control transistor for fast speed. A bad plastic E ring on the capstan spindle can cause variably slow or improper speed. Resolder the PC board around the drum stators when the capstan motor runs fast and shuts down occasionally. Replace the mode switch for fast speeds.

Check the electrolytics in the power supply and servo circuits when the cylinder motor runs too fast. A bad cylinder or capstan motor can produce fast speeds. Resolder all connections around the drum motor for fast-speed symptoms. Replace the motor driver IC for fast speeds. Check the 4.19-MHz crystal when the drum spins at a high speed. Check bypass capacitors (0.047 μF) in the servo circuits for capstan fast speeds.

Erratic functions and uneven speeds can result from a bad motor. Erratic speed can be caused by a defective motor driver IC. Check the voltage regulator IC when the capstan rotates erratically. Check for defective capacitors in the voltage supply of the capstan motor. Suspect all electrolytics in the voltage supply source, and check them out with an ESR meter. Do not overlook the 3.58-MHz clock data when the capstan and cylinder motor are erratic (Fig. 11-8).

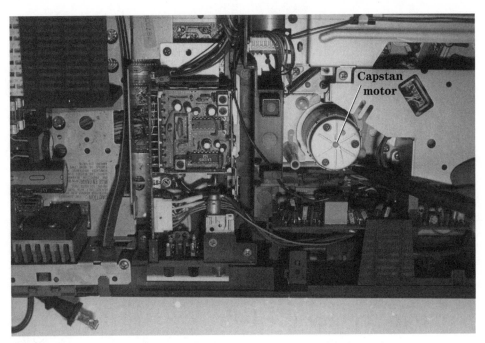

11-8 Check the capstan motor for erratic or intermittent rotation of the tape.

Wow and flutter

A bad or oily belt can cause a wow and flutter condition. Check the capstan motor when wow and flutter occur. Clean out the old grease from the capstan bearings, which can provide a drag, wow, and flutter sound. A bad clutch assembly and idler can cause wow and flutter. Replace the bad clutch assembly in a Hitachi VT1310A VCR for wow, flutter, and audio drags.

Dead, no operation

Check for an open or blown fuse when a VCR is dead. Shorted or leaky diodes can cause a dead, no-operation symptom. Defective diodes and transistors within the power supply regulator system can cause a dead chassis. A defective system control and servo IC can cause a dead chassis. Poorly soldered connections can be found with an ESR meter. Suspect blown parts or PC wiring traces due to lightning or a power surge when the chassis is dead. Leaky silicon and zener diodes in the power supply can result in a dead chassis.

The dead, no-function symptom can result from a bad sensor lamp. Check the mode switch when the chassis is dead. Suspect a defective side switch for a dead loading operation. No power-up, no display, or the tape loads and plays and then ejects can be caused by a bad connection at PMT01 on the main PC board. Check for bad fusible resistors in the power supply when the chassis is dead and there is no display or function operation. A bad power transformer can cause a dead, no-function symptom. Check all electrolytics for a dead power supply.

Check for a bad silicon diode on the 5-V line when there are no functions and no tuner action but the display lights up. No power-up and no function display may result from a faulty 4-MHz crystal resonator. No power-up and a flashing display can be caused by a faulty 4.19-MHz crystal on the tuner PC board. Suspect a faulty 220- to 270-μF electrolytic for a dead chassis and no display. A bad diode in the power supply can cause a dead, no-display, fuse-okay symptom. Suspect small open resistors in the power supply source for a dead, no-rotation symptom.

DEAD, NO POWER-UP, NO OPERATION

A Magnavox VR9040 VCR came in with a dead, will-not-power-up symptom. Very little voltage was found on the system control microprocessor (IC6001). Tracing through the different connections and wiring to the power supply revealed that a 1.5-A fuse was open. At first, the fuse looked and tested good. Replacing the open fuse in the 5-V power source solved the problem (Fig. 11-9).

Time to replace the belt

Loose, worn, or broken drive belts can cause slow or improper speeds in VCR functions. Check all belts for shiny surfaces, cracks, or broken areas. A shiny belt indicates slippage around a capstan motor pulley. Check the drive belts and wheels for erratic or intermittent movements. Clean the belts when cleaning the tape heads.

Replace the belts with original part numbers, when available. Universal belts and drive wheels can be ordered from local or mail-order firms. Sometimes exact drive wheels, idler pulleys, pinch rollers, and VCR gear assemblies can be obtained from large mail-order firms. Several VCR belt kits from different manufacturers offer convenience, economy, and a wide range of replacement parts. Pick up a belt gauge for a quick, accurate method of measuring the thickness, width, and inside circumference of broken

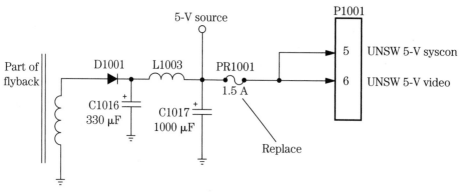

11-9 An open fuse caused a dead chassis, no-power-up symptom in a Magnavox VR9040 in the standby power supply.

11-10 A loose or dirty belt or oil on the belt of a loading or capstan motor can cause erratic or slow speeds.

or worn belts. Clean the motor pulley, cylinder, and belt in a Hitachi VTM1371 VCR for poor speeds and slow rotation (Fig. 11-10).

The tape eater

When the tape spills out and the take-up reel slows down, suspect a loose belt, a dry take-up reel, or a frozen bearing. A dirty tape head or a sticky roller or spindle can cause the tape to spill out. A bad or sluggish reel motor can cause a tape-eating symptom. Check the pinch roller bracket when tape pulls out in the playback mode. Replace the reel brake assembly when the VCR eats tape with a new idler, belts, and pressure roller. In the fast-forward mode, if the VCR eats tape and the fast-forward rewind is intermittent, replace the clutch-idler assembly. Chewed-up tape and intermittent symptoms can result from a bad brake assembly. A bad clutch-bearing binding can result in chewed-up tape.

Check the capstan motor and a possible cracked pulley when the VCR eats tape and does not eject the cassette. Replace the capstan motor when the motor runs backward and pulls out tape. A bad cylinder motor stator assembly can cause the VCR to eat tape in all modes.

Improper reel rotation can be caused by a bad driver reel motor IC and result in tape spilling out. The bad motor driver IC in take-up mode can spill out excess tape.

A defective loading IC on the take-up reel in an RCA VLT900 VCR shuts the unit down in a few minutes and causes the unit to eat tape and the reel to remain stationary.

Replace a cam stopper when the tape pulls out and the power goes off. A stripped motor pulley can pull out the tape. Replace a broken idler gear spring with a clicking noise when the unit eats tape. Check for a bad capstan motor stator assembly when the tape pulls out in all modes of operation. Cut off the excess end of the spring near the take-up reel when the unit shuts down after running for several hours and begins to eat tape. Suspect a bad idler when the take-up reel is inoperative, there is no fast forward or rewind, or the tape pulls out. A bad carriage assembly that is slightly out of line can produce a rubbing sound heard in fast forward or rewind and result in the unit eating the tape. Pulling of tape can be caused by a bad friction gear and transmitting arm assembly.

Replace the idler, belts, and capstan motor when the unit shows intermittent operation and then spills out tape. Replace the capstan motor, idler, and belts when the VCR eats tape intermittently. The tape was damaged in a Zenith VR1820 VCR as a result of a bad tension spring (809-1992).

Cylinder and drum motor problems

The cylinder motor drives the video heads at the required speed, and servo problems affect the quality of the picture. Check voltages on the cylinder drive motor IC and applied to the motor terminals. Locate the motor driving the video heads or drum assembly at the bottom of the unit, and take continuity test measurements.

A bad relay can cause no rotation of the cylinder or capstan motors. Resolder all cylinder and capstan PC board connections if the motors are not working or are intermittent, and then replace the drum stators. Check for improper voltage to the cylinder motor circuits caused by a shorted electrolytic in the power supply. Replace both end sensors that are open after the tape loads, the cylinder does not rotate, and then the unit shuts down. Replace the lower half of the cylinder when the tape loads, retracts, and then the cylinder does not rotate. Check for bad connections at the drum motor when the tape loads and there is no drum rotation. Suspect open small resistors in the voltage source when the cylinder motor does not rotate (Fig. 11-11).

A bad cylinder motor IC can cause the capstan motor to stop rotating. Replace the cylinder motor IC when the VCR powers up, the cylinder spins continually, and the unit shuts off. Check for a bad servo IC when the drum is inoperative. Replace the cylinder or drum motor IC when the cylinder does not rotate. Check the voltage regulator IC when there is no cylinder movement.

An open transistor voltage regulator in the power supply source can cause the cylinder motor to remain stationary. Check for badly soldered joints at the transistor voltage regulator when the capstan does not rotate. A bad zener diode regulator can cause the cylinder motor at power-up to shut down in play mode. Check for a defective diode in the cylinder circuits when the cylinder motor becomes intermittent and stops in play mode. The cylinder or capstan motor may not run with a bad end lamp.

11-11   Check for a bad servo IC when the drum or cylinder does not rotate.

When the cylinder motor runs fast and then shuts down, suspect poorly soldered joints on the motor stators. Replace a drum that shows irregular speeds. Check for a bad 100-μF electrolytic in the power source when the cylinder runs too fast. Suspect a defective electrolytic in the servo circuits when the drum motor runs fast all the time. Check all the electrolytics in the power source when the cylinder runs all the time with no other functions. Test all diodes in the power source when the cylinder runs too fast.

A bad servo IC can make the cylinder run too fast. An intermittent voltage regulator IC can cause erratic speed of the cylinder motor. Replace the 4.19-MHz crystal when the drum spins at a high rate of speed, there is no display, the LED lights up, and the cassette is inoperative.

SPEED JUMPS, COLOR FLASHES, PHILCO

The symptoms in a Philco VT3010A01 VCR were speed jumps, color flashes, and a sound like that caused by a bad cylinder bearing. The cylinder motor voltage source that feeds the cylinder assembly and driver IC checked okay. The 5- and 4-V sources were low. On checking the power source components within the power supply, the electrolytics were tested with an ESR meter. All diodes checked okay with the diode tester of a digital multimeter (DMM). C1017 (1000 μF) in the 5-V source and C1013 in the 14-V source were replaced (Fig. 11-12).

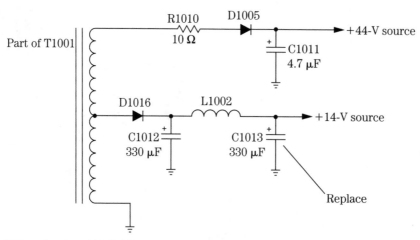

Part of T1001

R1010 10 Ω

D1005

+ C1011 4.7 μF

+44-V source

D1016

L1002

+ C1012 330 μF

+ C1013 330 μF

+14-V source

Replace

11-12   Replace C1012 (330 μF) and C1013 (330 μF) for speed jumps, color flashes, and a sound that resembles a bad cylinder bearing in a Philco VT3010AT01 VCR.

Capstan motor problems

The capstan motor provides tape motion in play, record, and fast-forward modes. The capstan motor is driven by a drive speed and system control IC. No tape motion can be caused by a defective capstan motor, drive belt, driver, or servo IC. Check for supply voltage at the driver and system control IC. Measure the voltage applied to the motor terminals during play mode.

Suspect a bad capstan motor when it will not rotate and voltage is found at the motor terminal. A dead VCR with a blown fuse can be caused by a defective capstan motor. A bad capstan motor will not operate in playback, record, fast-forward, or rewind modes with a blown fuse. The shorted capstan motor can cause parts in the power supply to become hot and smoke. Check for bad capstan motor bearings when the capstan motor will not rotate and the motor appears frozen. A bad capstan motor can cause the tape to load halfway and then eject, and the reel table will not turn. Replace D108, D109, and D110 in a GE VG2010 VCR when the capstan motor will not run. A defective end lamp can keep the capstan motor from operating. An open 2.2-ohm resistor feeding the capstan motor keeps the motor from operating.

Check for an open drive motor IC when there is no rotation. Check for a bad fuse and servo IC when the capstan motor will not rotate. Suspect a bad transistor voltage regulator feeding the drum motor IC. Replace the IC voltage regulator when the capstan is inoperative in play mode and shuts down. The capstan motor will shut down with badly soldered connections on the transistor voltage regulator.

A bad capstan motor can result in no tape rotation and no frequency generator (FG) pulse. Replace the capstan motor when there is no FG pulse from the motor and no motor rotation. A bad capstan motor can cause the cassette to get stuck inside, the unit to not play or eject the cassette, or the unit to not load or eject. Replace a defective capstan motor when there is a jerky motion in play mode and it looks like a bad bearing. A bad capstan motor also will not let the VCR rewind.

When the capstan motor will not rotate, check for a bad motor or a defective Hall sensor in the motor assembly. A defective capstan motor can cause the tape to load halfway and then eject and the reel table to remain stationary. The capstan motor in a Sony VCR has no rotation and eats tape, and this was caused by resistor R506 (1 kilohm) from CN1-4 to IC502-72 on the bottom of the PC board.

CAPSTAN RUNS FAST

Check for a defective diode in the power source with the diode tester of a DMM when the capstan motor keeps rotating very fast (Fig. 11-13). Check for bad diodes in the power source when the capstan motor is intermittent in start/run mode. A bad capstan motor also can operate at fast speeds. A bad capstan motor can cause the tape to run at a fast rate of speed in fast-forward, record, and playback modes with the FG pulse missing. The tape running at a higher speed on the audio control (AC) head can be caused by a glazed pinch roller. Repair the loose capstan magnet when the capstan motor speed is very fast.

Check all electrolytic capacitors (10 to 33 μF, 35 V) with an ESR meter when the capstan runs at full speed. Change the 330- to 1000-μF electrolytics when there is a low 5-V source that makes the capstan motor run fast in play mode. Check for defective bypass capacitors in the servo IC when the capstan motor runs at fast speed. Suspect the 0.47-μF capacitor off of the servo IC when the capstan runs at full speed, especially in playback mode.

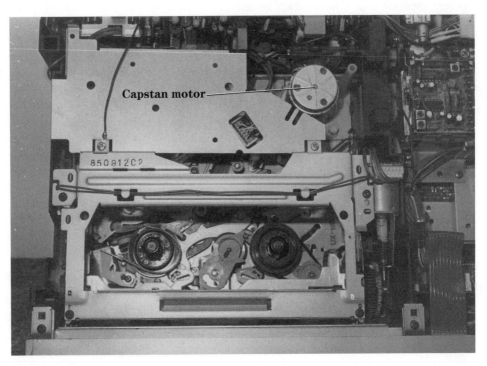

11-13 A bad capstan motor can operate at fast speeds.

Check the cylinder motor IC when there is erratic speed. The servo or capstan motor driver IC can cause the capstan motor to run at full speed in the play mode. Check the servo IC when there is intermittent capstan motor and drum rotation. A bad voltage regulator IC or transistor can cause erratic capstan motor rotation in fast-forward mode. Replace the capstan control transistor when the speed runs fast and fast forward shuts off after a few minutes. Badly soldered joints on the voltage regulator IC or transistors can cause erratic capstan motor rotation in fast-forward mode. A leaky voltage regulator IC can cause no or intermittent capstan motor rotation and can eject the tape when loaded. A bad capstan motor can cause a jerky play mode that might look like a bad motor bearing.

CAPSTAN RUNS SLOWLY

Check all small-ohm resistors in the voltage source when the capstan motor runs slowly. Replace the voltage regulator transistor when the system runs slowly, and the pause is inoperative. A bad IC regulator can cause the capstan motor to run slowly and result in a jittery motion. Check for a bad IC on the servo PC board when the capstan motor runs slowly and sometimes stops. Both the capstan and cylinder motors may run slowly when there is a bad signal at the 3.58-MHz crystal.

Clean out all grease in the capstan motor bearings and relube when there is a slow capstan speed or drag and wow conditions. Clean the capstan brake pads when the capstan motor is not running fast enough.

The capstan motor runs constantly with no on/off switching when there is a low-voltage regulator source at the capstan drive IC. A bad capstan motor can cause the motor to run constantly with the power turned off.

NO CAPSTAN MOTOR, JVC

Sometimes the capstan motor rotation was intermittent and at other times it was dead in a JVC HRD-170 VCR. At first, the capstan motor was suspected of being defective. The motor voltage was monitored at the terminals and was intermittent. Tracing the motor leads back to motor driver pins 3 and 7, the voltage was still intermittent. Both the 17- and 12-V supply voltages were normal at the capstan motor drive IC604. Replacing the defective IC604 cured the intermittent capstan rotation (Fig. 11-14).

STARTS TO PLAY, THEN STOPS

Check the capstan driver or control IC when the VCR starts to play and then shuts down. Inspect and clean the mechanism position switch in some models. This problem can be caused by a defective gear drive assembly or bracket assembly. Suspect a defective capstan motor if the VCR shuts down instantly.

A bad capstan motor can shut down the rotation immediately. When the capstan motor stops after a few minutes of operation, replace the motor. Check for badly sol-

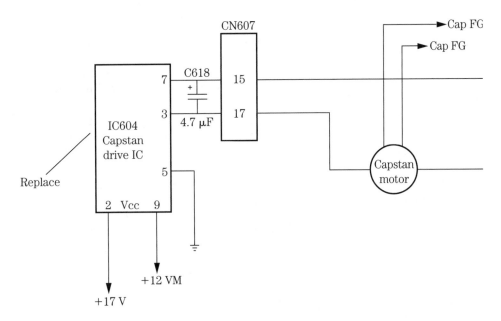

11-14   Replace a defective capstan driver IC for dead or intermittent rotation in a JVC HRD-170 VCR.

dered joints on voltage-dropping or isolation resistors in the power supply when the capstan runs fast and then shuts down.

Check the loading motor IC when the VCR shuts down after a few minutes of rotating and possibly pulls out tape. Suspect a capstan drive IC for intermittent and shutdown symptoms. A bad capstan motor driver IC can shut down after 5 seconds with no capstan rotation. The defective capstan motor IC may operate up to 3 hours and then shut down. Suspect a motor voltage regulator IC when in play mode the unit goes to stop and then the capstan motor does not rotate. Check all transistors on the servo PC board when the capstan motor runs in the stop mode. Replace the voltage regulator transistor when the capstan motor begins to play and shuts down.

Fast-forward and rewind problems

Check the 1-A fuse when there are no rewind or fast-forward modes. A bad capstan motor may cause the fuse to blow in play, rewind, and fast forward. The capstan motor may have a flat spot (on the armature) causing no fast forward or rewind. A bad loading motor assembly may produce no rewind, fast-forward, or play modes. Clean the old grease off the idler gear assembly and relube when there is no fast forward or rewind and the tape occasionally gets stuck.

Replace the forward/reverse bracket and broken shifter when there are no rewind, fast-forward, and play modes. Check for a bad plate assembly for no fast forward or rewind or slow operation. No play or rewind occurs with a bad slide switch on the carriage assembly. Check for a bad main lever latch on the main cam gear assembly when there is no fast forward, rewind, or play. A bad lever holder assembly and main cam gear can cause no fast forward or rewind. Check for a broken spring on the loading motor assembly when there is no fast forward or rewind, and all other functions are okay. Check for a bad AY assembly when the VCR will not rewind or fast forward.

Check for a bad coil or resistor connection in the voltage source of a motor with no fast-forward or play modes. Suspect a system control IC for no fast-forward, rewind, play, or eject modes. Replace the switching regulator transistor when the VCR will not play, fast forward, or rewind.

A soft brake not engaging or failure to tighten the spring on the soft brake assembly can cause no reverse or fast-forward mode. Clean a dirty cam switch for no rewind, fast-forward, or play modes. Replace the brake solenoid switch for no rewind or fast forward, whereas other functions might be okay. Damaged teeth in a cam gear can cause no rewind or reverse-search operations. Replace the mode switch when the VCR goes into record, rewind, and fast forward by itself.

FAST FORWARD AND REWIND SHUT OFF

Replace a bad Sycon IC when the VCR powers up, goes into fast forward or rewind, and then shuts off. Besides replacing the Sycon IC, a defective transistor may need replacing on the capstan control when rewind and fast forward shut off in 1 or 2 minutes.

Change both reel sensors when the rewind slows in 7 to 10 seconds and goes to stop. Replace the reel motor assembly when rewind or fast forward stops in 2 to 5 seconds. Check for a bad end sensor on the cassette housing when the VCR stops or goes into rewind or fast forward. Intermittent shutdown caused by an end sensor may let the VCR go to stop and record and then shut down. Replace the take-up reel sensor when the VCR shuts down during play, rewind, or fast forward.

Clean a dirty cam switch when rewind and fast forward are inoperative, and the unit shuts off. The unit may try to load, there is no rewind, and then it shuts off, and this is caused by a bad loading motor cam assembly. Clean the mode switch for slow fast forward and rewind and if the unit plays awhile and then shuts off. A bad mode switch in a Sharp VC2230U VCR caused no eject or fast-forward operation.

Tape ejection problems

Look for defective basket gears when the tape will not eject or the VCR keeps ejecting the tape after loading. A defective lift arm assembly or worn loading belt can keep the tape from ejecting. Check the loading motor and control IC when the tape keeps ejecting. A defective reel motor may not unload the tape. Suspect a right end sensor when the tape loads, the reel rotates, and then the unit ejects the tape.

Suspect a beginning tape sensor when the tape loads and then ejects. Replace the rack front-loading gear assembly when cassette loading is incomplete, and the unit ejects the tape. Replace the gear assembly and side plate when the gears jump, and the cassette will not load or eject. Suspect a bad lift cam when the tape will not eject. A worn wheel unit will not allow ejection of the cassette.

Check the basket gear assembly when there is no ejection operation. A faulty loading or cassette belt can prevent the tape from ejecting. A bad basket gear will not allow ejection of the tape. Suspect a bad basket gear assembly that will not load the tape and may keep ejecting a tape that does load. A bad loading motor and worn cam gear can cause no eject operation. Look for a bad plate assembly, broken teeth, or a faulty loading arm when the tape will not eject.

Intermittent ejection of the tape can be caused by a bad capstan motor. Replace a bad clutch assembly for intermittent ejection. The right-side bracket assembly may cause the front-loading assembly to bind and result in intermittent ejection. Replace the mode switch for no tape ejecting or loading. A bad carriage gear assembly also can cause intermittent tape ejection. A bad connection on the ribbon cable can result in the cassette loading, the capstan motor running, and then intermittent ejection of the tape.

A bad inverter transistor in the power supply can cause slow ejection, a stuck tape, or very little torque on the capstan motor. Check for an open resistor or change in resistance in the switching regulator IC for a unit that does not eject, the tape sticks, or the unit powers up and shuts down in a few seconds. Suspect a loading motor IC or capstan servo circuit IC for no tape ejection. Check for a defective diode in the power source on the main PC board when the cassette will not load or will not eject. A defective loading motor IC (IC215) and servo circuit IC (IC206) resulted in no ejection operation in a GE VG7500 VCR.

A bad bracket assembly can cause no loading or no ejection of the tape. Replace the loading bracket assembly when the tape jams in ejection mode. Suspect a bad capstan motor when the tape loads halfway, the reels will not turn, and then the tape ejects. Look for a defective capstan motor or a cracked pulley when the unit eats tape and will not eject. Replace the loading motor belt when the tape sticks or is intermittent and there is no ejection of the tape. A bad right end sensor may allow the unit to load the tape and the reels to turn, but the unit ejects the tape. Check for a loose ground screw on terminal deck PCB51 when there is no or intermittent ejection in a Zenith VR1820 VCR.

Record and playback problems

The main components within the playback and recording circuits are the video tape heads. In record mode, the video heads apply the signal to the magnetic heads and revolving tape. The magnetic video heads provide video recording and playback as well as audio and control information. In playback, the video heads pick up the recorded signal and amplify it to the playback circuits (Fig. 11-15).

Check the capstan motor IC when the capstan motor will not turn in play and record modes. Rewind, fast forward, and play do not occur when there is no drum rotation as a result of a faulty control IC. Suspect the voltage regulator IC when there is no fast forward, play, or record. Check the analog-to-digital (A/D) control IC when there is no play, fast forward, or rewind. Replace the capstan motor driver IC when the unit does not play, the take-up reel does not turn, and the capstan motor coils get hot.

Look for an open 1-A fuse when there is no rewind in fast-forward and play modes. A bad slide switch on the right side of the carriage assembly can cause intermittent

Video head
on drum

11-15   In playback, the video heads on the drum or cylinder pick up the recorded signal from the tape and apply it to the playback circuits.

functions with no play and record modes. Replace the forward/reverse (F/R) bracket when there is no fast forward, rewind, or play. Check for a bad safety switch when the VCR will not record. Test all leaky electrolytics with an ESR meter on the video PC board for no playback and record in SP mode.

Playback problems

Make sure that the audio/record and erase heads are cleaned before trying to repair the various playback problems. No playback mode can be caused by a broken pin on the limited post lever assembly. No play can be caused by a bad threading motor. Replace all belts and the mode switch for no video in playback and when the VCR shuts down. A bad diode in the power supply and no UNSW (37 V) can cause no video and tuner playback.

Check for a bad electrolytic (4.7 μF) in the servo IC section when there is no video playback and the tuner is normal. Replace the video/audio process IC when there is no video playback, the screen is black, and the audio is normal. Replace the drum motor IC when there is no playback and the tape goes to full speed and then back to a half speed. No capstan motor rotation in play and record modes was caused by a defective capstan drive IC.

Check the voltage regulator transistor in the power supply for no playback, rewind, or fast forward or the tape gets stuck in the player. No video in the playback mode with normal fast-forward and rewind modes can be caused by low voltage and a defective transistor in the luma-chroma circuits. No playback was caused by a bad 12-V record switch transistor. Replace a leaky record/play switch transistor when there is no video in playback but normal playback audio.

Check the capstan control and driver IC for fast play and intermittent fast-forward speed. Intermittent fast speed can be caused by poorly soldered joints on the driver IC. No capstan FG pulse to the system control IC can cause a fast play-back speed. Check the bias transistor when there is high-speed playback in EP mode. Clean and insulate the silicon shock absorber conductive plate when there is intermittent speedup of the playback mode. Look for leaky or open electrolytics in the power sources for fast speed in playback. Replace a bad mode switch when the play mode runs too fast and looks like a fast-forward search. A bad capstan stator can make the play mode very jerky.

PLAYBACK SHUTS DOWN

Replace the reset IC when play mode operates by itself and then shuts down. Play mode will shut down with a defective loading motor IC. Check the wave-shaper IC on the tape deck PC board when the play stops in a few seconds in both EP and SP modes and then unloads. A bad diode in the power supply source can cause intermittent capstan rotation and stops in both play and record modes. Replace the capstan reel driver IC when the VCR shuts down in play mode. Sus-pect a bad voltage regulator IC when the capstan motor shuts down in play mode. Replace the capstan motor IC when the VCR shuts down after playing for one-half hour.

Replace all belts and the mode switch when there is no video in playback, the tape loads, and the unit shuts down in a few seconds. Check the mode switch when the unit shuts down in play, fast forward, and rewind. Look for a bad reel sensor under the take-up reel when the unit shuts down in play mode. A bad clutch reel assembly can cause the play mode to stop. Check for a bad mechanism position switch if play starts and then stops. A bad drive gear assembly can cause the unit to play and then shut down. A bad bracket assembly can cause the tape to start to play and then shut down. Respace the take-up reel when play mode shuts down in a few seconds. Suspect a bad photo interrupter when the VCR quits in the play mode.

NO PICTURE IN PLAYBACK

Clean the video heads with chamois and alcohol or a cleaning solvent. Trace the video head wires back to the prerecord amp. Check the universal replacement man-ual for the right replacement part as well as what stage the IC functions in. If the IC replacement is not listed, trace the signal with a scope. Check the IC number in a similar VCR chassis or schematic.

Check the supply voltage applied to all IC and transistor components within the playback mode. A poor picture in playback can be caused by a faulty prerecord amp IC or luma process IC. Check the voltages on all ICs and transistors to determine if a component is leaky. Inspect the PC board for burned or open resistors.

Snow and noise in the picture or a faint picture can be caused by a defective upper cylinder and head amp IC. Check the tuner-demodulator circuits if there is only half a picture. Resolder the pins on the head amp board and check the VHF block assembly for a fuzzy and noisy picture. Check for a missing control pulse when there is noise in the playback mode. Replacing IC components within the video circuits can solve most playback audio and picture problems.

Poor recordings

Clean the video, audio, and erase heads when there is poor or fuzzy recording. Keep the video heads clean. Does the VCR play back normal audio and video? Usually the same IC circuits are used in both record and playback modes. If the video does not record, suspect additional circuits such as the luma record/playback process IC, transistor record switches, the video record amp, or supply voltages (Fig. 11-16). Also check the safety switch, record subassembly, and lower drum assembly when there is no video recording.

Replace the erase head if old recordings are present, and the record mode is inoperative. Check for a broken or out-of-line record tab switch when there is no record mode. Suspect a bad safety switch when the unit will not record. Look for a bad recording mode assembly when there is intermittent recording. Leaky capacitors in the video PC board can produce no playback or rewind in SP mode. Replace the lower cylinder assembly when the player shuts down in play or record mode.

Check the voltage regulator IC if the unit shuts down in playback or record mode. Resolder all voltage regulator IC or transistor terminals for shutdown in record and lost drum synchronization. A bad servo IC can make snowy recordings in SP mode. If the power shuts off after one-half hour in play and record, suspect a defective 12-V IC regulator. Look for a bad transistor in the playback circuits when there is no record or playback. Check all capacitors and ICs when there is no color in record mode.

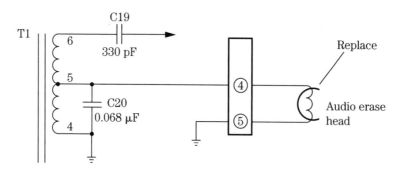

11-16 Replace the erase head when the old recording is not erased in a Zenith VR1820 VCR.

Suspect a bad control transistor when there is no tape movement in play or record and there are indifferent pauses. Look for a bad silicon diode in the power supply source when there is intermittent capstan rotation or the unit stops in play and record modes. Replace a bad surface-mounted switching transistor when there is no record mode.

POOR AUDIO RECORDINGS

Check large electrolytics in the power supply circuits when there is no audio in record mode. Test all electrolytics in the bias oscillator circuits with an ESR meter when there is no audio recording. Improper supply voltage in the oscillator circuits can cause no audio recording. A bad zener diode in the 9-V source on the main power supply PC board can result in no audio recording.

Replace small 1- to 5-μF electrolytics on the PB/REC preamp IC for no audio. Check for a bad PB/REC IC when there is no audio recording. A defective bias oscillator transistor can cause no audio in record mode. Replace a bad oscillator transistor when there is no audio recording, especially when there is old audio still on the tape. No audio record with normal playback can result from a bad bias oscillator transistor. Replace IC5AZ when there is a high-pitched audio squeal in a Mitsubishi HS410UR VCR.

A bad AC head assembly can cause no audio recording. Suspect a bad mode switch when there is no audio or video recording. Open coils on the main PC board can result in no audio recording. Replace the full erase head when there is no audio recording. A defective full erase head can cause intermittent audio recording. A glazed pressure roller can result in no audio recording and may cause a muffled sound. Replace the audio bias transformer for poor audio recording and very poor erasing of the previous recording. Suspect a bad record safety switch when the audio cuts out during a recording.

NO AUDIO ERASE, RCA VPT200

No audio erase in record or full erase can be repaired with a factory audio parts kit (195685). Also check C521, Q503, and C503 when there is no audio erase in an RCA VPT200 VCR (Fig. 11-17).

Audio problems

VCR audio bias oscillator circuits are similar to those found in a cassette player. A bias signal is applied to the full erase, audio erase, and audio record/playback heads. A waveform test at either head will indicate if the bias oscillator is functioning.

Check the voltages on the bias oscillator transistor. Look for a small transistor near the bias transformer. Trace the bias lead from the full erase or audio erase head to locate the small bias transformer. Resolder all transformer board connections when there is intermittent erasing. Check for leaky capacitors across the bias transformer when there is no waveform at the full erase head.

Replace the 12-V regulator IC when there is no driver or capstan rotation and no tuner audio. Check for bad resistors on the IC audio amp when there is no audio in

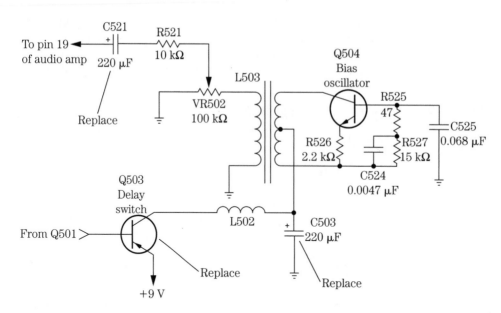

11-17 No erase occurred in an RCA VPT200 VCR with defective C521 (220 μF) and C503 (220 μF).

playback. No audio or video can be caused by a defective transistor in the PB 5-V SW circuits. Intermittent audio can result from a defective audio output IC. Suspect a transistor or IC regulator when there is low or no audio supply voltage. Intermittent audio and video can be caused by a defective IC near the video input circuits. Replace the servo IC for garbled audio in EP mode.

Distorted audio in playback can be caused by a 3300-μF electrolytic in the power supply circuits. Suspect a bad electrolytic in the servo phase error line for garbled audio and erratic operation. Check all electrolytics in the audio playback circuits with an ESR meter when there is no audio reproduction.

A shorted diode in the tuner 30-V line can cause no video or audio from the tuner. Replace the zener diode in the 9-V record source on the main PC board when there is no audio and normal video. A bad diode in the 6-V line can result in no video or audio.

Replace a bad clutch assembly for an audio wow and flutter symptom. A bad clutch gear can cause garbled audio and slack in the take-up action. When the VCR chews up tape and has garbled audio, suspect a bad take-up clutch bearing that is binding. Garbled audio can be caused by a bad capstan belt. A dirty or worn reel brake band also can cause garbled audio. A broken or loose tension band can result in distorted video and audio. A bad pinch roller bracket can cause the eating of tape and no audio and video. A defective upper cylinder motor bearing can cause garbled sound and a chattering capstan in reverse mode. Clean the audio switching relay when there is garbled audio in EP mode.

Clean the relay contacts for hum in the sound. Check for a broken post and a bad brake take-up reel when there is erratic take-up and quivering audio. Replace a bad brake band when the audio growls. Replace a bad lever assembly on top of the deck for motor boating in the sound.

POOR AUDIO/VIDEO PLAYBACK

Check for a defective IC on the capstan phase error line when there is no audio playback and normal video. A bad servo IC can cause no audio in playback mode. Check the 12-V regulator transistor for no audio and video in the playback mode. Replace a defective transistor on the Sycon PC board when there is no audio or video playback. No audio in playback can be caused by a regulator transistor in the audio 9-V line on the main PC board. Suspect a bad muting transistor when there is no audio in playback mode. Check the surface-mounted transistor playback switch on the video head assembly when there is no video or audio in playback.

Suspect electrolytic capacitors in the servo circuits when audio playback is garbled and erratic. Check all electrolytics on the head amp section when there is no video or audio in rewind or playback. Open capacitors in the audio section on the main PC board can cause no audio in playback with normal video. No audio or video in playback can be caused by a shorted electrolytic on the video head picture amp board. A bad electrolytic (330 μF) on the UNSW 5-V line can cause distorted video in playback. A shorted electrolytic on the audio/video board can cause no erase of the old audio.

A bad pinch roller bracket can result in no audio or video in playback. No audio in playback with a noisy right channel can be caused by a defective audio/CTL head assembly. Open coils in the head circuits will not play back the audio. A bad clutch assembly can cause a flutter in audio playback and record.

No audio erase can be caused by no voltage supply to the bias circuits with open resistors. Suspect a bad bias oscillator transistor when there is no erase of audio in record mode. A defective full erase head can cause low or no audio. A bad AC head or full erase head can cause a very low or no audio playback. Check the wiring on the heads for intermittent playback audio. Check for poorly soldered connections on the full erase head when the unit will not erase the old audio from a previous recording. No audio or full erase can be caused by a defective IC on the +9-V record bias circuit. Replace the audio bias transformer for poor audio recording and a very poor erase problem.

NO ERASE OF OLD AUDIO, RCA VR250

The bias oscillator tested normal in the erase head circuits of an RCA VR250 VCR. The voltage supplied to the bias oscillator was normal. Q501 and Q503 in the center-tapped circuit of the bias oscillator were checked with the diode tester of a digital multimeter (DMM). Q501 was open. Replacing Q501 with an ECG125 universal transistor solved the no-erase problem (Fig. 11-18).

Lines and noise

Very loose back tension can produce flagging at the top of the picture. A bad tension band can cause snow to come and go in the picture. Check all electrolytics for hum bars in the picture. Poorly soldered bad head amp connections can cause a snowy video in EP and SP modes and looks like a bad tape head. Poor adjustment of the cylinder can cause noise bars drifting through the picture. A bad upper cylinder can cause noisy lines in the picture in playback. A bad upper cylinder also can cause white lines in tape recording.

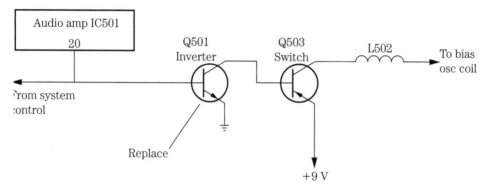

11-18 Q501 caused no erase of the old audio in an RCA VR250 VCR.

Check the system control IC when there is no video in playback or the picture jumps and flutters. A bad electrolytic on the system control IC can cause bad tracking, chipmunk speed, and lines of snow. A bad servo IC can cause lines through the video and looks like a bad drum motor. A bad servo IC can cause the upper cylinder to rotate more slowly than normal and produce lines in the picture (Fig. 11-19). A defective servo IC can produce a snow bank moving in the picture with audio wow and flutter. Replace the capstan control IC for noise bars in review mode. Check for a defective transistor on the preamp PC board when there are hash marks, no audio in record, and normal audio recording.

When the unit will not load all the way or there are lines in the top of the picture, lubricate the loading ring assembly, and tighten the back spring slightly. Check for a loose head relay switch when there are noise lines in record and play mode. Realign and service tape guides for lines in the video during playback. Lines in half the picture can result from a dirty or stuck switching relay.

Replace bad filter capacitors in the power supply for lines in the picture, bad color, and the appearance of a bad tape head. Bars through the lower part of the picture can be caused by a defective 1-μF electrolytic on the drum pulse generator line in the servo section of the PC board. Open electrolytics on the drum or cylinder drive PC board can cause horizontal lines in the picture at all speeds. Replace the 3.3-μF surface-mounted capacitor on the lower cylinder for lines in playback and the appearance of a bad tape head.

Noisy conditions

Replace the mode switch when the unit makes a grinding noise with no picture in reverse, pulls out tape when ejecting, and may have intermittent operation. A bad worm gear in the loading assembly can cause a clicking sound. A bad trigger gear assembly can cause a mechanical grinding noise. A loose tape guide post can cause a tracking noise. A high-pitched squeal on play and record can be caused by a bad static discharge arm on the upper cylinder assembly. Noise in the picture can be caused by a bad upper cylinder. Replace the left loading link lever when there is a noisy band at the top of the picture. Replace bad drum stators when there is noise in the picture. A bad clutch wheel can result in a loud grinding noise.

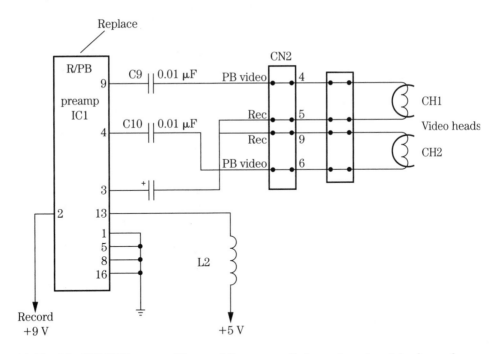

11-19 A bad PB/REC preamp IC caused the upper cylinder or drum to rotate slow and produced lines in the picture of a JVC HR-D1700 VCR.

Check all electrolytics on the power supply board when there is a squeal in the power supply. A bad power transformer on the switching block can cause a low-level whine or squeal. A bad electrolytic in the regulator circuits can cause noise in the picture when the capstan motor stops. A shorted diode in the 30-V line can cause a hissing noise.

When the loading belt screeches, check the rack and bracket assembly. A bad switching audio relay can cause noise in the audio. A bad relay also can cause a cracking and static noise in the audio playback mode. Check the carriage arm on the side of the cassette loading mechanism for a grinding noise. Replace the bushings in the capstan motor bearing holder for a vibrating noise in fast forward or rewind or during mode changes.

Dew circuit operations

Resolder the pin terminals on the servo control IC when there is no dew light. Check for a foil or trace break on the servo control IC terminals if there is no dew light. Replace the sensor lamp when there is no dew light, the tape loads, and there are no functions. Resolder the transistors or ICs for no operation of the dew light. Check for open connections on the servo IC when there are functions and the dew light is flashing. Check all traces and PC wiring breaks with an ESR meter. Replace a bad dew light sensor when the flashing stop light is on. Suspect a bad transistor in the dew light circuits when the dew light flashes.

Table 11-1. VCR service schedule (Radio Shack).

Ref. no.	Part	1,000 h	2,000 h	3,000 h	4,000 h
B2	Cylinder assembly	○ (check)	● (change)	○	●
B3	Loading motor			●	
B6	Pinch roller arm assembly		●		●
B8	Pulley assembly		●		●
B21	Belt LDG		●		●
B26	Clutch block assembly		●		●
B27	Band break assembly		●		●
B28	Main brake S assembly		●		●
B29	Main brake T assembly		●		●
B30	T break arm assembly		●		●
B31	AC head assembly			●	
B32	Reel assembly			●	
B37	Capstan motor		●		●
B52	Belt FWD		●		●
B54	Ground brush assembly			●	
B73	Full erase head		●		
B86	F break assembly		●		●

Service schedule of components

Clean all parts involved in tape transport (upper drum and video head/pinch roller/audio control head/full erase head) using 90 percent isopropyl alcohol. Check Table 11-1 for the suggested maintenance for VCR components. A listing of VCR numbers keyed to VCR manufacturers is shown in Table 11-2.

TV/VCR combo repairs

Most of the mechanical problems found in a regular VCR can exist in a TV/VCR combo. Troubles found in any TV chassis also can be found in the TV/VCR combo. You may find an AM/FM radio in the same TV/VCR chassis. The big difference in the TV/VCR combo is the power supplies. You may find two different types of power supplies, one in the TV and another in the VCR circuits. In a Panasonic PV-M2021 TV/VCR, a switching power supply powers the VCR circuits, whereas a line voltage regulator power supply supplies voltages to the TV circuits. Only one power supply may power both the VCR and TV in some models. Two different power supplies are found in the early battery-powered and power-line portable TV/VCRs. The same test equipment can be used to service a TV/VCR combo as found on any service bench (Fig. 11-20).

TV/VCR part layout

Usually the VCR is located at the front of the chassis within a shielded container. The VCR chassis can be removed by taking out several screws that hold moving and loading components. A shielded bottom piece can be removed to get at the TV PC board.

Table 11-2. UL listing number to VCR manufacturer (unofficial)
(courtesy of *Electronic Servicing & Technology* magazine).

UL number	Manufacturer	Brand names
146C	Goldstar	
153L	NEC	
16M4	Samsung	Supra, Multitech, Unitech, Tote Vision, Cybrex, GE, RCA, Sears
174Y	Toshiba	Sears
238Z	Hitachi	RCA, GE, JC Penney, Pentax
270C	Sony	
277C	JVC	
282B	Sharp	
289X	Emerson	
333Z	Symphonic	Teac, KTO, Realistic, Multitech, Funai, Porta Video, Dynatech, TMK
336H	RCA	
347H	NAP	
43K3	Kawasho	
403Y	Fisher/Sanyo	Realistic, Sears
436L	Quasar	
439F	JVC	Zenith, Kenwood, Sansui
444H	Zenith	
44L6	TMK	Emerson, Lloyds, Broksonic
504F	Sharp	Wards, KMC
51K8	Portavideo	
536Y	Mitsubishi	Emerson, Video Concepts, MGA
540B	GE	
570F	Sony	Zenith
623J	Sampo	
628E	Samsung	MTC, Tote Vision
679F	Panasonic	RCA, GE, Magnavox, Quasar, Canon, Philco
723L	Sanyo	
727H	Hitachi	
74K6	Funai	
781Y	NEC	Dumont, Video Concepts, Vector, Sears
828B	Panasonic	Olympus
843T	Magnavox	
86B0	Goldstar	Realistic, JC Penney, Tote Vision, Shinron, Sears, Memorex
873G	Mitsubishi	
41K4	Portland	

Follow the removal procedures found with any TV/VCR combo service literature if the exact service information is not available. Most of TV parts can be seen on the side and toward the rear of the TV chassis. For those who service TVs as a warranty station for certain brands, several different manufacturer cables and plug-in connections can be obtained for easy servicing.

11-20 The TV circuits are located around the side and rear with the VCR in the front in a Panasonic TV/VCR.

The power supply

A standard raw dc power line supply may contain a bridge rectifier circuit with small-ohm resistors and a large electrolytic capacitor filter network. The switching low-voltage power supply is the same as the switching supply in a regular TV circuit. You may find a raw dc power supply and a switching power supply in one TV/VCR chassis. A Panasonic PV-M1347 TV/VCR contains a switching power supply in the low-voltage circuits and secondary voltages from the flyback for both the TV and VCR circuits. A switch-mode power supply (SMPS) is found in an RCA 13TV701/19TVR60 unit. All ac power supplies should be serviced with the TV/VCR plugged into an isolation transformer.

In the early 9- and 13-in TV/VCR combos that operate from batteries and a switching power supply, a separate switching control IC, transformer, and silicon diodes provide a +117-V dc operating voltage. The power line switching power supply provides +117- and +33-V sources for the TV circuits. The battery power supply may furnish a +117-V source to the cathode-ray tube (CRT) and high-voltage (HV) circuits, with 6 and 12 V to both the TV and the VCR. Silicon diodes are placed in the voltage sources so that when in ac operation the voltage does not feed back into the battery circuits (Fig. 11-21).

Separate 5-, 6-, and 9-V regulator transistors are found in the output voltage circuits. The 5-V regulator is fed from the +6-V source, and the +6-V source is fed from the +12-V line with a voltage regulator transistor. The +117-V source feeds the CRT

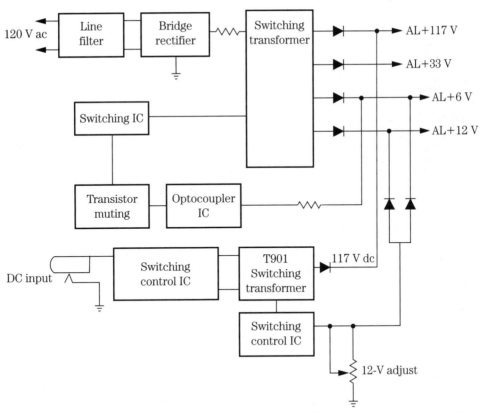

11-21 Block diagram of a battery-operated and ac switching power supply in a portable TV/VCR.

and HV circuits. The +33-V source provides power to the tuner and servo system control circuits. The tuner, system control, and TV micon circuits are supplied by the +5-V source. The +6-V source feeds the VCR servo system control circuits. The +12-V source provides voltage to the audio output and capstan control circuits. The tuner, video, and chroma circuits are powered by the +9-V source.

Extra care must be exercised when servicing the switching power supplies to take critical voltage measurements within the hot and cold systems so as not to damage other components and also to obtain the correct voltage reading. Critical voltage measurements without a schematic can be compared with those of another TV/VCR chassis. A slip of the test probe, however, can damage other parts in various TV/VCR circuits.

The raw dc power circuits

The raw dc power supply is like that found in a regular power line voltage source. The most common problems in the power line supply are a blown fuse, burned low-ohm resistors, shorted or leaky silicon diodes, open or leaky transistor or IC voltage regulators, and leaky or open filter capacitors (220 to 680 μF, 250 V). You may find a defective relay that turns on the TV or the degaussing circuits. Often a leaky or

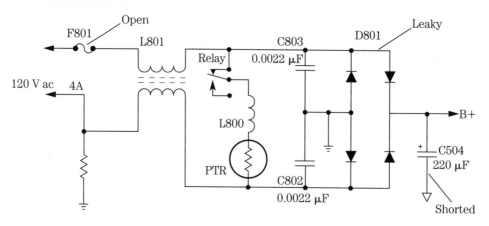

11-22 A shorted C504 (220 μF, 250 V) damaged D801 and F801, resulting in a dead chassis in a Daewoo DVQ-19H2FC TV/VCR.

shorted silicon diode blows the main fuse. A shorted electrolytic filter can blow the main fuse or lose capacity, which can shut down or prevent startup of the TV chassis. A bad ac relay may cause intermittent startup or prevent startup of the chassis. Service the power line circuits with extreme care (Fig. 11-22).

Relay problems

A relay may turn the TV on or off or operate in the degaussing circuits. Check the relay for an open solenoid, bad solenoid connections, bad relay driver transistor, or poor ac connections. When a relay does not click and there is a dead chassis, check for an open relay coil, a bad electrolytic in the relay circuit, or a damaged relay caused by lightning or a power line surge. A defective relay with stuck points, a defective relay driver transistor, or a shorted diode across the solenoid can produce a no shutoff symptom. Check the relay for bad switching points, a defective relay transistor, or badly soldered joints on the ac side of the solenoid.

When a relay chatters, check for a defective electrolytic in the relay voltage source, a faulty main filter capacitor, and poorly soldered terminals on the small resistors in the relay circuits (Fig. 11-23). The relay would not close in a Magnavox CRN200AT01 TV/VCR, and this was caused by a faulty diode (D008).

Switching transistor circuits

Notice that the input circuits from the raw dc power supply have a hot ground. The secondary side of the switching transformer has a cold common ground. Take the hot ground measurements from the main filter capacitor ground for accurate input readings. Most service problems found in switching power supply circuits are a leaky or shorted switching transistor, a bad switching IC, a blown line fuse, a defective IC regulator, a leaky voltage transistor regulator, and faulty silicon diodes. Service the switching transistor like you would the horizontal output transistor in the horizontal output circuits.

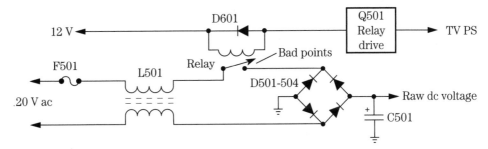

11-23 Bad relay points in an Emerson VT1920 TV/VCR caused an intermittent power-up symptom.

In a Panasonic PV-M2021 TV/VCR, the switching power supply receives the raw dc voltage from a bridge rectifier (D1001) and a main filter capacitor C1004 (120 μF, 200 V). The switching transistor (Q1001) and switching control transistor (Q1002) operate in a hot ground primary winding of the switching transformer (T1001). Five different voltage sources are found in the secondary side of the switching transformer. The secondary voltage sources have a common ground (Fig. 11-24).

Filter capacitor problems

Defective filter capacitors can cause dead, intermittent, poor power-up, and chassis shutdown problems. A dried-up filter capacitor can cause a low output voltage source, a shrunken picture on both sides, severe weaving of the picture, hum bars in the raster and hum in the speaker, an hourglass pattern, very light horizontal bars moving up the screen, and intermittent shutdown symptoms. Check for accurate dc voltage across the main filter capacitor. A shorted or leaky filter electrolytic can be checked on the low-ohm scale of a DMM. Quickly check the condition of the filter capacitor on an ESR meter.

Before taking ohmmeter and ESR meter tests, discharge the electrolytic filter capacitor so as not to damage the meter. Clip another electrolytic across the suspected one with the power turned off so as not to damage other semiconductor components. Notice if the hum or black lines are gone when the chassis is started up. Make sure that the test electrolytic has the same or higher voltage and capacity as the suspected capacitor.

Switch-mode power supply

The switch-mode power supply (SMPS) receives a raw dc voltage from the bridge rectifier power supply input circuits. A power line dc voltage is fed directly to the switching transformer with an output transistor and an IC regulator in the primary circuits. The feedback transistor is in the lower primary winding with an isolator-coupler separating the input and output circuits. All the hot grounds are found in the primary side of the switching transformer. The common grounds are found in the secondary circuits (Fig. 11-25).

Most switch-mode power supplies can be serviced as the switching power supply. A quick voltage test across the main filter capacitor indicates if raw voltage is

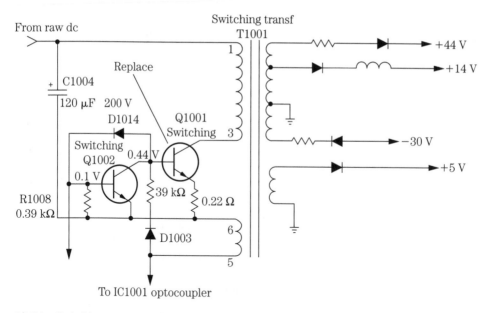

11-24 Switching transistor Q1001 became shorted and blew the main fuse F1001 (0.6 A) in a Panasonic PV-M2021 TV/VCR.

11-25 The switching transformer is found on the rear chassis of a Panasonic 13-in TV/VCR.

applied from the bridge rectifier circuits. Check the various voltage sources after the silicon diodes or across filter capacitors in the secondary circuits of the switching transformer. Dried-up electrolytics can lower the dc voltage sources.

Test the isolator IC with the diode tester of a DMM. The primary or hot side of the IC can check the light-emitting diode (LED) inside the isolator on terminals 1 and 2. Replace the isolator IC if a low ohm measurement is found on the transistor side. Check the transistor side with the diode tester of a DMM. A low resistance measurement with reversed test leads indicates a leakage between the collector and emitter terminals (3 and 4). A GE 13TVRU1 TV/VCR squealed when plugged into the power line and would not turn on due to a defective C810 (470 µF, 16 V) electrolytic in the power supply.

Dead chassis

A high dc voltage was found on the collector terminal of the switching transistor in an RCA TV/VCR chassis. TP01 tested normal with the in-circuit diode tester of a DMM. The emitter resistor RP20 (0.15 ohm) was fairly normal. All diodes within the primary winding of the switching transformer tested good. Voltage measurements on the IC regulator (IP01) were way off. The regulator IC (IP01) operates in the hot ground side of the switching transformer (LP03). Replacing the bad regulator IC (IP01) solved the dead chassis symptom (Fig. 11-26).

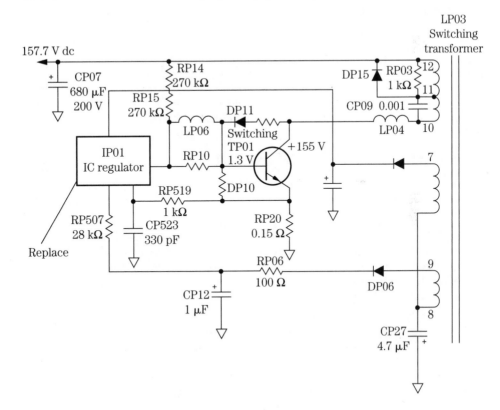

11-26   Replace the IC regulator (IP01) in an RCA 13TVR60 TV/VCR with a dead symptom.

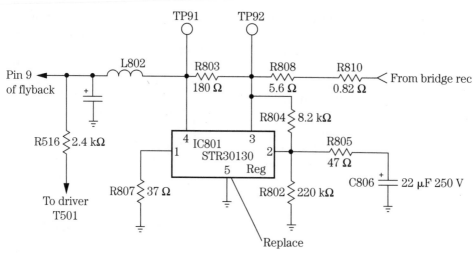

11-27 Replace a defective line voltage regulator (IC801) in a Panasonic PV-M2021 TV/VCR with a dead chassis and clicking of the relay.

Power line regulators

Besides a switching power supply, power line regulator circuits are found in a Panasonic PV-M2021 TV/VCR. The raw dc voltage feeds +132 V to power line regulator IC801. A defective power line regulator can cause no startup and shutdown problems. The defective power line regulator also can cause the TV to shut down when the power line voltage is over 100 V ac. Suspect the power line regulator when the TV becomes intermittent. A narrow picture may result from a defective line voltage regulator (Fig. 11-27).

No startup and shutdown

The startup and shutdown problems in a TV/VCR combo are the same as in a regular TV chassis. No startup in a TV/VCR was caused by a defective 33-µF electrolytic that lowered the dc voltage to 15 V from a 112-V source. Large electrolytics and diodes also can cause intermittent startup and shutdown symptoms. A dead TV/VCR might try to start up, and this may be caused by a defective line voltage regulator IC in the power supply circuits. A 470-µF electrolytic can cause intermittent startup and a ticking sound. A defective switching IC in the power supply can cause a dead chassis that tries to start up and has a pulsating noise in the speaker (Fig. 11-28). Intermittent startup with the power light on in a Samsung VT1921 TV/VCR was caused by a badly soldered joint on the 12-V secondary IC regulator in the power supply.

When a TV/VCR chassis shuts down, the results are the same as those found in a regular TV chassis. When the high voltage comes up and then shuts down immediately in a Symphonic TVCR13E1, this is caused by a bad 6-V regulator on the power supply board. In a Symphonic TVCR13F1 TV/VCR, the TV was dead, and when the voltage was raised to +112 V, the set came on and then shut off; this was caused by a bad 6-V regulator. A Zenith SMV-1315 TV/VCR came on momentarily with high voltage and then

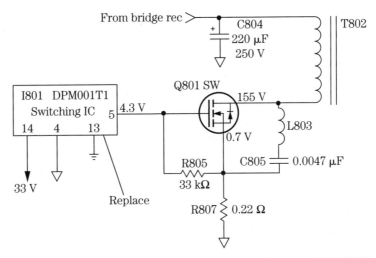

11-28 Replace the switching IC (I801) in a Daewoo DUN2056N TV/VCR when the chassis is dead, it tries to start up, and a pulsating noise is heard in the speaker.

shut off; this was caused by defective IC1003, IC1004, and IC1502. A Zenith SMV-19415 TV/VCR cycled on and off when plugged in, and this resulted from a badly soldered joint on the horizontal output transistor (Q9002).

Defective IC602 and IC603 in a Magnavox CCT13AT02 TV/VCR caused the TV to come on with high voltage and then shut off. The chassis in a Sylvania SSCO90 TV/VCR would shut down as a result of a defective IC602. A Symphonic SC3919 TV/VCR came on and then shut off immediately and would come on again for 35 minutes or so, and this was caused by a defective 6.8-V zener diode (D224) off of pin 2 of IC201.

EEPROM troubles

The TV/VCR chassis has the same EEPROM problems as a normal TV chassis. A TV/VCR with no vertical or horizontal sync, no color or audio, and no vertical or horizontal sync when a tape was played was the result of a defective EEPROM. No tuner action and raster with normal on-screen display was the result of a bad EEPROM IC. Many scanning lines with no tuner action was found in a TV/VCR to be the result of a defective EEPROM. A defective EEPROM (IC202) in a Sharp 27VSG300 TV/VCR caused a white raster with retrace lines, no tuner action, and no audio. When playing a tape in the same TV/VCR, there was no video with normal audio, and this also was caused by a defective EEPROM. Again, in the same Sharp TV/VCR chassis, there was no tuner action except an on-screen display, and when a tape was played, there was video and normal audio.

Horizontal sweep problems

Horizontal sweep problems in TV/VCR chassis are the same as those in a regular TV chassis. A dead Daewoo DVN20FGN TV/VCR tried to start up and had a pulsating

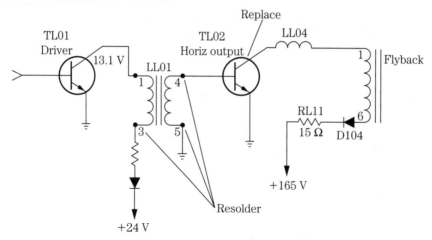

11-29 Resoldering connections on the horizontal driver transformer and replacing the horizontal output transistor (TL02) revived a dead chassis in an RCA 13TVR60 TV/VCR.

noise in the speaker caused by a faulty T803. After turn-on, a Funai F3809K TV/VCR worked okay for 15 to 20 minutes and then the raster became bright with retrace lines; this was caused by a defective flyback. The TV was dead but the VCR worked okay with an open F802 fuse in an MGN NVR9500 TV/VCR, and this was caused by a shorted horizontal output transistor (Q402) with a badly soldered joint on the horizontal driver transformer.

A bad flyback in a Symphonic SC319A TV/VCR produced a dark raster and poor focus when the screen control was turned up. The bad horizontal output transistor (Q402) in a White-West WTV-11911 TV/VCR caused a dead symptom with a normal fuse symptom. A Funai F19TRB1C TV/VCR was dead; when it was plugged in, a squealing noise was heard, and this was caused by a bad horizontal output transistor. In a Zenith SRV-1300S TV/VCR, the TV was dead but a noise heard inside the set, and this was caused by a defective horizontal driver transformer (T571) with an okay fuse symptom (Fig. 11-29).

A NARROW PICTURE

A narrow picture or a picture with both sides pulled inward can be caused by defective components within the power supply or horizontal circuits. A defective power line voltage regulator IC can produce a narrow picture. Poor connections on the horizontal driver transformer can cause an intermittent raster with the sides pulled in. Be aware of the narrow picture problems found in the power supply and horizontal circuits of a regular TV chassis and compare them with those in the TV/VCR chassis.

In a Panasonic CT-13R20K TV/VCR, the fuse was okay, and the narrow picture was caused by a defective IC801 (STR30130) power line regulator. Replace IC3001 in a Panasonic PV-M2021A TV/VCR that has a loss of video after the set is on for 15 minutes, and when it is turned on again, the picture is pulled to the right. This is caused by a faulty C553 capacitor (0.002 μF). Check C12 (4.7 μF, 250 V) and C231 (100 μF, 160 V) in a Zenith SLV-1940S TV/VCR when the picture shrinks on both sides and then the raster fades out after the TV is on for 2 to 5 minutes. Reduced width in a Memorex 16-411 TV/VCR was caused by a badly soldered joint on Q1008.

Vertical problems

The vertical problems found in the TV/VCR chassis are about the same as those in a regular TV chassis. No vertical sweep in an Emerson TV0952 TV/VCR was caused by a defective resistor (R467). Intermittent loss of vertical sweep was found in an Emerson VT3110 TV/VCR and resulted from badly soldered joints on IC402. Loss of vertical sync and flag waving at the top of the picture in a Magnavox CRN200AT01 TV/VCR were caused by a bad IC1 module. No raster was noted and, when the screen control was turned up, there was no vertical sweep in an RCA T25003BC TV/VCR, and this was caused by faulty IC501, R511 (3.9 ohm), and C508 (100 μF, 35 V). In another chassis with the same model number, loss of vertical sweep after operation for 5 minutes was caused by a badly soldered joint on IC501.

After replacing the IC602 (K1A7806) regulator for a dead chassis, a Symphonic TVCR9E1 TV/VCR had no vertical sweep, and this was caused by a defective vertical output IC (LA7837). No vertical sweep in a Zenith TVSA1320 TV/VCR resulted from a bad IC541 (LA7837) (Fig. 11-30). Loss of 1 in of vertical sweep at turn-on in a Funai SLV-19405 TV/VCR after warmup was caused by a faulty C01 electrolytic (100 μF, 16 V).

VERTICAL RETRACE LINES

Check the vertical sweep circuits for vertical retrace lines in a TV/VCR, the same as in a regular TV chassis. A white raster with retrace lines was cause by a bad EEPROM

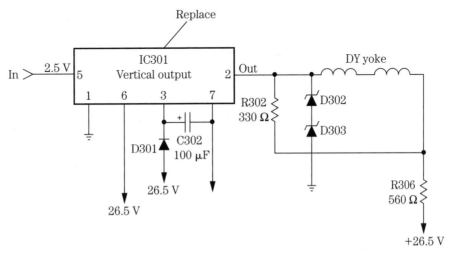

11-30 A normal vertical drive waveform was found on pin 5 of the vertical output IC301 with no vertical sweep and was caused by a bad output IC in a Daewoo CN-071 chassis.

in a Sharp 27VSG300 TV/VCR chassis. A Funai F3809A TV/VCR worked okay for 15 to 30 minutes, and then the raster became bright with retrace lines; this resulted from poorly soldered connections around the vertical output IC. Replace IC401 (LA7837) when two white lines appear in the bottom 2 in of the screen with horizontal tearing in a JVC TV-C2026 TV/VCR. At turn-on there was a bright screen and retrace lines and then no vertical sweep in a Symphonic TVCR13D1 TV/VCR, and this was caused by faulty C9012 (1 µF, 250 V) located near the flyback.

Poor or no video conditions

The same video problems found in today's TVs are the same as those found in TV/VCR chassis. After replacing C07 (220 µF, 6.3 V) in a Funai G19TRB1C TV/VCR for a dead symptom, the raster appeared extrermely bright and could not be turned down. This resulted from defective electrolytics C09 (4.7 µF, 160 V) and C12 (4.7 µF, 250 V). A washed out and smeared picture in a Funai FC1300T TV/VCR was caused by a leaky electrolytic C9003 (4.7 µF, 250 V). Wavy black lines with dark shading in the picture on both the TV and the VCR resulted from two defective 330-µF, 6.3-V electrolytics.

A bad EEPROM produced a white raster with retrace lines in a Sharp 27VSG300 TV/VCR. Excessive brightness in a Panasonic PV-M2021 TV/VCR was caused by a defective IC1 located near the VCR board. Defective IC02 in a Zenith SLV-1940S TV/VCR resulted in no video from either the tuner or the VCR but a normal raster and audio. Replace IC1 in a Quasar UV1220A TV/VCR that shows no video and a bright raster in VCR operation.

TV/VCR color problems

The color problems found in both TV/VCR chassis and regular TV chassis are the same, except a few symptoms may be different in the TV/VCR. A defective IC201 provided a blue screen with no audio and retrace lines in a Sharp 13VTF40M TV/VCR.

Intermittent color was found in a Panasonic PV-M2021 TV/VCR with a bad color IC. The luminance and chrominance main signal processor was contained in IC301. The supply voltage on pin 16 was very low and should have been around 12 V. No color waveform was normal off the color crystal on pin 19. No correct waveforms were found off pin terminals 9, 10, and 11 with very low voltage. Suspected IC301 was replaced, and this solved the intermittent and no color symptom (Fig. 11-31).

TV/VCR audio problems

Most small-screen TV/VCR combos employ a single audio output IC. Most audio problems are the same in a TV/VCR chassis as in a regular TV chassis, except those found in the VCR section. Check for a badly soldered joint on IC502 in an Emerson VT1321 TV/VCR that has no audio or raster. Replace IC801 (STR3220) and R805 (10 kilohms) when there is a white raster and no audio in a Panasonic ADP208 TV/VCR

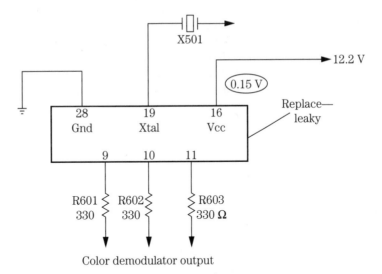

11-31 Intermittence and no color were found in the luminance and chrominance IC301 in a Panasonic PV-M2021 TV/VCR.

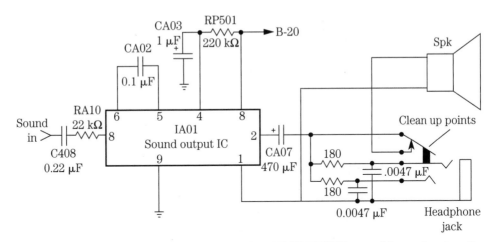

11-32 Dirty headphone plug points in an RCA 13TV70 TV/VCR caused intermittent audio reception.

chassis. No audio in a Protec TVCR13131 TV/VCR was caused by a faulty IC01 (AN5265). The surface-mounted IC3001 in a Panasonic PV-M2048 caused a no-audio symptom.

Suspect C4202 (220 μF) on the tuner board when there is no audio in a Funai F13TRB1 TV/VCR. Hum in the speaker, plus hum bars moving up the screen, resulted from a defective electrolytic C455 (2.2 μF, 315 V) on the TV board and C1102 (3300 μF,

35 V) on the VCR board in a Samsung VM7003 TV/VCR. Badly soldered joints on C801 (330 μF, 16 V) and C803 (10 μF, 16 V) caused no audio from the TV when a tape was played in a Symphonic TVCR13F1 TV/VCR (Fig. 11-32).

A bad speaker in a Panasonic PV-M2559 TV/VCR indicated a distorted-audio symptom. A defective head phone jack with dirty contacts in a Symphonic TVCR961 caused a no-audio symptom in the TV or VCR mode. A cracked board where the earphone jack was connected produced intermittent audio in a Zenith SMV-19415A TV/VCR.

12
CHAPTER

Repairing remote control units

Today, remote controls are found with virtually every electronic product available on the market. Early mechanical remotes turned the TV off and on and had control of the volume in the audio circuits. Along came the radiofrequency (RF) remote for operating TVs. The RF remote sometimes would turn on the neighbor's TV if you lived in an apartment house. Later, the infrared remote controlled every function on the TV chassis. The recent remote control transmitter not only operates the TV but also controls the VCR. Now the infrared remote operates the TV/VCR, CD player, home theater, DVD player, AM/FM stereo/CD combo system, auto CD trunk player, liquid-crystal display (LCD) TV/video monitor found in automobiles, and digital satellite system (DSS) dish receiver (Fig. 12-1).

The remote control transmitter can be damaged by rough treatment. Although children may be able to operate the remote control better than some adults, little ones can cause a lot of damage. They carry the remote around and drop it on the floor, pour soda pop into it, stick sharp objects into the buttons, and sometimes use it as a hammer. Besides hiding under newspapers, blankets, and rugs, a remote control can be stepped on easily. Actually, a remote should be left in a safe place so that the little ones cannot play with it.

A mistreated or dropped remote cannot operate with dislodged batteries, a broken case and PC boards, or sprung battery contacts. This can result in intermittent or no operation. Most remote control transmitters operate on two or three AAA or AA batteries.

Remote control functions

The infrared signal from a remote control device is picked up by a sensor in the unit to be controlled. The signal is picked up and applied to a system control integrated

12-1 Early remote transmitters used RF control until the infrared remote was introduced.

circuit (IC) that provides the various functions. Today, a TV remote can operate most of the features of both a TV chassis and a VCR. The DSS/TV/VCR remote can control the TV channels, up and down channels, single-channel button control, DSS cable, display, rewind, record, fast forward, menu processing, clear, picture in picture (PIP), move PIP, freeze PIP, skip, swap PIP, reset, antenna, volume up and down, and mute.

DSS remote control features may include power, set, TV, VCR, DVD, mute, up and down volume, channels up and down, exit, menu, quick, guide, WHO, audio, back, freeze, pushbuttons for each channel, OK, play, stop, record, pause, rewind, and fast forward for VCR operation.

A deluxe A/V receiver remote control may operate the TV, VCR, CD, tape, and receiver functions. Some of the buttons found on the face of the A/V remote control include on and off, TV, VCR, tuner, tape, CD, DVD, record skip, fast skip, play, reverse, TV channels up and down, volume up and down, independent TV channel buttons, tape deck, and muting controls.

RF remote control

The supersonic RF remote control transmitter sends out an RF signal that is collected by an RF indicator and sent to a system control processor. The RF control generates an electronic signal to control certain frequencies to be radiated through a speaker or transmitter. The frequency chosen was between 44.75 and 47 kHz. The frequency avoids most erroneous signals that might trigger the remote control receiver. Of course, garage door openers, door and telephone bells, and other supersonic signals can trigger some TVs. Erroneous signals from various electrical and RF-generating devices can trigger remote control receivers.

Infrared remotes

You must point an infrared remote transmitter directly at the unit to be controlled or no function will occur. If someone or something is in front of the infrared detector, the infrared signal will not trigger the infrared receiver. Nothing happens when a person walks in front of a TV as the remote is being used to select a station. The infrared remote will not trigger another remote electronic product found in a different room.

The infrared remote sends out an infrared signal from one or two transmitting light-emitting diodes (LEDs). This infrared signal is picked up by an infrared photo transistor sensor, amplified by transistors or one IC component, and fed to a system control IC. The system control may consist of a microprocessor, a central processing unit (CPU), or a system control IC in the latest TV circuits. The processor chip sends the remote control messages to the various circuits to be operated (Fig. 12-2). In the sound circuits of some remote-controlled receivers, a small motor also rotates the volume control up and down.

TV remote   **Disc remote**   **DVD remote**

12-2   A remote transmitter may operate a VCR, CD player, and DVD player besides the TV.

The dropped remote

One of the biggest difficulties when using a remote control is dropping the remote on a hard floor. Sometimes the remote becomes tangled up with newspapers and magazines and is dropped accidentally on the hard floor. You try to turn the TV or receiver on, and there is no operation. The dropped remote can come apart, or the batteries can pop out from the plastic case. Simply remove all the batteries and reinstall them with the correct polarity.

When these small batteries are installed backward, the remote will not function. Check the positive and negative terminal signs inside the remote when installing new or dropped-case batteries. Make sure that the battery end terminals are not bent out of line. Take a small screwdriver or knife blade and bend the battery terminals toward the batteries to make a good contact. Sometimes if the remote is dropped on a cement floor, the PC board can be cracked or the wire traces can be broken.

Dead, no operation

A dead remote may be caused by defective, corroded, or weak batteries. Most remote control units operate with AAA or AA batteries. Two or more batteries (3 to 6 V and 9 V) are connected in series to power some remote control transmitters. Some units are operated with only two AA (1.5-V) batteries. The high-powered DSS control may operate with four AA (6-V) batteries. A remote transmitter will not function with dislodged batteries, improper polarity, bad end-contact strips, or weak batteries.

The remote transmitter may not operate the TV or receiver when it is left for a long time and not used. Remote transmitters left in a cold room may not control the TV the next time the on button is pressed, may become weak from many hours of operation, or may not work if the battery voltage falls to 1 V. Sometimes one battery becomes defective and causes the remote to malfunction. If the remote will only turn on the TV when it is within 2 or 3 ft of the TV, suspect weak batteries (Fig. 12-3).

Preferably, check the batteries with a battery tester. If one is not available, check each battery under load with the low-voltage range of a digital multimeter (DMM). Hold down one button, and test the voltage across each battery. Replace all batteries that test 1 V or less. Weak batteries can cause intermittent or weak operation. It is best to replace all batteries when one or more are found to be weak or dead. Always replace all the batteries at the same time. Replace remote control batteries with high-energy batteries.

Heavy-duty power cells or ultra-alkaline batteries have a relative service life of from 2 to 10 times longer than ordinary carbon batteries. Wipe the new battery contacts by rubbing against a cloth, towel, or pant leg before installing. Make sure that the positive (+) terminal is inserted properly. Most battery holders have marked positive and negative terminals. Double-check the battery polarity.

Aim at the target

You must aim the remote transmitter directly at the receiver's infrared indicator or the electronic product will not operate. Aim the remote high above any object that might be in front of you and the electronic product to be operated. The infrared remote operates silently, and the only indication that the remote is working is when

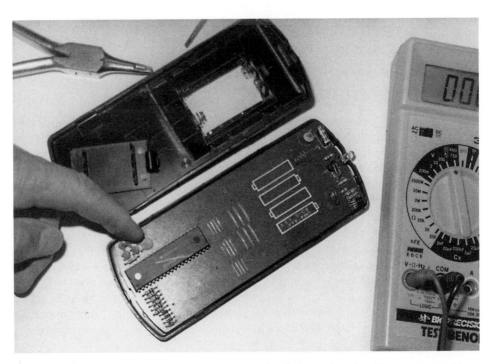

12-3 Suspect weak batteries when remote operation is intermittent or the remote must be close to the electronic product to function.

the light appears on the remote transmitter. Bring the remote close to the TV to see if it will function. Check for weak or dislodged batteries if the remote fails to turn on the TV. Suspect weak batteries when the remote will only operate a few feet from the TV and will not function while you are sitting in an easy chair.

A quick method to determine if the remote is not functioning or the infrared indictor within the electronic product is defective is to push each button of the remote next to a battery-operated portable radio. Each time a button is pressed, you can hear a gurgling sound in the portable radio. If a button produces an erratic gurgling noise, this particular button assembly is dirty and not making a good contact. However, this test does not indicate the strength of the remote infrared signal.

Another method is to use an infrared indicator card in front of the remote transmitter. The infrared card will change to a different color if the transmitter is working. These indicator cards can be obtained through TV and parts distributors. The RCA infrared indicator card for checking the output of remote control transmitters is stock number 153093. Such infrared cards will only indicate if the infrared signal is present but not how strong or weak the signal is. They only indicate that the remote control is operating.

Try another one

Simply try another remote to see if it will turn on the TV set, VCR, CD player, or DVD player. Usually, there are several different remotes in the house. Most remotes will have a function that will operate another receiver in the TV or VCR. If the subbed remote operates one or two functions, you can assume that the sensor and receiver are operating. Then check the defective remote transmitter.

Tap that remote

Just tapping the remote may make it work, indicating weak batteries, poor battery contacts, or batteries that are not seated properly. An intermittent remote can irritate most anyone. Remove the battery cover and test each battery on the battery tester or under load with the voltage tester of a digital multimeter (DMM). Clean the battery tab contacts. Make sure that these contacts are not corroded. Sometimes, if a battery begins to leak, it can corrode the metal battery connection. Sandpaper or a finger nail strip can easily clean a damaged battery clip.

Loose components or cracked printed circuit (PC) boards can cause a remote to operate intermittently. Inspect the PC boards under a strong light. Check all the coils, transistors, and capacitors for loose or broken leads. Resolder all connections on the PC board. Make sure that the wires to the battery terminals are intact and soldered. When large defective components are found inside a remote transmitter, send it in for repair, or if it is under warranty, exchange it at the manufacturer's warranty or tuner depot. Replace the remote if the unit is out of warranty because you can find them everywhere at a very low price.

Infrared power meter

The infrared power meter that is used to check the laser optical assembly in a CD player also can be used to test out infrared remote control transmitters. Start with the lowest scale of the power meter (0 to 0.3 mW). If the results are poor, move up to the next scale. The laser power meter may have a 0.3-, 1-, or 3-mW range with switchable wavelength settings of 633 and 750 to 820 nm. A laser power meter that can be used in such infrared tests is a Tenma number 72-670 (Fig. 12-4). In addition, a Leader laser power meter number 70-510 can be ordered by mail from MCM Electronics, 650 Congress Park Drive, Centerville, OH 45459-4072.

Place the probe 3 in from the pickup probe, and register the measurement on the power meter. Likewise, check the power of the remote control unit, and write down the measurement. By making comparison tests with new remotes, defective or weak remotes can be discovered. Move the remote back and forth to acquire the best reading. A weak or a dead remote will have a low or no measurement compared with a normal remote. Then adjust the batteries and contact terminals to obtain a higher measurement. Remote comparison tests with the laser power meter can indicate if the remote or receiver is defective.

Weak reception

Weak, intermittent, or erratic remote control operation results from weak batteries, bad battery terminals, and dislodged batteries. Try the remote out as you approach the TV or receiver to determine the distance at which the remote operates. Replace

12-4   A laser power meter can quickly test a remote transmitter.

weak batteries when some of the functions of the remote will perform only extremely close to the electronic product. Inspect the end terminals where the batteries make contact. Clean the contacts with cleaning fluid. Try another remote to determine if the remote in question is defective or the problem is in the infrared indicator in the receiver.

Infrared remote circuits

The infrared remote transmitter may operate with one or two infrared LEDs in the output. These LEDs are usually driven by one or two transistors in a series operation. The driver transistor has a code output terminal from the remote encoder IC. Several different pushbuttons provide key in, decoder in, and scan out terminal connections from the encoder IC. Two AA or AAA batteries provide voltage to the remote transmitter circuits.

In a JVC VCR-225 VCR, the remote encoder IC1 drives an NPN and a PNP transistor in a series circuit with two infrared LEDs at the output. The first transistor (Q01) acts as a infrared driver, and the second transistor (Q02) acts as a infrared amp output. Both infrared LEDs are fed in a parallel output circuit from collector Q02. The key in pin terminals of the remote decoder IC1 are fed from fast-forward, rewind, power, play, pause/still, stop, TV/video, channel next, and channel back switch keys. The record (REC) button is tied to decoder in terminals 9 and 10.

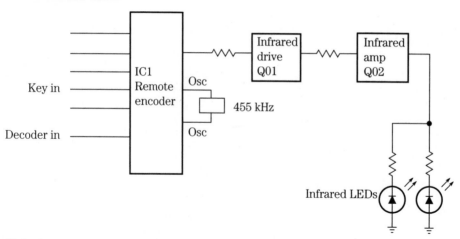

12-5   Block diagram of a JVC VCR-258 remote control transmitter.

A 455-kHz oscillator network provides oscillator action on pins 2 and 3 of IC1. The infrared code input is taken from pin 17 to the base of Q01 with a low-ohm resistor tied in series with the collector of Q01 to the base terminal of Q02. The two infrared LEDs are tied to the collector terminals of infrared amp Q02. Two AA 1.5-V batteries power the remote control transmitter (Fig. 12-5).

In a Panasonic PV-M2021 TV/VCR, the remote transmitter has only one emitting infrared LED, an infrared driver transistor (Q6801), and a multifunction remote control IC6801. A crystal oscillator is found on pin terminals 20 and 21. The key in circuits from the different pushbuttons tie into pins 3 through 10. The scan-key circuits are tied to pins 12 through 18, whereas pin 19 goes to the base terminal of Q6801. A 3-V battery dc source powers the remote transmitter (Fig. 12-6).

Remote control problems

One of the biggest problems with remote controls is dropping the remote on a hard surface and dislodging the batteries. A weak battery can cause intermittent or no operation of the remote transmitter. Oozing batteries can corrode the battery contacts. Dirty or poor battery contacts can cause no or intermittent operation of the remote. Replace the batteries when the remote will work one time and not the next. If the remote control will only operate up close to the TV, replace the weak batteries. Reseat the batteries when the plastic battery lid will not close properly.

If the batteries are replaced and the remote still does not function, remove the top or bottom cover. A stiff knife blade between the two lids can pop open the plastic case. Check each transistor with the transistor tester or the diode tester of a DMM. Test the infrared diode as you would any silicon diode. The infrared diode can be checked on the diode tester of a DMM with a higher reading than a normal silicon diode. Each pushbutton can be tested with low-ohm resistance tests on a DMM. Clip the DMM leads across the pushbutton terminals, and press the button down for a shorted measurement. Replacing the large encoder IC, however, may

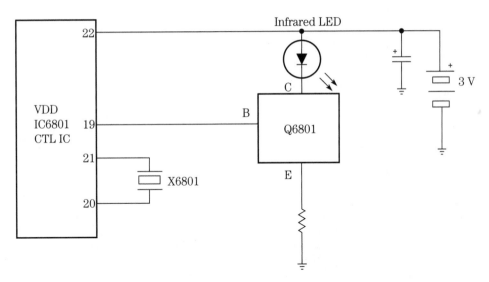

12-6 The infrared wireless remote transmitter block diagram found with a Panasonic PV-M2021 TV/VCR.

cost as much as a new remote. Generally, it is easier to replace the remote with a new one because replacements are fairly cheap and can be purchased anywhere (Fig. 12-7).

More than one

A remote control transmitter can operate most TVs, VCRs, CD players, DVD players, and deluxe receivers. The General Electric RRC500 does the work of three remotes, whereas model RRC600 does the work of four remotes. RRC600 controls up to four infrared audio/video products, with over 200 key combinations, program sequencing, LCD display, and a low-battery indicator.

This remote, like all other universal remote control units, can be programmed to work on virtually all electronic products. These remote control units can be purchased at TV dealers, hardware stores, mall stores, and Radio Shack (Fig. 12-8). The universal remote operates on batteries and can be checked with the infrared tester like other remotes. The RCA DSS receiver can be operated with some universal remote control transmitters (e.g., Radio Shack 15-2116).

Inside the remote

The plastic top or bottom plate of a remote control can be removed by inserting a screwdriver or knife blade in the broken seam. The top and bottom plastic case parts are held together by plastic clips on each side and at the ends. Remove the battery cover or lid and batteries before trying to open the plastic case. Very few components are found on the inside of a remote control (Fig. 12-9).

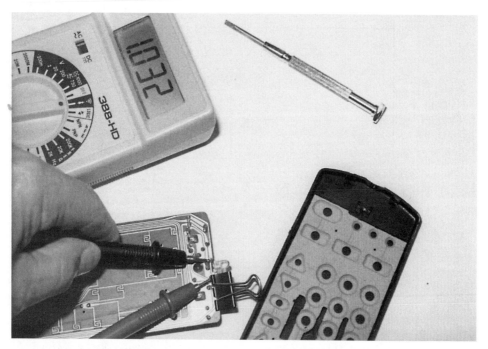

12-7   Quickly check the infrared LED with the diode tester of a DMM.

Batteries for TV set

Batteries for VCR and disc player

12-8   Inside view of a remote control with separate buttons for the TV and VCR/CD player.

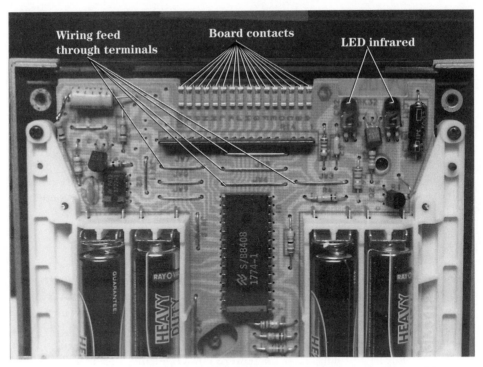

Wiring feed through terminals　Board contacts　LED infrared

12-9　The infrared IC, transistor, and LEDs found in a dual remote control transmitter.

If the PC board is cracked and cannot be repaired, replace the remote control. Check the emitter LED with the diode tester of a DMM; it should read about twice the resistance of a silicon diode in only one direction. Likewise, the LED light can be checked in the same manner. Check the transistors with the diode tester of a DMM. Often the pushbutton terminals are found in the PC wiring side of the remote.

Remote receiver problems

When the remote has been tested or subbed with another unit and the electronic product still does not respond, the trouble can be in the infrared receiver or standby power supply circuits. In today's TV receivers, the standby voltage is on all the time, even when the TV is turned off. The remote receiver must be alive to receive the commands of the remote control to turn the electronic product on.

The remote control receiver sensor is found in the front panel of any electronic product. This infrared sensor is actually an infrared phototransistor that picks up the infrared signal and connects it to the remote receiver in the TV, VCR, or CD player. The infrared sensor is usually placed in a shielded container to prevent outside noise and unwanted signals from striking it.

In earlier infrared receivers, small transistors were used to amplify the weak signal to the system control IC. You may now find an IC component between the infrared sensor and the decoder. One large IC may be found with a transistor amp before feeding the remote signal to the system control IC. In turn, the system control IC operates the various functions of the remote transmitter.

TV RECEIVER REMOTE CIRCUITS

The TV remote circuits consist of an infrared detector, a remote receiver, a driver transistor, and a system control IC or microprocessor. The standby power supply must be on all the time to provide voltage to the infrared receiver and system control circuits. The infrared detector picks up the remote transmitter's signal, which is amplified and decoded within the IC receiver circuits. Sometimes the infrared detector or phototransistor and IC provide signal directly to the IC microprocessor or system control IC. This infrared signal may go through several different connectors and boards before it is applied to the system control IC.

JVC INFRARED IC RECEIVER

In early infrared IC receivers, the infrared LED detector signal was amplified by several transistors. Today, the infrared detector has an IC component in the receiver circuits. The infrared receiver (IC1) receives the infrared signal from the detector, and the output is fed to an operational board or to the control IC in a JVC VCR-255 VCR. The infrared IC1 receiver signal is fed to an operation and monitor board and then on to the mechanical control IC601. IC1 is powered by a regulated 12-V standby voltage source. A standby voltage source is on all the time as long as the TV is plugged into an ac receptacle (Fig. 12-10).

RCA CTC167 INFRARED RECEIVER

The infrared preamp IC (U3401) is found in the tuner control schematic with CR3401 as the LED detector that feeds into pin 2 of U3401 of an RCA CTC167 chassis. The output signal at pin 8 feeds to an infrared amp transistor (Q3401) and to input termi-

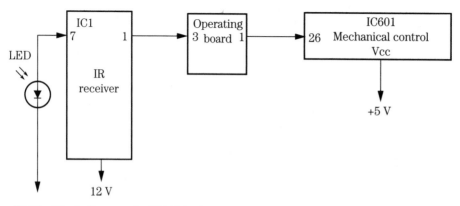

12-10   Block diagram of a JVC VCR-255 infrared receiver.

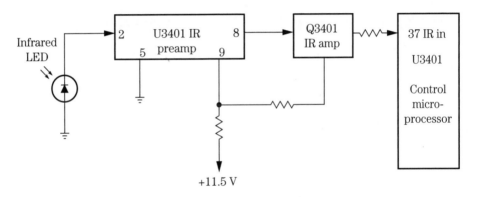

12-11   Block diagram of an RCA CTC167 infrared receiver and remote circuits.

nal 37 of the control microprocessor (U3401). The infrared receiver IC is powered by an 11.5-V source from the secondary voltage of the flyback (T4401). A standby voltage source feeds a 12- and 5-V source to power control microprocessor U3101. U3101 provides the remote signal to the various control functions of the TV (Fig. 12-11).

PANASONIC REMOTE RECEIVER

The infrared remote receiver signal is picked up in a Panasonic CT-27S18S/CS TV by the infrared remote receiver (RM001) and fed to a remote switch transistor (Q007). The NPN remote switch transistor feeds the infrared signal from the collector terminal to the remote receiver at pin terminal 36 of the microprocessor (IC001). The RM001 remote receiver is powered from a 5.2-V source to most terminals on the microprocessor IC001. The 5.2-V source is fed from a separate standby transformer source. The manual control button functions are found on the terminals of the microprocessor (IC001) (Fig. 12-12).

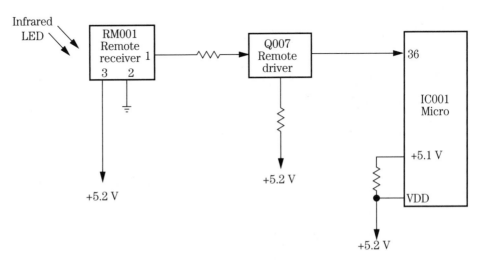

12-12   Block diagram of the infrared receiver in a Panasonic CT-27S18S/CS TV.

Troubleshooting the TV infrared receiver circuits

To determine if the infrared remote or the electronic product receiving circuits are defective, try another remote. Usually there is more than one in the house. If neither remote control operates the TV, VCR, CD player, or DVD player, suspect a defective component inside the infrared receiving unit. Notice if the TV works with the manual controls but not with the remote. Locate the infrared sensor, and trace it to the preamp IC. Some infrared receivers have a scope test point at the output of the infrared IC. Check the supply voltage on one pin terminal of the infrared IC. Usually the supply voltage is either 5 or 12 V dc. Go directly to the standby power supply if very low or no supply voltage is measured at the infrared IC and driver transistor.

Sometimes loose or broken cable connections produce intermittent remote control functions in the receiver circuits. Scope the remote signal, and take accurate voltage and resistance measurements on the IC and transistor components to locate the defective component. Test diodes and transistors in the receiving circuits with the diode tester of a DMM. Check for a badly soldered joint on the infrared sensor where it connects to the PC board when there is intermittent remote operation in a Magnavox 19PRC-0121 TV.

NO RECEIVER ACTION, RCA CTC167

Although manual control operated all functions in an RCA TV, remote operations were dead. However, the same remote control operated the VCR player. This meant that the trouble had to be in the infrared preamp section of the receiver. The defective component had to be in the infrared preamp IC (U3401). A quick voltage measurement on terminal pins of the infrared IC (U3401) showed practically nothing (0.57 V).

At first, the standby power supply was suspected of providing no or very little supply voltage to pin 9 of the infrared preamp IC. CR3402 and CR3401 checked normal with the diode tester of a DMM. A low resistance measurement was found from pin 9 to ground. After removing pin 9 from the PC wiring with solder wick and an iron, the 5.1-V supply source showed up. The infrared preamp (U3401) was replaced because a leakage test indicated that the IC was causing the low supply voltage source (Fig. 12-13).

NO INFRARED RECEIVER RESPONSE, GE CTC146

Although the manual buttons on the front of the TV worked, no remote functions occurred with the infrared remote in a GE CTC146 TV. A quick voltage test was quite normal. A continuity test from the infrared circuits to pin 8 of U3401 showed an open circuit. Resoldering the trace PC wiring restored the remote control functions at pin 8 of U3401 (Fig. 12-14).

Infrared receiver case histories

A leaky channel up switch caused the channels to scan continuously with reception only on channels 3, 4, and 83 with no remote action in an Orion TV1928 TV. No re-

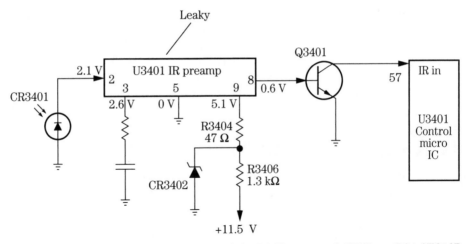

12-13 A leaky preamp IC (U3401) reduced the 5.1-V source to 0.15 V in an RCA CTC167 infrared remote receiver.

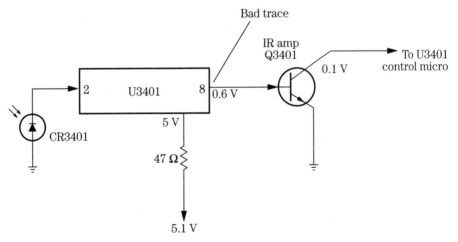

12-14 A badly soldered joint at pin 8 of U3401 caused no remote functions in a GE CTC146 TV.

mote control action in a Panasonic CTZ-2042R TV was caused by a bad M004 remote receiver (part number EUR37234). The defective IC003 remote receiver in a Panasonic CT-27G24A TV produced no remote control action. A corroded resistor R139 (100 ohms, ⅛ W) located at the front of an Emerson TC1966D TV caused the remote not to function. The auto programming would not lock in on cable channels 7 to 13, 16 to 18, and 19 to 37, and when the remote was used, the picture would tear horizontally as a result of a bad electrolytic C052 (100 µF, 16 V) in a KTV-19TEC TV.

In an Admiral JSJ-12674 TV, channel 8 would not change and there was no remote action as a result of a bad EEPROM memory chip (IC2701). Replace the MM001 remote receiver when there are no remote functions in a Panasonic CT-27G31U TV. Check and replace the microprocessor (IC2001) in a Sharp 25G-M100 TV when the remote will not operate. A badly soldered joint on pin 5 (ground) of connector J4 of the ribbon cable from the infrared receiver to the main board in a Magnavox 25G1-03 TV caused intermittent loss of remote control action. Replace the IC (CX20160K) in the remote receiver for no remote operation in a Sony KV-13TR24 TV. No remote functions occurred in a Zenith SS1915N8 TV as a result of a badly soldered joint connection from the infrared receiver where it joins the main PC board.

RCA CTC145 REMOTE RECEIVER

In this TV chassis, the remote sensor is fed into a transistor infrared amp and a second, third, and fourth infrared amp in series. All the infrared amp transistors and sensors are enclosed in a shielded area. The top shield cover must be in place before the remote will function (Fig. 12-15). A +12-V source is fed to the infrared receiver and analog microprocessor (U3300).

When the infrared receiver does not function, check the dc voltage (+12 V) to the infrared stages. Test the infrared detector (CR3404) with the diode tester of a DMM. Take an in-circuit transistor test of each transistor. A scope test at the fourth infrared amp output or pin 36 of U3300 should indicate whether the remote receiver is functioning. The infrared test point is found at FB3304 and FB3305. Suspect IC U3300 if the infrared receiver is functioning but there is no remote control. Check for a standby +5 V on the microprocessor.

TV standby circuits

The standby circuits in the low-voltage power supply provide an operating voltage source to the infrared receiver and system control or microprocessor circuits. The standby power supply might have a separate low-voltage power transformer regulated supply in some TV circuits. The latest standby power supplies are taken off the secondary winding of the switching transformer. The standby power supply provides a 5- or 12-V source to the infrared receiver circuits.

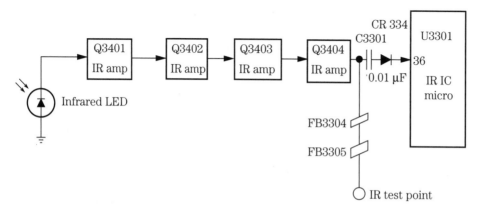

12-15 In the early RCA CTC145 TV remote receiver, transistors Q3401 through Q3404 amplified the infrared signal to the system control IC (U3301).

A simple low-voltage power transformer supply may consist of a low-voltage power transformer, a half-wave silicon diode, filter capacitors, and a zener diode or transistor regulated circuit. The power transformer is connected after the main ac fuse and is connected directly to the primary winding (120 V ac) of the power transformer. The secondary winding of the standby power transformer is connected to a single silicon diode and common ground terminal. Several resistors, diodes, and electrolytic capacitors are found in the 5.2-V standby source. No remote action in a Sylvania 20B1-03 TV was caused by a defective C2 capacitor (3.3 μF, 50 V) inside the remote control receiver.

Dead remote operation

A Panasonic CT-31S18S/CS TV would not turn on with the remote control. The remote transmitter operated the VCR but not the TV. This meant that the trouble had to lie in the infrared receiver circuits. The remote receiver was located at the front of the TV chassis, and no supply voltage was found at pin 3 of the system control IC. Since most remote infrared receivers feed into a system control IC or microprocessor, the voltage source was traced back to the standby power supply.

A separate power transformer (T001) was found on the TV chassis with D001 rectifying the standby voltage. No standby voltage source was noted at the collector terminal of D001 or at the positive terminal of the main filter capacitor C045 (2200 μF). Since the secondary voltage of most low-voltage power supplies is wound with heavier copper wire than the primary winding, a quick continuity check of the primary winding was taken.

The primary winding should have a resistance of around 500 to 850 ohms, and the winding was open. Usually, power transformers are damaged by shorted silicon diodes or electrolytic capacitors. D001, D002, D003, D004, and D006 were normal on the diode tester of a DMM. Replacing T001 resolved the dead remote power source. These small low-voltage transformers can be found at most electronic outlets or Radio Shack (Fig. 12-16).

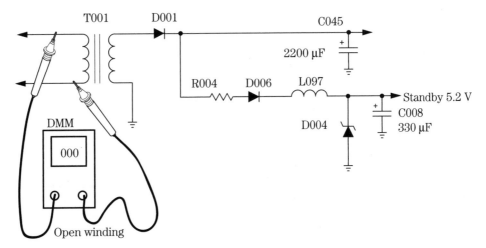

12-16 An open primary winding of the power transformer (T001) caused no 5.2-V standby voltage in a Panasonic CT-31S18S/CS TV.

Switching transformer

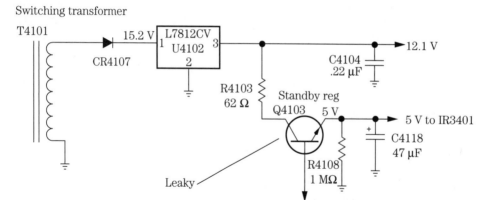

12-17 A leaky regulator (Q4103) caused low standby voltage (5 V) to the IR3401 remote receiver.

Switching transformers

In the latest infrared remote power circuits, the standby power supply is taken from a switching transformer winding. The switching transformer is alive all the time the TV is plugged into an ac outlet. A silicon diode, transistor, and IC voltage regulators are found in a regulated standby power source. You may find a 5- or 12-V source feeding the infrared remote receiver and system control IC so that the remote transmitter can trigger the remote control system (Fig. 12-17). The power switch (SW2501) was found to be open in a Sharp 25J-M100 TV, and this resulted in no remote control action.

Intermittent standby voltage

The infrared remote receiver in an RCA CTC166 TV consisted of only the IR4301 remote receiver module that picks up the infrared signal and feeds it into pin 1 of the system control microprocessor (U3101). The 5-V standby power source is fed from a +5-V standby regulator (Q4103), which receives a 12-V input voltage from the 12-V standby regulator IC (U4102).

The 15.2-V source that feeds U4102 comes from the secondary winding transformer T4101. CR4101 rectifies the dc voltage applied to IC (U4102) and the 5-V source from a +5-V standby regulator transistor (Q4103). The same +5-V source feeds most system control functions of U3101.

The intermittent +5-V source was traced to the low-voltage power source. The 5-V source appeared erratic or intermittent when the DMM was used to monitor it at the emitter terminal of Q4103. Sometimes the voltage was okay for a few hours, and at other times the TV could only be operated by the pushbutton on/off switch, volume up and volume down, and channel up and down functions. The 12-V source feeding the collector of Q4103 was normal at all times. Q4103 was replaced with part number 223704, and this resolved the intermittent remote control operation (Fig. 12-18).

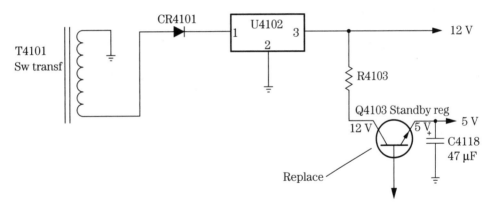

12-18   Replace Q4103 for intermittent +5-V standby voltage in an RCA CTC166 TV.

Open 5-V regulator, no remote function

In an RCA TX82N TV, the remote would not operate the remote control functions, but the front panel controls were okay. Another remote was tried with the same result. This indicated that the defective component had to be in the infrared receiver or standby power supply. A quick voltage measurement on preamp IR02 indicated very little supply voltage. The standby voltage source was traced to the low-voltage power supply circuits. A quick method to locate the voltage supply source is to take an ohmmeter measurement from the supply pin of IR02 to the low-voltage circuits.

The 5-V source was traced back to CR24 (470 μF) and DR01. No voltage was measured on DR01 or on the emitter terminal of the 5-V standby regulator transistor (TR03). A dc voltage was found on the collector and base terminals of TR03. Both DR05 and DR01 were checked with the diode tester of a DMM. TR03 was checked in the circuit with the diode tester of a DMM, and the emitter terminal appeared open. TR03 was replaced with a universal NTE123AP transistor, and this resolved the missing 5-V power source to the remote control receiver circuits (Fig. 12-19). Also, defective TR03 and DR05 can cause intermittent startup and shutdown symptoms in this same RCA chassis.

Servicing standby power circuits

When low or no dc voltage is found at the IC or transistor preamp within the infrared receiver, go directly to the standby voltage source. Carefully look over the TV chassis or electronic product for a small power transformer that supplies ac voltage to the standby power circuits. Determine if the standby voltage is fed to a low-voltage regulator source from a switching power transformer in switching power supply circuits if no single low-voltage transformer is found on the TV chassis. You may find that the standby power source is in the switching power supply of many of the latest TV chassis. If in doubt, check a similar TV chassis or a schematic from the same manufacturer.

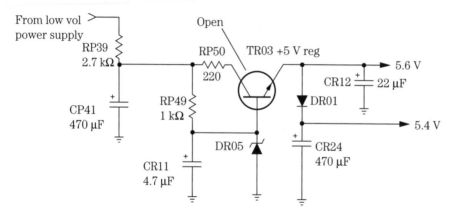

12-19 An open regulator transistor (TR03) in the 5.6-V source caused no standby voltage to the preamp IC (IR02) in an RCA TX82N TV.

Quickly check the dc low voltage on each diode in the secondary of the switching transformer circuits for no or a low-voltage source. Next, check the dc voltage on each IC or transistor regulator for a 12- or 5-V source. The standby output voltage will be at the emitter terminal of the transistor regulator. Another method is to check the voltage on each electrolytic capacitor in the secondary voltage sources. The regulator that has no or very low voltage can be the suspected power source.

Now trace the missing voltage source to a diode, transistor, or IC voltage regulator. Check each transistor in the low-voltage sources with the diode tester of a DMM. Check the dc voltage on the collector and base terminals of the suspected transistor regulator. The dc voltage may be a little higher on the base and collector terminals when there is a defective transistor regulator. A shorted transistor regulator may have very low voltage on all three terminals.

The input terminal of an IC voltage regulator will be high with no or low voltage out of the output terminal when there is an open regulator transistor. A leaky IC regulator may have low voltage on the input terminal and very little voltage on the output terminal (Fig. 12-20). Since the IC regulator has only three terminals and one terminal is grounded, low input and output voltage can indicate a defective IC regulator.

Intermittent remote operation

Sometimes the remote function would operate and at other times not in a Panasonic PV-M2021 TV/VCR. Trying another remote control transmitter did not help. A quick voltage measurement (+5 V) on the infrared sensor (U7501) was normal for several hours. When the wiring connector P7551 was touched, the remote system appeared normal and then went into the intermittent mode. The connecting wire to pin 3 was resoldered, and the remote began to function. The badly soldered connection at P7551 was the cause of the intermittent remote operation (Fig. 12-21).

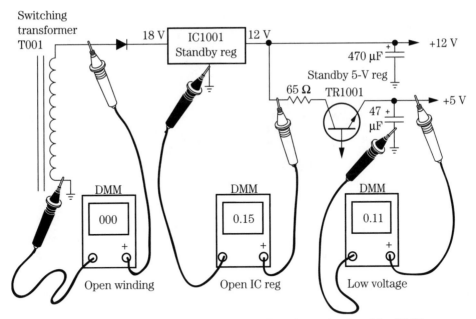

12-20 Check the various circuits within the standby voltage circuits with a DMM.

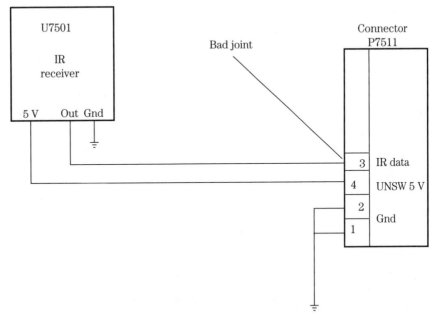

12-21 Intermittent remote action as a result of a poorly soldered joint at pin 3 of the connector (P7551) in a Panasonic PV-M2021 TV/VCR.

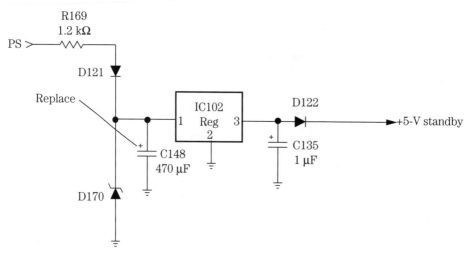

12-22 Sometimes no channel up or channel down operation was caused by electrolytic C148 (470 μF) in an Emerson TC1965D TV.

No channel up and down

The manual up and down buttons operated okay, but the remote control would not function in an Emerson TC1965D TV. The remote transmitter seemed okay. Sometimes the remote up and down would function and at other times not. The 5-V standby source was taken off the external power supply and fed to the remote receiver circuits. The TV remote would operate for a whole day every time it was used.

Sometimes the +5-V source appeared low and at other times normal. The input voltage on pin 1 would vary on the remote IC. IC102 was suspected but tested okay. Since a bad electrolytic filter can cause low voltage in a power supply source, C148 was checked with an equivalent series resistance (ESR) meter. When C148 was shunted with another 470-μF electrolytic, the remote worked for days. Replacing the C148 (470-μF) electrolytic filter capacitor resolved the no channel up and down symptom (Fig. 12-22).

No remote receiver action

The remote control would not turn on a Sharp 19SB60R portable TV. The remote was tested and seemed normal. Very little voltage was found on receiver terminal pin 2. A voltage test was made on the +5-V standby voltage and was fairly normal. Resistor R1016 (100 ohms) appeared quite warm (Fig. 12-23). A low resistance measurement was made at pin 2, and the receiver shielded area showed only 27 ohms to ground.

Pin 2 was removed from the remote receiver, and the voltage was still low. On checking the wiring, a zener diode appeared burned and quite warm. Replacing leaky diode D1052 solved the problem.

No remote, dead TV

A dead RCA CTC157 TV receiver with no remote operation was brought in. Because the picture tube had a blank raster and there was no remote action, the symptoms

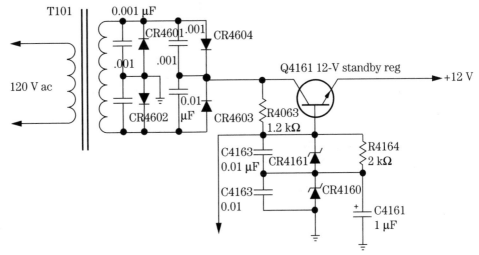

12-23 Leaky D1052 in a Sharp 19SB60R portable TV resulted in a dead remote control receiver.

12-24 Replace both CR4161 and CR4160 in the standby circuits when the remote will not function in a dead RCA CTC157 TV.

were traced to the low-voltage standby circuits. The 12-V voltage source was found in the standby regulator power supply circuits.

The standby transformer was located on the chassis, and the ac voltage was measured to the bridge rectifier circuits. A reading of 24.7 V was found on the collector terminal of the 12-V standby regulator (Q4161). Very low voltage was found at the emitter terminal (Fig. 12-24). Q4161 tested good with an in-circuit transistor test. CR4161 and CR4160 were found to be leaky and were replaced. The 12-V standby voltage was now measured on the emitter terminal of Q4161 regulator. CR4161 and CR4160 were replaced with universal ECG5071A zener diodes.

Check Table 12-1 on p. 410 for troubleshooting remote controls, infrared receivers, and standby circuits.

Table 12-1. Remote control troubleshooting.

Symptom	Repair
Dead remote control	Check and test the batteries. Check the remote against the indicator card. Test the remote with a laser power meter. Sub another remote.
Intermittent remote operation	Clean the battery terminals. Check for loose batteries. Remove the batteries, and clean the end terminals with sandpaper. Bend the terminals out for a tight fit. Suspect the remote has been dropped several times. Inspect the PC board for cracks or loose parts.
Weak reception	Check batteries—replace all the batteries when one is found bad. Test the remote on an indicator or power meter.
Remote normal—no action	Check the voltage source to the IC or transistor amps. Take waveform tests. Test each transistor in the circuit. Take voltage tests on the IC amp or receiver. Test the sensor unit with a diode test of the DMM. Suspect the standby voltage supply.
Standby power supply problems	Check the output voltage supplied to the IR receiver. Test the output voltage on the standby supply. When low or no voltage is found at the IR regulators, test the transistors and zener diode regulators with a diode test of the DMM. Remember the standby voltage must be present or the remote won't work.

13
CHAPTER

DVD player repairs

A DVD player may appear in a separate cabinet, by itself, or with other electronic products. A DVD/CD player combination provides digital video and audio music on separate discs. You can enjoy a stunning picture with a DVD and play your favorite music on a CD. Some DVD players can play DVD video, MP3s, CD-Rs, and CDR-Ws. A DVD/CD combo player can be found in home theater systems and even in boom-box players (Fig. 13-1).

A DVD/CD/VCR combination player can provide digital video in the TV chassis or VCR and your favorite music with a CD player. Some of these DVD players also can play back your MP3s. A DVD/CD/VCR combo player may appear in a single cabinet or within a TV set. You can record your own discs on the DVD/VCR recorder.

DVD/CD RW combo drive units appear in portable 2004 notebooks. Portable DVD notebooks can play a DVD, a CD-R/RW, a video CD, or an MP3 on a 5-in or 6.8-in liquid-crystal display (LCD) screen. An in-dash or overhead DVD/CD player is now produced for mobile video viewing within an automobile. You can play many hours of recordings on a DVD/CD three- or five-disc changer.

CD and DVD comparison

The CD provides a method of storage for video, audio, or computer data. The original CD provided high-fidelity music with a typical capacity of 650 MB, whereas the DVD offers a minimum 4.7 GB by having a shorter laser wavelength, greater pit density, narrow track spacing, faster rotation speeds, and faster servo systems.

The DVD laser can produce a shorter-wavelength beam that is measured in nanometers. A shorter wavelength provides greater precision in operation. Two different data layers provide accurate focusing systems. The second DVD layer must have a laser beam that focuses deeper. The laser beam in a DVD player must be stronger so as

13-1 Loading a DVD in a DVD/CD player.

to read both the first and second layers. As in a CD player, the laser beam strikes the bottom side of the disc, opposite the label side, which is mounted upward.

Safety precautions

Before a DVD player is returned to the customer, make cold and hot leakage tests. Simply connect a jumper wire between the unplugged ac two-prong plug of the ac cord. For a cold leakage tests, take the resistance measurement between the jumped ac plug and the outside of the metal cabinet, screw-head connectors, control shafts, and exposed metal parts. The ohmmeter measurement should be somewhere between 1 and 5 megohms.

For a hot leakage test, plug the DVD player into an ac outlet. Connect a 1.5-kilohm, 10-W resistor in parallel with a 0.15-μF capacitor between an exposed metal part of the DVD player and a good ground (Fig. 13-2). Measure the ac voltage across the resistor and capacitor. The ac voltage should not be over 0.75 V root mean squared (RMS). Check for possible shock hazard in the DVD player if the ac voltage reading is higher. Recheck and repair the equipment before it is returned for easy viewing and high-fidelity listening.

Electronically sensitive devices

Some semiconductors can be damaged by electricity or continuous handling of electronically sensitive (ES) devices. Integrated circuits (ICs), field-effect transistors,

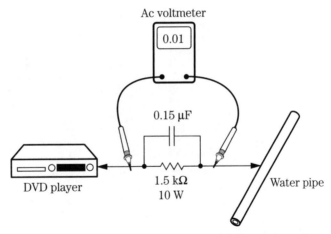

13-2 Check the ac voltage across a 1.5-kilohm resistor for a hot current leakage test.

and semiconductor chip components can be damaged by electrostatic discharge (ESD). Always wear an ESD wrist band or strap to drain off electrostatic charges from your body. Do not place ES devices on a conductive surface such as metal or foil. Make sure that optical assembly is grounded at all times.

Use only an antistatic soldering iron, or ground the metal point of the soldering iron. Use only an antistatic solder removal device. Do not use Freon or chemicals that can generate electric charges that can damage ES devices. Remove the replacement ES device only when it is ready to be mounted. Remove the electrical leads before mounting. Touch the unit to the chassis in which the ES device is installed. Make sure that no power is applied to the DVD player when installing ES devices.

Optical pickup

As in a CD player, the optical pickup assembly in a DVD player consists of a laser, a mirror, and a photo diode required to receive the information stored on the DVD/CD and change it into electrical energy. The spindle or disc motor rotates the disc while the DVD and CD data are picked up by the optical laser assembly. The objective lens focuses the beam onto the disc and collects the reflected light. The reflected light directs the laser beam toward a photodetector.

The optical mechanical assembly consists of the spindle or turntable motor, the SLED motor, focus and tracking coils, the clamper, and the servo control. The loading motor pushes out a loading drawer or tray and takes it back into the spindle motor to be loaded. A clamper device holds the disc in position on the spindle or disc platform (Fig. 13-3).

Do not take a peek

Remember, an invisible laser radiation is emitted from the optical assembly. You cannot see the laser light. Do not look directly at the pickup lens assembly. You can damage your eyes if you stare at the bare optical lens assembly while the player is

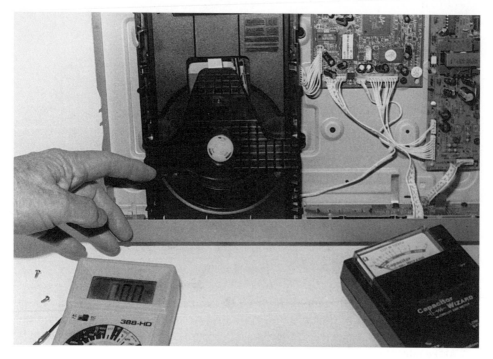

13-3   The laser pickup assembly found within a DVD/CD player.

operating. Usually a laser beam warning is found attached to the side of the optical assembly. Always keep a disc loaded on the spindle when you are servicing a DVD/CD
player. Remember, the laser beam is not visible like that of a light-emitting diode
(LED) or pilot light.

Electrostatic grounding

The laser diode of the optical lens assembly can be damaged by static electricity built
up on your clothes or your body. Make sure that your body is grounded with a antistatic wrist strap. Place a conductive material on the work bench where the DVD/CD
player is placed, and the conductive sheet should be grounded. Do not let your
clothes touch the traverse or optical assembly when installing a new optical pickup
assembly. Be sure to remove the shorting clip from the flexible ribbon connecting
the optical unit to the DVD/CD player after installation.

Block diagram of a DVD/CD optical
assembly

The optical pickup assembly of a DVD/CD player may consist of a high-frequency modulator (HFM), a DVD modulator (DVDM), and a CD modulator (CDM). The linear drive
(LD) switch is found between the front-end processor and LD in of the high-frequency

Optical pickup

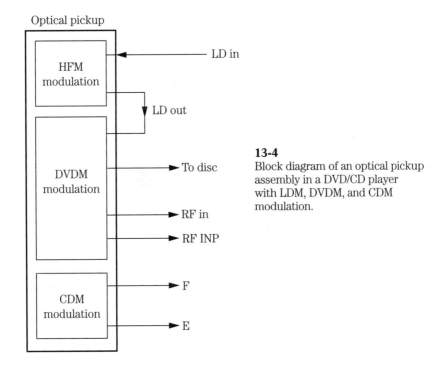

13-4
Block diagram of an optical pickup
assembly in a DVD/CD player
with LDM, DVDM, and CDM
modulation.

modulator. The LD out of the high-frequency modulator is fed to the LD in of the DVD modulator. The SW signal is fed to the DVD modulator and onto a disc servo control (DSC) processor (IC2001).

The tracking and focus signals from the DVD modulator are fed to the front-end sensor (IC5201). RFN and RFP are also fed to IC5201. A CD supply switch is fed into the LD terminals of the CD modulator. The focus, error, and monitor signals are also fed into the front-end processor (Fig. 13-4).

The HFM optical pickup assembly consists of an IC module and surface-mounted components. The DVDM optical assembly contains many optical integrated circuits (ICs), transistors, and diode components. The CDM assembly may consist of several optical ICs, diodes, and transistors within the CDM block of the optical pickup unit.

Block diagram of a DVD/CD/VCR optical assembly

The DVD traverse unit in a DVD/CD/VCR player may consist of an optical pickup assembly that is fed into a front-end processor (FEP), and the output of the FEP is fed to a DSC processor. The traverse motor is fed from a motor driver IC that is controlled by the DSC processor. Likewise, the spindle motor, focus motor, and tracking motor are controlled by the motor drive IC with signal from the DSC processor. The DSC IC signal is fed to the optical disc control processor and then fed to the decoder IC (Fig. 13-5).

Traverse unit

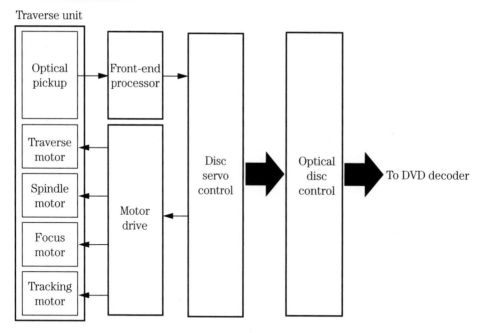

13-5   Block diagram of the traverse unit found in a Panasonic DVD/CD/VCR player.

Always use a variable-voltage isolation transformer when servicing a DVD/CD/VCR chassis. The isolation transformer can prevent accidents resulting in personal injury from a possible electric shock. The isolation transformer also protects the DVD/CD/VCR player from being damaged by accidental shorting that may occur during servicing. Besides protecting the electronic technician and the DVD/CD/VCR player, expensive test equipment also is protected from extreme damage by the isolation transformer. The isolation transformer can be used to vary the power line voltage for erratic and intermittent operation of the DVD player.

The DVD/CD player

A DVD/CD player may have a high-speed scan, Dolby digital/data transmission system (DTS) output, ultra surround sound, and remote operation. A DVD/CD player may have DVD-R/CD-R/RW/WMA/MP3 playback. A DVD/CD player may have twin lasers for DVD/CD playback with an on-screen component video and RCA video output. A DVD/CD player has loading, spindle, and disc motors and a SLED motor. Besides the loading and spindle motors, a DVD/CD changer has stepping motors, as well as focus and tracking coils within the optical pickup assembly.

The DVD/CD/VCR player

A DVD/CD/VCR player may contain a DVD/four-head hi-fi stereo VCR combo with Dolby digital/DTS, DVD/DVD-R/VCD/CD/CD-R/RW/MP3 playback that is operated from a remote control transmitter. A DVD/CD/VCR player may have Color Stream component video outputs, icon-based on-screen display, and Toslink for optical connectivity. The slim-design DVD recorder and player may have progressive scan, Dolby digital/DTS outputs, play list playback, and up to 12 hours of extended recording. A DVD recorder can play DVD video, DVD RAM, DVD-R, video CD, CD, CD-R, and CD-RW.

A DVD/CD/VCR player may contain DVD/CD loading, spindle, and turntable motors and a SLED motor. A VCR section contains capstan, loading, cylinder, and drum motors. The capstan, loading, and cylinder motors are controlled by a capstan/cylinder/loading motor driver IC.

The optical pickup assembly within a DVD/CD/VCR player may contain traverse, spindle, focus, and tracking motors. A motor drive IC controls the traverse, spindle, focus, and tracking motors, whereas a front-end processor receives the laser optical pickup signal. The DSC IC controls the motor driver IC and also receives the front-end processor's signal. The DSC IC signal is fed to the optical disc control processor and on into the DVD decoder microprocessor.

Look me over

After removing the top cabinet shell, take a peek at the various components of the DVD/CD player chassis. The power supply components should be located near the large main filter capacitor and switching transformer. Likewise, all transistors and IC regulators are found nearby. Simply trace from the ac cord into the DVD/CD player chassis to locate the low-voltage bridge rectifier and switching control IC or transistor. They will have the highest dc voltage on their terminals. The two different power supplies found in a DVD/CD/VCR player will have two switching transformers, one feeding the DVD/CD circuits and another supplying power to the VCR circuits. Only one switching transformer is found in a DVD/CD player (Fig. 13-6).

Disassembly of a Panasonic DVD/CD changer

Remove the four screws that hold the top cover of the DVD/CD changer, two on each side of the unit. Remove the two screws at the top and rear for top cover removal. To remove the front panel, remove the three metal screws at the bottom edge and the two screws, one on each side, of the plastic front cover.

Remove the four screws that hold the tray assembly at the rear panel. Remove two screws, three connectors, and four FFCs from the top of the tray assembly. A closed lock gear on the left side of the tray assembly must be pressed to release the plastic tray assembly. Pull the tray assembly toward the front, and remove all connecting cables. Push and release the four claws, and then remove the tray assembly.

Disassemble the rear panel by the removing 10 screws at the rear of the DVD/CD/VCR player. Remove the clamper plate assembly by removing the two screws, one on each side, mounted on the plastic clamper. Push out the claw clamp at the corner, and the clamper plate assembly can be raised upward. To disassemble the fixed plate, magnet, and clamper, release the three claws inside the round clamper plate assembly. Extreme care should be used when removing the loading mechanism, back-end

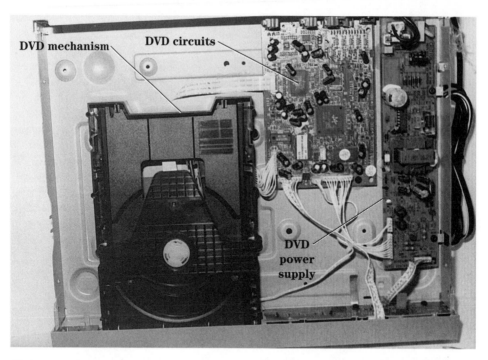

13-6 Try to locate the various components on the topside of a DVD/CD player.

module, and middle chassis. Make a diagram or screw-mounting drawing and lay the parts in line so that the components can be replaced without any difficulty and without any service literature.

After replacing the DVD/CD/VCR changer, the traverse unit must be left in a standby position. Press the open/close button to close the loading tray. Press the power button to turn the power off. Now disconnect the power plug from the ac outlet. Do not disconnect the power plug from the ac receptacle with the tray open. If you try to close the tray manually, the traverse unit will not go to the upper (standby) position, and the player cannot be handled or transported.

The focus and tracking coils

The purpose of the focus servo circuits is to keep the laser beam focused correctly on the pits of the disc surface. The focus circuits detect the focus error signal and are used with the FOK circuit to determine the focus adjustment timing. The focus circuits shift the objective lens up and down to find the correct focus point. A signal from the processor or system control IC controls the focus driver IC or transistors that are tied to the focus coil. All the latest focus coil circuits are controlled by a motor driver IC. The focus and tracking coils operate the same as in early CD players.

The purpose of the tracking servo system is to place the laser beam directly in the center of the pit track both laterally and horizontally. Tracking coil assembly movement

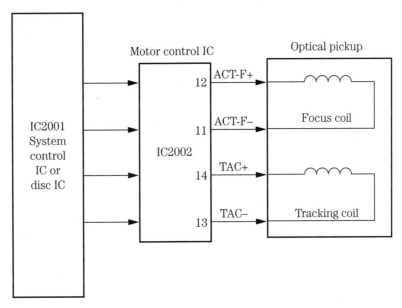

13-7 The tracking and focus coils are driven by a motor driver IC that is controlled by a system control IC or microprocessor.

is horizontal, where the focus coil assembly moves closer or further. The tracking error signal from the pickup optical assembly goes to the servo IC, which drives a tracking coil driver IC (Fig. 13-7). The tracking driver IC provides voltage to the tracking and focus coil assembly located in the optical pickup assembly.

In a Panasonic DVD/CD/VCR player, the focus and tracking motors are driven by a motor driver IC that is controlled by a DSC microprocessor. The spindle and traverse motors are driven by the same motor driver IC. Separate motors drive the focus, tracking, spindle, and traverse operations instead of coils in DVD/CD/VCR players (Fig. 13-8).

Low-voltage power supplies

Most low-voltage power supplies found in DVD/CD players consist of transformer switching circuits. The normal 120-V ac input is like most low-voltage power supply TV circuits. A bridge rectifier produces a high dc voltage to the switching transformer and switch-drive transistor. The switching transistor provides a switching off and on path in the primary side of the switching transformer that produces a magnetic field to create a voltage in the secondary winding of the transformer.

A switching power supply or a switch-mode power supply (SMPS) is found in DVD/CD players (Fig. 13-9). These same switching circuits are found in present-day TV circuits. No on/off switch is found in the input circuits of the switching power supply.

Several half- and full-wave silicon diode circuits are found in the various voltage sources in the secondary winding of a switching transformer. A photocoupler or isolator provides feedback from the secondary winding source to the primary winding of the

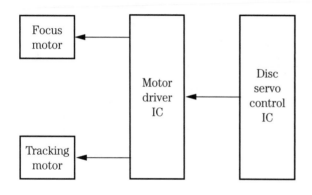

13-8
The focus and tracking motors are driven by a motor driver IC from a DSC microprocessor in a DVD/CD/VCR player.

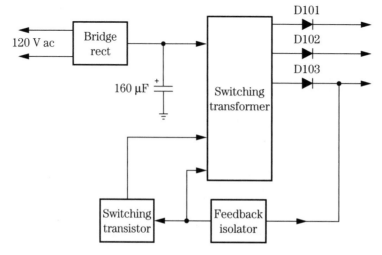

13-9 Block diagram of a switching power supply found in a DVD/CD player.

switching transformer. Low-pass-filter (LPF) transistors may be found in the different voltage sources. Some DVD/CD player power supplies have only transistor voltage regulators within the various voltage sources. IC voltage regulators are found in the larger voltage sources. A switching IC, instead of a switching transistor, may be located in more recent switching power supplies.

The switch-mode power supply (SMPS) input circuits are above the ground that is called a hot ground. When taking voltage measurements on a switching transistor or IC component, make sure that the ground clip of the voltmeter is connected to the hot ground or a bad voltage measurement will occur. Remember, the main low-voltage input is at the hot ground side. All components tied to the primary winding of the switching transformer are at hot ground potential. Clip the ground or black lead of the voltmeter to the main filter capacitor hot ground for accurate voltage measurements within SMPSs or switching power supplies.

A signal processor or secondary windings of the switching transformer operate at a cold or common ground. All voltage measurements within the secondary windings of a

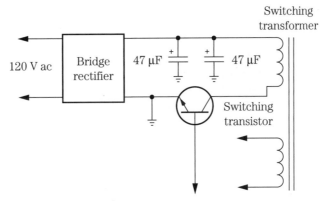

13-10 Block diagram of a DVD/CD changer low-voltage power supply with a common input ground.

switching transformer can be taken at the common or cold ground. Again, use the main filter capacitor negative terminal as the common ground side. Some low-voltage power supplies with a switching transistor and transformer have a common ground on the secondary side (Fig. 13-10).

The main filter capacitor in a DVD/CD player is not as large as that found in a TV chassis. A TV main filter capacitor may be 470 to 680 μF and 150 to 200 V, whereas two 47-μF electrolytics are found in the parallel filter circuit of a DVD/CD player. Determine if the main filter capacitor has a common or hot ground before taking critical voltage measurements.

The secondary voltages in a DVD/CD player are usually 5-, 8-, and 12-V sources. A higher 20 to 24 V dc is fed to the display circuits. The DVD/CD/VCR deck has UNSW +14-V, audio +15.7-V, and UNSW +33-V sources. The regulated output voltages in a DVD/CD changer are at 2.5, 5, and 9 V. You may find several voltage control switch and current-limiter switch transistors within the DVD/CD output circuits. Transistor and IC voltage regulators are found throughout the secondary voltage sources.

Transistor regulators

Although most of the latest DVD/CD players contain IC voltage regulators, transistors are found in switching, LPN, and voltage regulator circuits. The positive voltage source is taken from the emitter terminal of the transistor and filtered with electrolytics in the output voltage source. A transistor voltage regulator works as a series regulator, whereas a zener diode appears in a shunt-type regulator. The regulator circuit stabilizes the rectified-filter power supply, keeps the voltage source stable, and does not vary with a change in load (Fig. 13-11).

A defective transistor regulator can appear intermittent, open, or leaky. An intermittent voltage measured at the regulator output emitter terminal can produce intermittent operation at the voltage regulator source. An open transistor regulator has no or very little voltage at the output voltage source and produces a dead symptom. A shorted transistor regulator can produce a lower dc voltage output, and the voltage on all three terminals of the transistor may indicate a common low voltage. Critical voltage

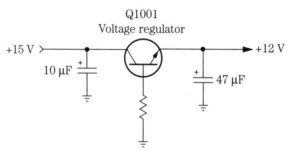

13-11 The output supply voltage is taken from the emitter terminal of a transistor regulator.

input and output tests on the transistor voltage regulator can indicate a defective transistor voltage regulator. Take a transistor in-circuit test with the diode tester of a digital multimeter (DMM) or transistor tester if in doubt. Also check the switching transistors with the diode tester of a DMM.

IC voltage regulators

The latest IC voltage regulators are found in DVD/CD player chassis. A regular IC voltage regulator has input, output, and ground terminals. The shunt IC regulator may appear as an internal zener diode or a multiterminal voltage regulator (Fig. 13-12). An adjustable IC voltage regulator might have four or five connecting terminals. IC shunt, digital voltage IC, and motor IC regulators (9 V) are found in several DVD/CD players.

An open IC voltage regulator will have very little or no voltage at the output terminal. The shorted or leaky IC voltage regulator may have a low voltage at the output terminal. Suspect an intermittent IC regulator when there is a varying voltage on the output and a normal voltage on the input terminals. Take critical input and output voltage measurements with a DMM on the suspected IC to determine if the IC regulator is defective.

Photocoupler or isolator

The switching transformer power supply circuits have a photocoupler or isolator IC that provides feedback from the output secondary voltage to the primary side of the transformer. An optical isolator IC provides the same results in switch-mode power supply (SMPS) circuits. The optical isolator contains an internal light-emitting diode (LED) and a photodiode or transistor in one package. The optical isolator with the switching transformer isolates the secondary (cold) ground from the primary (hot) ground. An LED is found on the secondary voltage side, and the photodiode or transistor is at the input circuits (Fig. 13-13).

Test both sides of the voltage-controlled photocoupler or isolator with the diode tester of a DMM. A diode measurement on the LED side will be at least twice the resistance measurement of a normal silicon diode in only one direction. If terminals 1 and 2

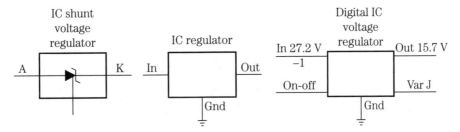

13-12 The different types of IC voltage regulators found in today's low-voltage power sources.

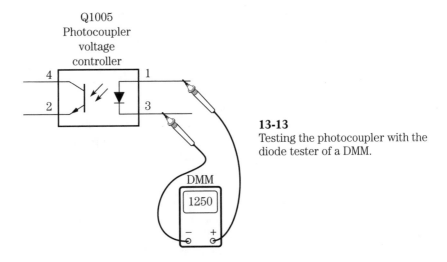

13-13
Testing the photocoupler with the diode tester of a DMM.

show a resistance measurement in both directions with the diode tester of a DMM, suspect a leaky coupler or isolator. Pin terminals 3 and 4 should show no resistance with reversed test leads.

Troubleshooting low-voltage circuits

Troubleshoot the low-voltage circuits in a DVD/CD player as in a TV chassis. Quickly check the dc voltage across the main filter capacitor. No voltage here may indicate an open fuse, isolation resistor, or bridge rectifier. Most DVD/CD players employ a bridge rectifier circuit instead of single silicon diodes. The low-voltage bridge rectifiers work directly off the 120-V ac power line as in a TV chassis.

Low dc voltage across the main filter capacitor may indicate a leaky or shorted diode or a defective electrolytic capacitor. Unplug the DVD/CD player from the ac outlet, and check each diode within the bridge component with the diode tester of a DMM. Suspect open diodes within the bridge circuit when there is no or very little output voltage if the power line voltage is at the input terminals of the bridge rectifier.

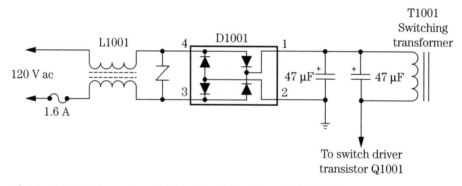

13-14 A typical low-voltage bridge circuit found in a DVD/CD player.

Sometimes a defective electrolytic filter capacitor can have lower capacity and cause a lower dc output voltage from the power supply. Check for an equivalent series resistance (ESR) loss and different capacitance of the main filter capacitor with an ESR meter. If the capacitor shows ESR capabilities or a loss of capacity, replace it. A defective main filter capacitor may cause the chassis to not start up or to start up and then shut down. Usually, a lower dc voltage is measured across the main filter capacitor if the component is leaky or defective. Remember, the output voltage on the main filter capacitor is much higher than that found in secondary voltage sources.

If an ESR meter is not handy and low voltage is measured at the main filter capacitor, pull the ac cord out of the wall receptacle. Clip a higher or same capacity and voltage electrolytic across the suspected capacitor. Now power the unit up and take another voltage measurement. Always observe the correct polarity of the electrolytic capacitor. The dc voltage should rise to normal when the original capacitor has lost its capacity or has ESR problems. Replace the defective electrolytic with one of the same or higher capacity and working voltage (Fig. 13-14).

Troubleshooting the switching power supply

Servicing the switching power supply in a DVD/CD player is the same as troubleshooting a TV switching power supply. Check the dc voltage across the main filter capacitor. If no or low dc voltage is found across the main filter capacitor, check the low-voltage circuits. Check all isolation resistors, bridge rectifiers, and electrolytic capacitors.

Go to the switching IC or transistor when high dc voltage is found across the main filter capacitor and there are no secondary voltage sources. Take critical voltage measurements on the switching transistor. Very high voltage on the switching transistor collector terminal indicates an open transistor or emitter resistor if one is found between the emitter leg and hot ground.

Most switching transformer power supply problems are caused by the switching transistor or IC. This switching transistor can appear leaky, shorted, or open. A shorted switching transistor may open the main power line fuse. Do not overlook an open or leaky voltage control switch transistor within the base circuit of the switching transistor.

Sometimes a switching transformer is found to be bad, but very seldom. The switching transformer should be replaced with the original part, although some switching transformers are now available. Check for open or burned resistors within the switching transistor or IC circuits.

Check the secondary voltage on the silicon diodes within the switching transformer sources. When a certain section of a DVD/CD player does not function, always check the dc voltage source. Check the voltage across each secondary filter electrolytic network with a quick voltage measurement. The output voltages in the secondary sources will vary form 3.3 to 12 V.

Simply measure the electrolytic working voltage on the capacitor when a schematic is not available. The measured voltage will not exceed the working voltage found marked on the electrolytic capacitor. By quickly measuring the dc voltage on each silicon diode or corresponding filter capacitor, you can locate a defective low-voltage source (Fig. 13-15). Double-check the electrolytic with an ESR meter.

When all the secondary voltages are found across all filter capacitors, go to the input and output terminals of the IC regulator. Measure the input voltage on the collector terminal of the transistor regulator. The normal output dc voltage on the emitter terminal will be less than that found on the collector terminal. Again, check the working voltage marked on the electrolytic capacitor at the output voltage source.

For instance, a filter output capacitor with a 1000-µF, 25-V rating works in a 13- to 15-V output source. The 10- to 47-µF, 6.3-V electrolytic may be located in a 5-V source.

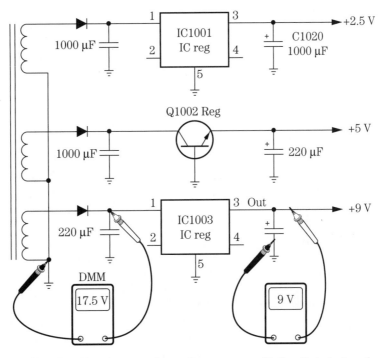

13-15   Checking the secondary voltage source with the diode tester of a DMM.

A 10- to 47-μF, 50-V electrolytic filter may be located within a +33-V source. A 220-μF, 16-V electrolytic filter may provide a smooth filter action for a 9-V motor source. Remember, the working voltage of an electrolytic filter capacitor will always be higher than the actual voltage source. Check all electrolytic capacitors within the low-voltage and switching power supplies with an ESR meter.

The DVD/CD/VCR player power supply

The low-voltage power supply found in a DVD/CD/VCR player may be composed of two different power supplies. The two switching power supplies have a switching transformer and transistor operating within the switching circuits, whereas the other power supply may have a switching IC component. Both switching power supplies may be protected by the same 3-A fuse. A bridge rectifier in each power supply may contain a bridge rectifier or separate silicon diodes and be filtered by an 82- to 150-μF electrolytic capacitor. The input switching power supply circuits have a hot ground, whereas the secondary circuits of the switching transformer have a cold ground (Fig. 13-16).

A feedback isolator or voltage error detector IC provides feedback from the secondary voltage source to a separate primary winding. The hot and cold grounds are isolated by the switching transformer and the error switching detector or isolator IC. When taking voltage measurements within the switching input circuits, take the measurement across the main filter capacitor, and use the hot ground of the capacitor for other voltage measurements in the input circuits. All secondary transformer voltage sources can be taken from the common (cold) ground side, or use one of the negative ground terminals of a filter capacitor as common ground.

A shunt regulator IC is found between common ground and the isolator error voltage detector in a DVD/CD/VCR player (Fig. 13-17). The shunt regulator (IC1003) is tied to the LED photodiode within the error voltage detector (IC1005). A diode test with the diode tester of a DMM can indicate a good diode assembly with a low ohm measurement in one direction.

Several secondary windings within the switching transformer have several silicon diodes rectifying the different voltage sources. Fairly large electrolytic capacitors (47 to 100 μF) are found in the secondary voltage sources. Transistor and IC voltage regulators may be found in these voltage sources. A +12- and +5-V source may be connected to the DVD circuits. A separate +5-V source may feed the analog and digital circuits. Most of the processing ICs operate from the 5-V source.

The second switching power supply provides a voltage source to the VCR circuits. The 3.3- and 3.6-V sources are developed from an IC voltage regulator in the 5-V line. A 3.3-V source may feed the SD RAM IC. The 3.6-V source may feed the flash-memory circuits. A 5-V source feeds the Syscon, video, and LED circuits, whereas a 12-V source supplies voltage to the switching power on +12-V and power on/off transistors. The +14-V source applies voltage to the UNSW14 Advance processor IC, and the 15.7-V source feeds the audio VCR circuits. You can service the switching IC power supply like the switching transformer circuits in a TV chassis. If a schematic is handy, troubleshoot the waveforms with an oscilloscope.

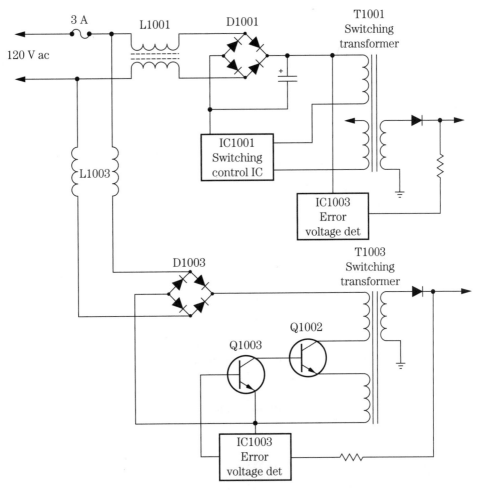

13-16 Block diagram of a switching transformer circuit of a dual power supply in a DVD/CD/VCR player.

Switching transistors

Many different switching transistors are found in a DVD/CD player. The current limiter switching transistor operates in the low-voltage primary circuits of the switching transformer. Several voltage-control switching transistors are found and operate within the secondary winding of the switching transformer. A couple of switching transistors may be found in the switch power on/off of the analog +5-V circuits. Two +9-V motor-switching transistors operate the 9-V motor circuits. Two switching transistors are found in the muting audio output jack circuits.

Switching transistors are located in the power supply that feeds the VCR and turn-on circuits within the DVD/CD/VCR player. Switching power-on +12-V and switching power-off transistors are found in the 12-V source (Fig. 13-18). The switching power-on

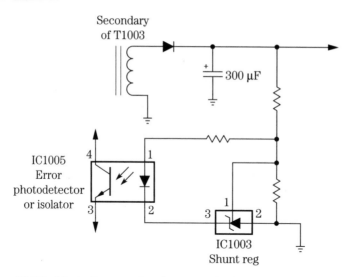

13-17 Block diagram of a shunt coupler between the isolator (IC1005) and common ground.

+5-V transistor is located within the video SW 5-V source. A switching rec on transistor and switch delay rec on-on transistor are found in the system control (servo) circuitry of the VCR section. Suspect a defective in-circuit switching transistor when a certain operation will not function. Check the suspected switching transistor with the diode tester of a DMM. Do not overlook switching transistors located within the hi-fi section of a DVD/CD/VCR player.

Chips and more chips

There are many surface-mounted device (SMD) chips found within DVD/CD/VCR player circuits. There are many resistor and capacitor chips in virtually every DVD circuit. Besides SMD transistors and IC components, several chip inductors and chip filters are found in a DVD player chassis. Check the resistance of the chip resistors and inductors with the low-ohm range of a DMM. A chip inductor will have very little resistance. Test the chip capacitors with an ESR meter.

Check each chip transistor with the diode tester of a DMM. Compare the input signal with the output signal of a suspected chip IC or microprocessor. Take a critical voltage measurement on the Vcc or Vdd terminal of a suspected chip IC. Be very careful when checking voltage on a chip IC or microprocessor. Use sharp-pointed test probes on the test equipment so as not to short out a terminal nearby (Fig. 13-19).

The DVD test disc

A DVD test disc provides signals for tilt, luminance level, and color-bar adjustments. A DVD test disc may provide a split-field National Television System Committee/Electronic Industries Alliance (NTSC/EIA) code-bar pattern with a 1-kHz audio tone. The

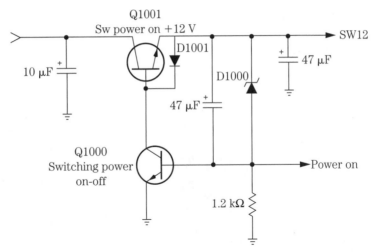

13-18 Switching transistors within the switching power supply circuits of a DVD/CD player.

13-19 There are SMD transistors, ICs, inductors, and filters within a DVD/CD player.

full-field NTSC/EIA color-bar pattern with 20 Hz to 20 kHz to 20 Hz at −10 dB is found on the DVD test disc. A 10-bar gray-staircase test pattern with 440 Hz (musical A note) at −10 dB is found on the test disc. The DVD test disc may have a multiburst bar sweep signal from 0 (carrier) to 4.5 MHz with white noise at −10 dB. You also may

13-20 A DVD test disc can check the color-bar pattern, full-field pattern, grey-staircase test pattern, multiburst bar sweep, cross-hatch, and white, red, blue, and green rasters.

find a cross-hatch pattern with a 250-Hz tone alternating left to right every 10 seconds on the DVD test disc. Most selections are continuous. Silence is used to test audio systems for generated noise, ground loops, and interference (Fig. 13-20). A DVD test disc (32-16660) can be purchased at MCM Electronics, 650 Congress Park Drive, Centerville, OH 45459-4072.

There are other DVD test discs that can be purchased at most electronics parts distributors. Panasonic has a DVD test disc (DVDT-S01) for electrical adjustments. Check with the service warranty station that you service for its test discs.

Tray motor operations

The tray motor rotates the rotating tray that might hold three or five discs in a DVD/CD changer. The tray motor drives a belt that turns a tray gear that turns the large tray holding the different DVDs. The tray motor may operate from a +9-V source from the power supply or a voltage regulator IC. The +9 V is fed from the power source to pin terminals 7 and 9 of a motor driver IC. The motor driver voltage from motor driver IC500 is found on terminals 3 and 10, which feed voltage to the motor terminals (Fig. 13-21).

Check the dc voltage across the tray motor terminals to determine if the motor driver IC500 is supplying voltage to the tray motor. If the motor does not operate with a dc voltage found on the motor terminals, take a continuity ohmmeter test across the

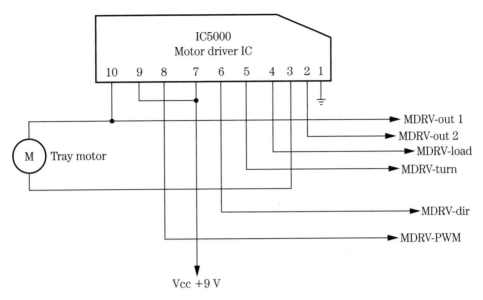

13-21   The tray motor with a motor driver IC in a DVD/CD changer.

motor terminals. An open tray motor will have no resistance, whereas a normal measurement has a low continuity reading. No voltage across the motor terminals may indicate a defective motor driver IC. Measure the supply voltage at the Vcc terminals of the motor (7 and 9). Suspect a leaky motor driver IC if no or low voltage is found at the supply terminals. If the voltage is low, suspect a leaky motor driver IC or improper supply voltage.

Measure the voltage at the supply terminals (Vcc) on the motor driver IC. Go directly to the low-voltage power supply for the +9-V source. A defective IC voltage regulator can provide no, low, or intermittent supply voltage. Monitor the supply voltage at the output terminal of the tray motor for erratic or intermittent rotation of the tray motor. A defective tray motor also can cause intermittent tray rotation.

Loading motor operation

The loading motor may be controlled by a voltage from the motor driver IC500 that also drives the tray motor or from a separate motor IC. The loading motor in some DVD/CD/VCR units is controlled by a signal from the system control microcontroller to a loading motor driver IC. The microcontroller processor controls the forward, reverse, and stop rotations.

The capstan motor, cylinder, and loading motor may be controlled by the same capstan/cylinder/loading motor driver IC. The loading motor load and unload terminals are LDM out 2 and LDM out 1. The loading motor loads and unloads the disc in DVD/CD/VCR players (Fig. 13-22). The UNSW +14-V source is fed from the secondary voltage source to the capstan/cylinder/loading motor driver IC.

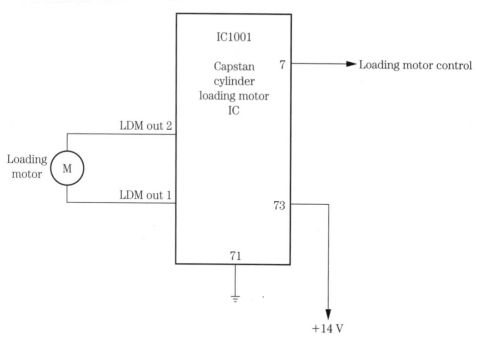

13-22   The loading motor circuits and capstan/cylinder loading motor driver IC.

Check the voltage source applied to the loading motor in either load or unload modes. No or low dc motor voltage may indicate a defective motor driver IC or a bad loading motor. Remove one motor terminal, and take another voltage measurement. Suspect a bad motor if the supply voltage to the loading motor is higher than normal. A bad motor is indicated by normal drive voltage and no motor rotation. Check the continuity of the motor with the low-ohm scale of a DMM. When motor rotation is erratic or intermittent, monitor the dc voltage at the motor terminals to determine if the driver IC or the motor is defective.

Loading motor belt problems

Most loading and capstan motors have a drive belt. A worn, loose, or stretched motor belt can cause intermittent or erratic loading. The loading motor belt with grease or oil on it may result in erratic loading. A worn and stretched loading motor belt can come off the pulley into the gear area, resulting in no loading. A cracked or broken belt can cause the disc not to load. Check all belts when the disc will not load or a cassette will not load in a VCR player.

Clean the old, hardened grease or oil off the capstan motor and drive belt when the capstan motor speed is slow or the belt drags and produces a wow and flutter tape speed. A worn or loose capstan belt can cause a garbled audio sound. A dry or frozen capstan bearing can cause the capstan belt to slip. Loss of playback and recording functions can be caused by a cracked or broken capstan drive belt. The broken capstan belt

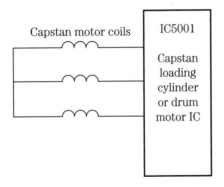

Capstan motor coils

IC5001

Capstan
loading
cylinder
or drum
motor IC

13-23
Block diagram of the capstan motor driver IC in a DVD/CD/VCR player.

results in no rotation of the tape in the VCR. Realign the new belt if it is too loose or keeps coming off the motor pulley.

Capstan motor circuits

The capstan motor circuits within a DVD/CD/VCR player may operate from the same driver IC as the loading and cylinder motors. A capstan motor assembly may include the capstan motor Hall elements, MR head, cylinder, Hall IC, capstan motor, and the motor control IC. Troubleshoot the capstan motor as in any VCR (Fig. 13-23). The capstan motor may operate from a 12- or 14-V source.

A defective capstan motor or frozen motor bearings can cause the tape to become stuck in rotation. Improper or low dc voltage applied to the capstan motor can cause a dead rotation or slow speeds. Sometimes the bad capstan motor can open the low-ohm resistors in series with the motor and supply voltage source. A flat armature or a dead spot with no rotation can occur in a defective capstan motor. Try rotating the motor pulley by hand when the capstan motor is dead and then takes off. Simply replace the defective capstan motor.

Take a low-voltage measurement across the capstan motor terminals. Suspect a bad motor if a normal dc voltage (9 to 14 V) is found at the motor terminals with no rotation. Check the capstan motor driver IC when no or low dc voltage is found at the motor terminals. Recheck the motor voltage source with one motor terminal removed from the motor driver IC. Usually, a defective motor driver IC also will cause no rotation in any motor that is connected to the same motor driver IC. Suspect a motor driver IC when none of the motors will operate that are connected to the same driver IC.

Cylinder or drum motor operation

The cylinder motor circuits may consist of a cylinder motor, Hall IC, cylinder or drum motor driver IC, a motor position detector, and a cylinder pulse generator (PG). Service the cylinder or drum motor as in any VCR player. No rotation of the drum motor may be caused by improper voltage to the motor, a poor or defective voltage regulator IC, or a faulty drum motor driver IC. A poor connection between the cylinder motor and the servo IC can cause no cylinder rotation. Check for bad PC board lower drum

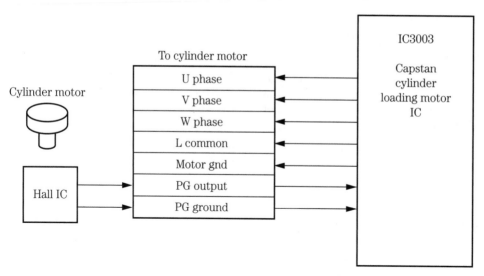

13-24   Block diagram of the cylinder or drum motor IC circuits in a DVD/CD/VCR player.

assembly connections. Open or low-ohm resistors in the voltage source of the cylinder motor can cause no motor rotation. Check for a badly soldered joint on the cylinder circuit board assembly (Fig. 13-24). The cylinder motor may operate from a 12- to 14-V source.

A defective drum or cylinder motor can cause the tape to load and then the motor to shut down. Bad tape end sensors can cause the tape to load with a motor shutdown symptom. A bad cylinder or drum motor can cause the motor to start and quit with motor shutdown symptoms. Suspect a bad cylinder motor driver or servo IC when the motor begins to run and then shuts down. Take critical voltage measurements on the cylinder motor driver IC and servo control processor. Check for an improper voltage source to the driver and servo ICs supplied from a defective IC or transistor voltage regulator source.

Stepping motors

Stepping motors are found in a DVD/CD (three- to five-disc) changer. The stepping motors help to load several discs and sometimes are called the up and down motors. Like all motors found in a DVD player, a motor driver IC drives the rotation of the stepping motors. The optical pickup assembly may be driven by the same motor driver IC2500 (Fig. 13-25). Check the stepping motors as you would any other motor within a DVD/CD player.

Disc or spin motor operation

The disc, spindle, or turntable motor starts to rotate the disc after it has been loaded in the DVD/CD player. A small platform is mounted at the end of the spindle motor's

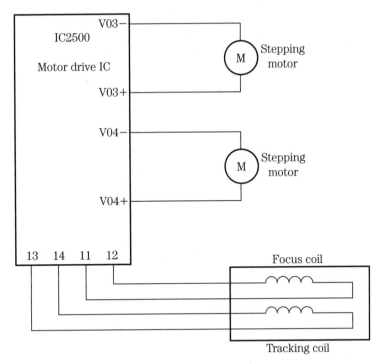

13-25 Block diagram of the stepping motor found in a DVD/CD changer.

shaft that spins the CD or DVD at a variable rate of speed. The spindle motor is usually located right under the clamper assembly. The disc motor starts out at 500 rpm and slows down to 200 rpm as the pickup assembly moves toward the outer rim of the CD in a CD player (Fig. 13-26).

The spindle motor can be controlled with a separate IC or combined with the SLED motor from a motor driver IC500. Check for a dead disc motor by starting at the motor circuits and working toward the servo or microcomputer IC. Check the motor voltage across the motor terminals. If the dc voltage is low or there is no supply voltage, check for a defective motor driver IC501. Measure the supply voltage on the motor driver IC. Low voltage on the motor driver IC may be caused by a leaky motor driver IC or a defective voltage supply source.

A defective spindle or turntable motor can cause no disc rotation. Suspect a defective motor driver IC when the spindle motor will not stop in the stop mode; a low dc voltage can be found on the disc motor terminals. Check for poor terminal pin connections on the motor driver IC if the spindle or disc motor starts at a high rate of speed. A defective disc motor can cause very high speed problems (Fig. 13-27).

SLED or slide motor operation

The SLED, slide, or feed motor in a CD player moves the optical assembly across the disc from the inside to the outside rim of the CD, keeping the optical lens constantly

13-26 The spindle or disc motor found in a DVD/CD player.

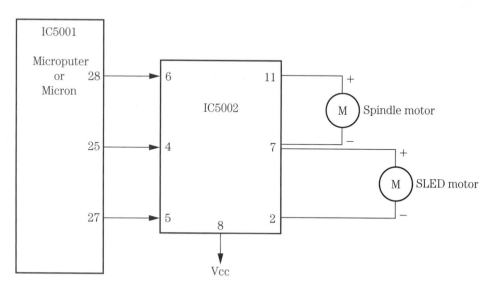

13-27 Block diagram of the motor driver IC to the spindle and SLED motors.

in line with the center of the optical axis. Usually the motor is driven by a rotating gear that moves the laser beam down one or two sliding bars. The slide motor may have fast-forward and rewind mode operations. A separate spindle and traverse motor may be driven from the same driver IC that also may operate the focus and tracking motor in a DVD/CD/VCR player.

Check for a gummed-up track or poor meshing of the pulley and gears when there is erratic or intermittent operation of the slide motor. Notice if the voltage on the slide motor terminals is intermittent. Monitor the voltage on the motor terminals. Erratic or intermittent voltage can be caused by a defective motor driver IC. A defective motor can cause erratic or intermittent rotation of the disc (Fig. 13-28).

You may find a small motor belt that drives a worm gear pulley to move the pickup assembly in some models. A broken or worn gear belt can result in no or erratic optical assembly movement. Clean the motor belt of oil or grease spots if movement is erratic. Replace the motor drive belt if it shows any signs of slippage.

Suspect a motor driver IC when the SLED or spindle motor does not rotate if they are controlled by the same motor driver IC. Simply scope the waveforms on and into the motor driver IC. Check the voltage from the motor driver IC to each motor. If there is no voltage, measure the dc supply voltage (Vcc) to the motor driver IC from the low-voltage source in the power supply. Scope the supply signal from the microcomputer or system control IC with no signal applied to the motor driver IC.

13-28 The SLED or slide motor moves the optical assembly from the inside to the outside rim of a DVD/CD.

Radiofrequency (RF) preamplifier

In early RCA and Zenith DVD/CD players, the output from the laser optical pickup was fed to the DVD A, B, C, D, DVD RF and CDA, and BEF signals. The RF signal processor is somewhat like the CD RF amplifier circuits. This RF signal is amplified and passed on to a CD/DVD DSP, and DVD servo IC205. The FE, TE, and DVD/CD RF is fed from the RF amplifier IC to IC205. IC205 provides demodulation and C1-C2 error-correction circuits.

The RF signal processor also may include internal RF automatic gain control (AGC) circuits, automatic phase control (APC), internal auto asymmetry, an internal disc effect detector, and an internal focus protect function against disc effects. Scope the output of the RF amplifier to determine if the optical pickup signal and the RF amplifier are normal.

In some models, the servo control phase duration modulation tracking and CD-ROM compatibility are fed from the RF amplifier IC to the servo IC. A RAM buffer or DRAM IC is connected to IC205. The demodulation, corrector, and DVD data are fed from IC205 to the block decoder IC301 (Fig. 13-29).

System control or microcontroller

Like the standard CD player system control IC, the microcontroller in a DVD/CD/VCR player controls the loading motor operations. The loading motor is controlled by IC8002 to a loading motor driver IC to the loading motor terminals. The system control IC also controls the mode switch positions, reset, power down, drive, and servo data. The take-up reel, take-up photo, and supply photo (TRL) LEDs are controlled by the system control IC.

The Xtal oscillator, safety tab broken, video delay, and PC and clock serial data 1 are controlled by IC8001. The system control IC also operates the DVD ready, DVD reset, host ready, DVD serial data, DVD serial data 1, and DVD serial clock function. IC8002 controls the progressive on, key data (1–5), VCR LED on, VCR rec LED on, DVD LED on, LED on, and progressive LED on operations. The infrared remote receiver signal is connected to pin 5 of the system control IC. A +5-V volt (Vdd) source powers the operations of the system microcontroller (IC8002) (Fig. 13-30).

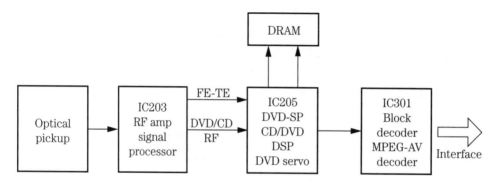

13-29   An early DVD/CD player with an RF amplifier signal processor that amplifies the RF signal from the optical pickup assembly.

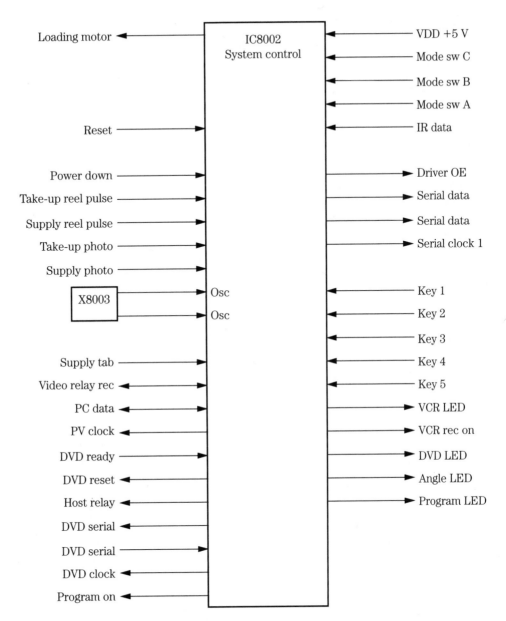

13-30 The system control microcontroller (IC8002) controls many different operations within a DVD/CD/VCR player.

Front-end processor

The front-end processor IC receives the DVD audio/video signal from the HFM, DVDM, and CDM optical pickup unit in a DVD/CD changer. In a DVD player, the optical pickup output is fed from the RF amplifier IC. The optical pickup signal from the traverse unit is fed to the front-end processor, and the output is fed to the disc servo and optical disc control process IC in a DVD/CD/VCR player. The front-end

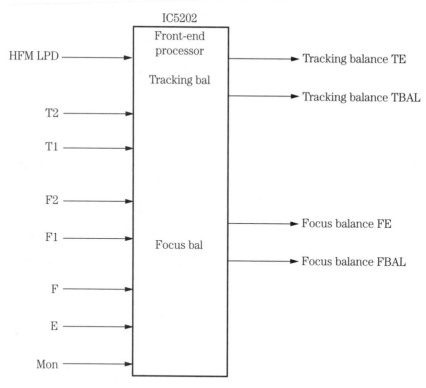

13-31   The DVD front-end processor found in a DVD/CD changer.

processor (FEP) receives the DVD modulation and CDM modulation signal (DVD audio/video signal line) from the optical pickup unit in a DVD/CD player (Fig. 13-31). The tracking balance, focus balance, and TE signals are fed to the decoder IC.

A +5-V source is fed to pin 1, and a 3.3-V source is applied to pin 2. Simply scope the input and output of the DVD audio/video signal lines of the front-end processor IC. Take critical voltage tests on the supply voltage source (Vcc) terminals of the IC front-end processor. The tracking and focus balance signal is sent from the front-end processor to the DVD decoder or disc servo control IC.

From the optical pickup assembly in a DVD/CD/VCR player, the optical pickup signal is fed to a front-end processor IC, a disc servo control IC, an optical disc control processor, and a decoder microprocessor (Fig. 13-32). The same DVD/CD/VCR player's optical pickup unit contains the traverse, spindle, focus, and tracking motors. Here the focus and tracking motors are found instead of coils.

DVD digital signal processor IC

The DVD digital signal processor converts the high-frequency input signal to digital with an 8-bit analog-to-digital (A/D) converter. The gain control is set by an AGC circuit for the best performance of the converter. The A/D converter–CLC circuits provide a clocking sync for the preceding operations. The amplifier and equalized RF

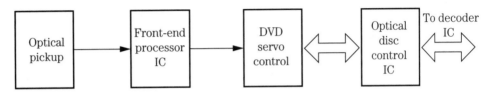

13-32 Block diagram of a DVD/CD/VCR player showing the DVD signal from the optical pickup assembly to the decoder IC.

signal that is fed into the DVD DSP circuits also includes a portion of the DSP signal that is sliced data, which functions as a bit decoder and provides error correction.

MPEG decoder

The DVD technologies depend on MPEG encoding and decoding to provide high-fidelity movies and other video programs. The MPEG audio/video decoder IC8002 in a DVD/CD/VCR player receives the host data (16 bits) from the optical disc control IC. The host IF signal is fed to a DVD decoder and error-correction circuits. IC8102 (64-MB SDRAM) provides clock, an SDRAM address bus, and an SDRAM data bus to and from the error-correction circuits. The MPEG decoder signal is fed to a video signal processing circuit feeding a YUV data (8-bit) signal to the NTSC error IC8201. The video process circuits also provide a horizontal and vertical output sync signal.

The DVD decoder circuit separates the video and audio signals and applies the audio signal to an audio decoder circuit that is fed to a D/A converter (IC8003). A reduced instruction set computing (RISC) processor circuit provides a memory address bus and a memory data bus to and from the flash memory IC8012. The output of the RISC processor is fed to a flip-flop IC8002. A regulated +3.6-V supply source is fed to the MPEG decoder IC8002 (Fig. 13-33).

NTSC encoder

The NTSC format has 525 scanning lines, whereas the standard phase alternate line (PAL) format has 625 scanning lines at 50 Hz instead of 60 Hz. A DVD player must play both the NTSC and the PAL standard. This means that additional video is included so that either system can be used.

The YUV data (8 bits) derive from the video signal processor circuits within the MPEG decoder IC. Also, the horizontal and vertical sync signals are fed into IC8201. The three outlets from the YUV decoder are fed to a sync mix and burst mix to the matrix circuits. The four D/A converter signals are fed from the matrix circuits to a transistor buffer and filters circuits, providing a DVDS-Y/Y, DVD S-C/PR, DVD video, and DVD PB to the video signal path circuits in a DVD/CD/VCR player (Fig. 13-34).

D/A audio converter

The D/A converter IC actually changes the digital signal from digital to analog (audio). An audio decoder from the MPEG decoder IC in a DVD/CD/VCR player is

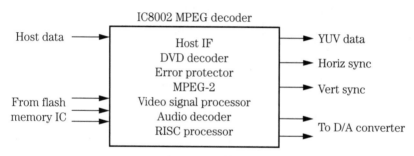

13-33   The MPEG decoder (IC8002) found in a DVD/CD/VCR player.

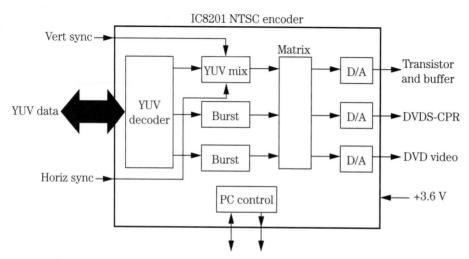

13-34   Block diagram of the NTSC encoder within a DVD/CD/VCR player.

fed into the inputs of the D/A converter IC. The flip-flop IC provides a mode control clock, mode control latch, mode control data, and a reset signal at the input of the D/A converter. The audio intermediate frequency (IF) signal is fed internally to the D/A converter stage, separating the audio into left and right audio signals (Fig. 13-35).

The D/A converter within a DVD/CD changer is fed SCK, data, LRCK, and BCK from the A/V decoder. The D/A audio output is tied into two different OP amps before it is fed to the RCA right and left output jacks.

The D/A converter circuits can be serviced in the same manner as within a CD player. Check the audio signal with the oscilloscope and at the left and right output jacks. The audio signal can be signal traced from the output of the D/A converter to the audio IC amp. Scope the left and right input audio into the OP amp and at the output terminals to the right and left output jacks. You also can check the audio from the D/A

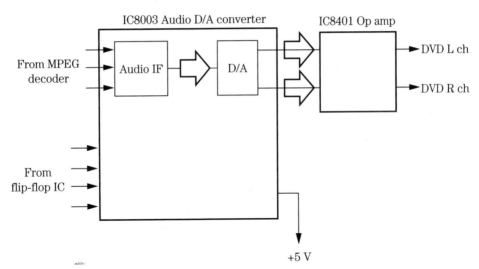

13-35 The audio D/A converter supplies right and left audio signals to the OP amplifier.

converter stage with the external audio amplifier. One or two audio OP amps may be found in other DVD/CD players after the D/A converter IC.

Scope the input terminals of the D/A converter IC when there is no left or right output signal. If the audio signal from the audio decoder is found into the D/A converter and not at the output, suspect a defective D/A IC or an improper voltage source. When a scope is not available, check the audio output from the D/A converter IC with an external audio amplifier.

Try to locate the +5-V source with a DMM. A very low voltage source may indicate a leaky D/A IC or an improper voltage source. Remove the voltage supply pin from the D/A converter from the PC wiring. Notice if the voltage returns close to the +5-V source. Suspect a defective voltage regulator source if the voltage is still low or there is no voltage. Replace the leaky D/A IC if the +5 V returns to normal. Most circuits found in a DVD/CD player can be serviced like those found in a CD player.

Transistor muting

Like most CD and DVD players, the audio output is muted before the audio output jacks. The audio signal is fed into the collector terminal of the right and left audio lines, providing muting of both audio channels. The base terminal of each muting transistor is controlled by a switching or system control IC. Muting of the sound is accomplished when mechanical operations occur and should not be heard within the audio channels.

Remove the collector terminal when one or both muting transistors have no sound in the audio output channels. If the sound output is intermittent, check the muting transistors. Test each transistor when one or both cut off the sound all the time. Suspect a switching transistor or system control IC if both muting transistors are normal (Fig. 13-36).

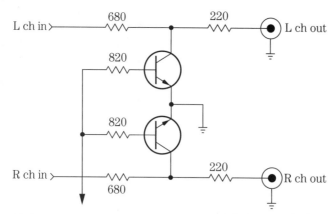

13-36 The left and right audio muting transistors in a DVD/CD player.

Remote control operations

Check the remote transmitter as discussed in Chap. 12. Suspect a defective infrared receiver when the remote control transmitter can operate certain functions within other electronic products. Usually the infrared signal from the infrared receiver is fed into the system control IC. The infrared receiver picks up the infrared signal from the remote transmitter and feeds it into the infrared data of the system control IC.

Check the supply voltage on the infrared receiver, which is fed by a +5-V source. Suspect a +5-V regulator IC when the voltage is low at the infrared receiver. Remove the +5-V pin terminal from the infrared receiver, and take another measurement. Check the IC components within the receiver when the +5-V source is normal. Scope the output terminal (3) or at the test point to prove that the infrared receiver is functioning (Fig. 13-37).

DVD troubleshooting chart

Most CD player problems are the same as those found in DVD/CD players. Check Table 13-1 for additional DVD/CD player service problems and solutions.

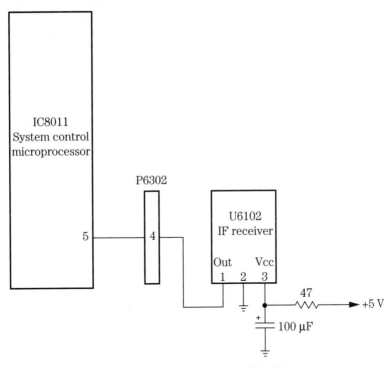

13-37 The infrared receiver circuits in a DVD/CD/VCR player.

Table 13-1. Troubleshooting DVD/CD player problems.

Problem	Procedure
Basic troubles	
No power	Check power off/on switch
	Check connecting cables
No power supply	Check fuse
	Check bridge rectifiers
	Check main filter capacitors with an ESR meter
	Check switching transistors and ICs
Mechanical problems	Check tray loading
	Check loading motor
	Check traverse unit
Optical pickup assembly	Check laser diode with laser meter
	Scope RF waveforms on RF amp
	Replace OPU if no function
Circuit problems	Check LD drive
	Servo system control, traverse, focus, and tracking disc servo
	Signal processing IC or microprocessor
	Flip-flop IC
	D/A converter

Table 13-1. Troubleshooting DVD/CD player problems (*continued*).

Problem	Procedure
Actual DVD/CD problems	
Dead, no power	Check ac fuse Check bridge rectifiers Test electrolytic capacitors
No play	Make sure disc label is up Recheck power cord at receptacle Check ac fuse When will not play outer disc, check ribbon cable to optical assembly
Play fails to start	Check deck for DVDs, video CDs, and CDs Dirty disc can cause reduced video and audio, clean disc
Disc starts to play and stops	Clean disc Try another disc Make sure label side is up
Plays only part of disc	Open resistor in carriage motor circuits Gummed-up rails Clean with alcohol and cloth
Disc spins slowly, will not play completely	Clean laser lens assembly with alcohol and Q-Tip Check voltage regulator IC to disc motor
Disc will not load	Check bad loading belt Check disc reverse switch Dirty mechanism Broken worm gear Tray binding, clean up Spray and clean loading switch Replace defective loading motor
Tray will not move back and forth	Check voltage to loading motor Check voltage at loading motor IC Test each electrolytic in power supply with an ESR meter Check loading switch Replace defective loading motor
Intermittent skipping	Clean lens assembly Worn receptacle, unstable clamper
Will not spin, skips, and stops	Check table height Bad worm gear assembly Replace disc motor
Improper tracking	Clean lens assembly Check tracking coil movement Test tracking driver motor IC

Table 13-1.   Troubleshooting DVD/CD player problems (*continued*).

Problem	Procedure
Actual DVD/CD problems (*continued*)	
Improper tracking (*continued*)	Check driver motor IC regulator Replace split laser, head gear
No video or audio	Clean lens assembly Clean timing mirror Check for scratched lens assembly Check flat ribbon cable to optical pickup
No picture	Make sure TV is on Check all connecting cables Check DVD and TV hookups Try another disc
Picture distorted during rapid advance	This is normal on most DVD players
Program cannot be viewed on TV	Check TV to DVD connections Check all cable connections Set VEN/TV to selector TV Turn the TV to the right channel
No audio in the left channel	Check audio from the D/A converter to audio output jacks Test audio from D/A converter with external audio amplifier Check D/A converter supply voltage Replace D/A converter for no audio Replace demodulator IC
No audio while disc is playing	Bad decoder IC Suspect leaky coupling capacitors Open resistors in audio output Check audio OP amp ICs Check defective muting transistor
No audio, dim display	Check voltage regulator source Replace defective regulator Solder bad terminals on decoder IC
Noisy audio	Replace D/A converter Check filter capacitors with an ESR meter Replace IC OP amp with static in both channels Replace optical pickup with static noise while playing Replace RAM IC for high-pitched noise in sound Replace RAM IC for ticking noise in audio Replace D/A converter for noise in background Replace crystal for low-level noise in the audio Check audio output jacks with the external audio amplifier to check output stages

Table 13-1. Troubleshooting DVD/CD player problems (*continued*).

Problem	Procedure
Actual DVD/CD problems (*continued*)	
Distorted or garbled audio	Replace D/A converter
	Replace RAM IC
	Replace defective OP amp IC
	Check for change in resistance in audio output circuits
	Check all coupling capacitors with an ESR meter
	Troubleshoot the audio output circuits with scope or external audio amp
No muting during program mode	Check muting transistors
	Check system control IC
Protection relay will not turn on or off speakers	Replace mechanism micro IC
	Check protection relay
	Clean points on relay
	Check for poorly soldered connections on the relay
	Replace defective zener diode
	Check silicon diode across relay coil
	Check all connections on regulator transistor
Disc changer will not play	Replace worm gear and level assembly

Index

Pages shown in **boldface** have illustrations on them.

About the Author

The late Homer L. Davidson wrote more than 45 books and over 1000 articles on technical-level electronics. His highly popular books include *Troubleshooting and Repairing Audio Equipment,* now in its Third Edition; *Troubleshooting and Repairing Compact Disc Players,* now in its Third Edition; *Troubleshooting and Repairing Camcorders,* now in its Second Edition; and *Troubleshooting and Repairing Solid-State TVs,* now in its Third Edition, all published by McGraw-Hill. At the end of his life he was the TV Servicing Consultant for *Electronic Servicing & Technology* magazine.